# The Evolutionary Biology of Viruses

# The Evolutionary Biology of Viruses

Editor

## Stephen S. Morse, Ph.D.

*The Rockefeller University*
*New York, New York*

Raven Press New York

**Raven Press, Ltd., 1185 Avenue of the Americas, New York, New York 10036**

Made in the United States of America

**Library of Congress Cataloging-in-Publication Data**
The evolutionary biology of viruses / editor, Stephen S. Morse.
    p.  cm.
   Includes bibliographical references and index.
   ISBN 0–7817–0119–8
   1. Viruses—Evolution.  I. Morse, Stephen S.
QR370.E96   1994
576′.64—dc20                        93–29820
                                       CIP

9 8 7 6 5 4 3 2 1

# Contents

## IV. DRIVING FORCES IN EVOLUTION II: NATURAL SELECTION

# Contributors

**R. Antia**  *Parasite Epidemiology Research Group, Imperial College, London SW7 2BB, United Kingdom*

**Patrick Argos**  *European Molecular Biology Laboratory, Postfach 10.2209, Meyerhofstrasse 1, D-69012 Heidelberg, Germany*

**A. J. Leigh Brown**  *Centre for HIV Research, Division of Biological Sciences, University of Edinburgh, Kings Bldgs, Edinburgh EH9 3JN, Scotland*

**Lin Chao**  *Department of Zoology, University of Maryland, College Park, Maryland 20742*

**Peter de Haan**  *Department of Virology, Wageningen Agricultural University, Binnenhaven 11, 6709 PD, Wageningen, The Netherlands*

**Esteban Domingo**  *Centro de Biología Molecular (CSIC-UAM), Universidad Autónoma de Madrid, Cantoblanco, 28049 Madrid, Spain*

**Thomas H. Eickbush**  *Department of Biology, University of Rochester, Rochester, New York 14627*

**Frank Fenner**  *John Curtin School of Medical Research, Australian National University, Mills Road, Canberra 2600, Australia*

**G. P. Garnett**  *Parasite Epidemiology Research Group, Imperial College, London SW7 2BB, United Kingdom*

**Rob Goldbach**  *Department of Virology, Wageningen Agricultural University, Binnenhaven 11, 6709 PD, Wageningen, The Netherlands*

**Jaap Heringa**  *European Molecular Biology Laboratory, Postfach 10.2209, Meyerhofstrasse 1, D-69012 Heidelberg, Germany*

**John J. Holland**  *Department of Biology and Institute of Molecular Genetics, University of California, San Diego, La Jolla, California 92093-0116*

**Peter J. Kerr**  *CSIRO Division of Wildlife and Ecology, Canberra 2912, Australia*

**Edwin D. Kilbourne**  *Department of Microbiology, New York Medical College, Basic Science Building, Room 315, Valhalla, New York 10595*

**Bette Korber**  *HIV Sequence Database and Analysis Project, Theoretical Division, Los Alamos National Laboratory, Los Alamos, New Mexico 87545*

**Varuni L. Mallampalli** *Department of Entomology and Center for Agricultural Biotechnology, University of Maryland, College Park, Maryland 20742*

**Ernst Mayr** *Museum of Comparative Zoology, Harvard University, 26 Oxford Street, Cambridge, Massachusetts 02138*

**Stephen S. Morse** *The Rockefeller University, 1230 York Avenue, Box 120, New York, New York 10021*

**Gerald Myers** *HIV Sequence Database and Analysis Project, Theoretical Division, Los Alamos National Laboratory, Los Alamos, New Mexico 87545*

**Thomas W. Scott** *Department of Entomology and Center for Agricultural Biotechnology, University of Maryland, College Park, Maryland 20742*

**Simon Wain-Hobson** *Unité de Rétrovirologie Moléculaire, Institut Pasteur, 28 Rue de Dr Roux, 75724 Paris, Cedex 15, France*

**Scott C. Weaver** *Department of Biology (C-0116), University of California, San Diego, La Jolla, California 92093*

# Preface

These are exciting times for the study of viral evolution. The increasing availability of molecular data and powerful molecular techniques, and the development of tools for phylogenetic analysis based on these data, have made it possible for the first time to address a variety of questions of profound evolutionary interest. The paucity of data that once plagued the study of viral evolution is giving way to something approaching a small flood of information.

There is still a need, however, to place this flood of information within a coherent framework. Two principal objectives of *The Evolutionary Biology of Viruses* are to help place viruses within a general evolutionary framework and to encourage more integrative thinking concerning viral evolution. This book provides a current and accessible introduction to viral evolution and evolutionary biology for scientists in virology, the biomedical sciences, and in other biological disciplines, including evolutionary biology. Its purpose is thus to bridge the gap between these disciplines by considering the evolution of viruses from a number of perspectives and attempting to present and integrate the various approaches. Viral genetic variation is well appreciated, but so far has been considered mostly as an end in itself. My personal bias is that most discussions of viral evolution have remained generally outside the "neo-Darwinian" synthesis, and the relationship between genetic variation and natural selection remains largely unexplored with respect to viruses. We need a clearer appreciation of how the two relate to each other, why they are not necessarily antithetical, and where the study of viral evolution might go from there. This book, therefore, was developed and organized largely along thematic lines, with the intention of elucidating basic themes, mechanisms, and principles, using appropriate examples as available, and avoiding an encyclopedic approach in favor of the illustrative. As a result, those viruses that have been more intensively studied receive more attention. No attempt has been made to represent every viral family; their sheer number alone would have made this a difficult task. Although some viruses receive less attention in this book than may be warranted by their importance (and less in some cases than I would have wished), I hope that there is a compensating advantage in breadth of approaches. A number of reviews, papers, and even one or two books dealing with viral variation are now available, as well as many on the molecular biology of individual viruses or viral families. References cited throughout the text should guide the interested reader to these appropriate specialized sources.

Three experiences were largely responsible for shaping this book. I had become interested in questions involving viral evolution as a result of my work on emerging viruses begun a few years earlier. In 1990, I was invited to lecture at the Seventh International School of Biological Sciences, held at Ischia in June 1990 under the auspices of the Naples Zoological Station and organized by Professors Mirko Grmek, Baruch Blumberg, and Bernardino Fantini. I decided to discuss our knowledge of viral evolution and the relationship between viral evolution and emerging viruses (see the closing section of this volume for a discussion). As I was organizing my thoughts for the Ischia lectures, Dr. Jan Geliebter at my own institution (The Rockefeller University) offered a course in Molecular Evolution, which afforded me both a systematic exposure to evolutionary theory and additional opportunities to speak informally and at length with those leaders in the field who had been invited as lecturers. These experiences were indispensable in developing both my views and the plan for this book, and I am grateful to the organizers of these courses for their catalytic role and their assistance. The opening chapter of this book, especially, springs from these experiences. Finally, as this book was nearing completion, I was privileged to attend the EMBO Workshop on Variation and Molecular Evolution of Viruses, organized by Drs. Tom Kirkwood, Charles Bangham, and Esteban Domingo, which allowed me to broaden my knowledge and further refine my thinking (if I have remained unrefined, it is, therefore, not for lack of opportunities for improvement).

It is a natural progression from these expressions of gratitude to additional formal acknowledgments. Any effort of this type must necessarily involve the help and goodwill of many people, beginning with the authors who contributed chapters to this book. I thank the many additional colleagues who have given invaluable assistance in many ways, including Roy Anderson, David Baltimore, Charles Bangham, D. H. L. Bishop, Baruch Blumberg, Andrea Branch, Charles Calisher, Merrill W. Chase, John Coffin, Seymour Cohen, Paul Edelson, Paul Ewald, Niles Eldredge, Mitchell Feigenbaum, Bernard Fields, Walter Fitch, Jan Geliebter, Irwin Gelman, Mirko Grmek, Hidesaburo Hanafusa, Lily Kay, Karla Kirkegaard, T. B. L. Kirkwood, Joshua Lederberg, Bruce Levin, Arnold Levine, Marcella McClure, Robert May, Peter Palese, Maria Ricchetti, Hugh Robertson, Bernard Roizman, Paul Sharp, Robert Shope, Eric Siggia, James Strauss, Howard Temin, David Thaler, Noel Tordo, and Eckhard Wimmer. They have given generously of their time and knowledge, and patiently clarified many uncertain points for me; those errors that remain are the result of my own obduracy.

I owe a great deal of thanks to Dr. Kerry Willis, Senior Science Editor at Raven Press during the preparation of this book, who entered into this project with enthusiasm and as an active collaborator. Her many valuable suggestions and advice earn her a large part in this book, as well as my lasting appreciation. My sincere thanks to Kathleen Lyons, Jill Barnowski, and Glynis Seaton at Raven Press for essential support to a sometimes overwhelmed volume editor. And of course a very special, and personal, word of thanks to my wife, Marilyn Gewirtz, who as always has been my valued coauthor. Her penetrating insights and questions brought clarity to many of my moments of deeper and murkier immersion in the primordial soup of evolution.

I am supported by the National Institutes of Health, U. S. Department of Health and Human Services (RR 03121 and RR 01180), and am grateful to the Comparative Medicine Program for support. Work on emerging viruses and their evolution was supported by the Division of Microbiology and Infectious Diseases (DMID), National Institute of Allergy and Infectious Diseases, National Institutes of Health, and by the Fogarty International Center of NIH. I am grateful to Dr. John R. La Montagne, Director, Division of Microbiology and Infectious Diseases (DMID), National Institute of Allergy and Infectious Diseases, National Institutes of Health, and Dr. Ann Schluederberg, Virology Branch Chief, for their support and encouragement.

*Stephen S. Morse*

# The Evolutionary Biology of Viruses

*The Evolutionary Biology of Viruses*,
edited by Stephen S. Morse.
Raven Press, Ltd., New York © 1994.

# 1

# Toward an Evolutionary Biology of Viruses

## Stephen S. Morse

*The Rockefeller University, New York, New York 10021*

Despite some interest in evolution on the part of virologists, viruses have not generally been considered from the same evolutionary vantage point as other organisms. In this chapter, I wish to consider some of the possible reasons for this, by tracing the historical development of ideas in viral evolution, and reviewing some current trends that are giving new insights into viral evolution. Several of the latter subjects are covered in greater detail in the other chapters of this book, to which I would refer the reader. I will then explore how a theoretical framework for viral evolution could be developed. Finding ways to develop this framework, in fact, is one of the main themes of this book, and will be further elaborated in the chapter by Mayr that follows this one.

In surveying recent work on viral evolution, one sees several trends and major themes: emphasis on viral variation and (especially in the past) speculations concerning viral origins (I will consider origins in the next section, before going on to the other issues). Interestingly, these themes were introduced fairly early, at least a half century ago.

Variation has long been a main theme, in fact the dominant theme, in thinking about viral evolution. As early as 1944, F. M. (later Sir Macfarlane) Burnet noted, "Variability is so characteristic of viruses that it is hard to find a paper which fails to mention some aspect of virus variation" (1). Therefore, it is not surprising that this theme has continued to be important. All major reviews within the past 15 years, including many that I would recommend (2–8), and almost all of the original literature, emphasize variation in the viral genome. This is appropriate, given the unique availability and value of molecular data; what is missing is a biological context or framework in which to interpret this information. Viral variation and viral evolution have essentially become synonymous. Correspondingly, there has also been a major change in the means of discourse, away from biological considerations. Early work, such as that referred to by Burnet, of necessity emphasized phenotype and phenotypic variation (and, therefore, perhaps was more oriented toward the broader biological and functional perspective), whereas more recent work (essentially since the development of molecular evolution in the 1960s, but beginning earlier) has emphasized genotypic variation, with little reference to phenotype and to the role of host or ecological factors, or to evolutionary constraints. In the development of evolutionary thought in virology, emphasis on variation per se thus became divorced from the observation that (as Burnet put it; 1), "The classical example of constancy of a disease type through the centuries is a virus disease." It is the nature of this distinction, and how these two observations might be reconciled, that I wish to address here. I suggest that viruses can be accommodated within the "evolutionary synthesis" of evolutionary biology. Natural selection, in particular, should be introduced into evolutionary thinking about viruses.

## VIRAL ORIGINS AND CLASSIFICATION

Let us consider first the meaning of the term *viral evolution*. In his widely used textbook (9), Futuyma defines evolutionary biology as

> consist[ing] of two principal endeavors: inferring the history of evolution and elucidating its mechanisms. In the study of evolutionary history, as in any historical study, inferences about past events are made from typically incomplete and often misleading data. Sometimes there are historical records (i.e., fossils); quite often there are not, and past events must be inferred from present patterns. . . . Although the historical existence of an event can often be inferred with some confidence, its causes are often much more difficult to elucidate, and frequently must remain a matter for speculation.

### Hypotheses on Viral Origins

However, in the past, viral evolution often meant specifically how viruses originated in the first place, rather than the other questions (such as relatedness and evolutionary mechanisms) defined by Futuyma, and the term still carries that meaning for many. Therefore, before considering these other aspects of evolutionary biology, we will consider viral origins, a question that has understandably fascinated many scientists, despite its highly speculative nature.

Three major hypotheses of viral origins were enumerated by Burnet in his 1944 lectures (1). With some appropriate modifications, they remain the major alternatives. These are the degenerative, or retrograde evolution theory, often called the Green–Laidlaw hypothesis (10,11), summarized by Burnet (1) as the view that "viruses represent the degenerate descendants of larger pathogenic micro-organisms"; viruses as vestiges of an ancient precellular or prebiotic world; and viruses as genetic elements that "escaped" from cells. The last is the most widely accepted hypothesis today.

Some years later, Andrewes, in a series of lectures from the mid-1960s, discusses the same hypotheses in assessing speculations about how viruses originated (12). He also offers there the "Galatea hypothesis," as a variant of the cellular origin theory: "Let us suppose that one of these bacteria . . . produced for its own ends a nucleoprotein. If the nucleoprotein should gain access to another organism and be capable of replication, it would, if not under strict control by the genome of its host, be automatically subject to the laws of natural selection and the debutante virus Galatea would be 'out'." He discusses evidence from bacteria and suggests that "as multicellular organisms appeared, their viruses could have evolved along with them."

As might be expected, the matter is still unresolved, although roughly since the 1970s there has been increasing consensus, largely as the result of molecular data that have become available. The major change between 1940 and now is that the view most widely accepted then, the degenerative, is the only one ruled out by subsequent data and, hence, is out of contention today. The other two hypotheses, which may have seemed far less likely in the 1940s, are more heavily favored today. Viruses as "a foot-loose gene" (Burnet's expression) originating as "pathologically active fragments from the cells of higher forms" is, in an elaborated form, the most widely accepted view today. The third hypothesis (in Burnet's words, "that viruses represent the surviving descendants of primitive precellular forms of life" which became extinct in other settings) has varied in popularity from the 1940s on, although rarely first choice.

The early acceptance of the Green–Laidlaw hypothesis is quite logical in retrospect. It was an attractive hypothesis that seemed in keeping with the unfolding biological knowledge of parasitic organisms. It best fit the data on viruses then available; in fact, as Burnet pointed out in 1944, at that time it was the only view for which any quantity of supporting data was available. It is also logical in terms of historical development of the virus concept. When viruses were first identified, bacteria had recently been recognized as major causes of disease. In the beginning, size was an important criterion for classification. Viruses were primarily defined by their ability to pass certain filters that retained bacteria. Reasoning by analogy, it was easy to envision viruses as being like other microbes that had recently been characterized, only smaller—hence, "degenerate" bacteria. By the same criteria of size and inability to replicate on inanimate culture media, the *Rickettsia* and *Chlamydia* genera were also initially classified with the viruses. These other organisms have clear bacterial affinities, and today it is reasonable to suggest that they were probably descended from bacterial ancestors. However, the retrograde hypothesis became less attractive when it was recognized that *Rickettsia* and *Chlamydia* were quite distinct from viruses, and it was gradually abandoned as more detailed molecular characterization began to elucidate major differences between viruses and bacteria.

One of the alternatives, viruses as prebiotic nucleic acid, was in abeyance for many years. Both Burnet and White (13) and Andrewes raise the valid objection that, as Andrewes put it (14), "Their [viruses'] intracellular parasitism is so much a part of them that their origin from something other than living cells seems difficult to imagine." But the notion of prebiotic origin has been revived in the last few years by recent findings of RNA enzymatic activity and self-splicing (15). Such data led Gilbert (16) to suggest that RNA might have predated DNA as genetic information, and that an "RNA world" may have flourished before DNA rose to dominance

(16,17). If so, it is conceivable that at least some RNA viruses could be vestiges of that RNA world, although this hypothesis is still controversial. It would, however, explain the puzzling findings that RNA genomes, found uniquely among viruses, represent a great variety of genomic types, including RNA types unlike cellular messenger RNA (mRNA; a conceivable alternative is that some of these RNAs could have originated from the "antisense" strand of corresponding DNA, or from RNAs serving functions other than as mRNA). The status of this view at the moment is perhaps best summarized by Holland (18), a leading researcher in viral evolution:

> Thus, today's biosphere is DNA-based, and the only known RNA "life forms" are RNA viruses. These depend on DNA-based host cells for their existence. Whether any existing RNA virus genomes contain sequence vestiges of the earliest RNA life forms is unknown (and probably unknowable), but it is clear that RNA viruses are ubiquitous, extremely successful cellular parasites of considerable disease importance.

The third view of viral origins is the endogenous theory, that viruses originated in some manner from cellular genetic sequences. In one form or another, this view apparently has had a fairly long history. Burnet and White (13) succinctly summed up the situation as it stood in 1971:

> In the early days when we knew virtually nothing of viruses this hypothesis was rather popular and the description of viruses as "footloose genes" was often used. Then it became so unfashionable that in the 1940s it was legitimate to entitle a book *Virus as Organism* and come down firmly on the side of parasitic degeneration. Now that we know a great deal particularly about bacterial viruses the wheel has come nearly full circle.

Writing in the 1960s, Waterson (19) also noted that this view had once again become acceptable. Waterson pointed to latency in bacteriophage as a parallel (the integration of retroviruses was yet to be discovered). Waterson might have been influenced by André Lwoff, who also considered this hypothesis in the context of bacteriophage latency. Reviewing the origins of viruses in 1957, using the prophage form of bacteriophage as an example, Lwoff (20) contrasted the two alternatives then prevalent, "(1) the prophage is the residue of the degradation of a parasitic bacterium or of a more or less primitive organism; (2) the prophage was born by genic or chromosomal mutations of the bacterium which thus became lysogenic." He then suggested a compromise, that the genetic material of *Escherichia coli* and of the prophage originated from the same primitive ancestral genetic material: "the genetic material of the bacteriophage and the genetic material of the bacterium have evolved from a common structure, the genetic material of a primitive bacterium. Whatever the origin of the genetic material of the prophage might have been, we know to-day that the prophage behaves as if it were a bacterial gene."

Stating that the supporters of an endogenous theory of viruses were accused of heresy, Lwoff defends the theory, declaring that it is not absurd. Support for this hypothesis has increased since then with the availability of molecular data for genetic comparisons, and this former "heresy" is the most widely accepted view today. Howard Temin has been the most important current proponent and elaborator of this view for animal viruses. The original formal expression by Temin was his protovirus theory (21), which "suggests that leukemia viruses do not preexist but arise from other elements, protoviruses, by genetic change." Temin later refined and extended the hypothesis, suggesting that retroviruses originated from cellular mobile genetic elements (22). In general terms, the concepts were succinctly stated by Lederberg (23): "the very essence of the virus is its fundamental entanglement with the genetic and metabolic machinery of the host."

For retroviruses, at least, the supporting data are extensive. In addition to cellular and retroviral homologues of oncogenes, numerous endogenous retroviruses (retroviral-like DNA sequences integrated into chromosomal DNA in the host germline) are known in mice, humans, and other mammalian species (24–27). Especially in mice, some of these, from time to time, have given rise to infectious retroviruses. In addition, smaller "retroelements" exist in a wide variety of other species (24,25, 28). The likely relationships among this large extended family of retroelements is addressed in the chapter by Eickbush in this volume. In an interesting twist, retroelements have recently been identified in bacteria (29), perhaps repaying in some sense the debt of animal virology to bacteriophage. Many other mobile genetic elements have been identified, including P (hybrid dysgenesis) elements in *Drosophila* (30). As suggested by the protovirus theory, such genetic elements can be seen as past or potential (prospective) viruses; Andrewes (12) makes a similar observation for bacteriophages.

Although Temin developed his original theory with retroviruses, oncogenes, and cellular genes controlling growth and differentiation, apparent similarities between viral genetic sequences and other types of mobile elements lend additional support to the view that this may be generalizable to most, and possibly all, viruses (31). (It is also my impression that, since the mid-1970s at least, Temin has felt that the cellular origin theory is generally applicable to other viruses.) Although an origin in cellular DNA seems straightforward for retroviruses and DNA viruses, other RNA viruses, which do not undergo a DNA phase in the cell, could be more problematic. Replication of RNA is a function not normally carried out by host cells. Therefore, for many RNA viruses, especially those termed *negative-strand* (viruses for which the genomic RNA is complementary to mRNA, or *antisense*), specialized RNA polymerases or replicases are usually required to copy or to use viral nucleic acid (32). Explaining the origins of such specialized viral polymerases, therefore, is essential to any unified theory of viral origins. Several chapters in this volume, including those by Heringa and Argos and by Goldbach and de Haan, address this important question in various ways. Studies showing similarities among viral nucleic acid polymerases (33–36), and between viral and cellular polymerases (37, 38), support the hypothesis of a common (cellular) origin even for viruses that appear only distantly related. Although one caution is the difficulty of distinguishing between true evolutionary relatedness and convergent evolution, always a complication for proteins that carry out essential functions, the origins of the specialized viral replicases may be explainable, at least in principle. At the moment, viral capsid proteins are harder to explain, as fewer cellular homologies are apparent, but may be susceptible to similar analysis. A few cellular homologues have been identified (39), and it seems likely that most capsid genes will prove to have cellular homologues. Among possible reasons that could be invoked to explain failure to identify cellular homologues for some viral genes are divergent evolution after acquisition of the genes, origins of ancestral viral genes from host genes that have not yet been identified, or origins of ancestral viruses in unstudied organisms that are phylogenetically distant from well-characterized hosts. Andrewes, for example (12), speculated that insects might be key organisms for the origin of viruses.

### Virus Superfamilies and Viral Taxonomy

Did viruses originate monophyletically (from a single ancestor, followed by subsequent evolutionary diversification) or polyphyletically (from more than one)? This has interesting implications; for one thing, the frequency with which new viruses have arisen in the past might allow some estimates of the probability of such events occurring in the future. Unfortunately, the data are now insufficient to resolve this question. The diversity of virus types (40) would appear to indicate a polyphyletic origin, and this is the traditional view. The question is also complicated by the likelihood of gene exchanges between viruses and host cells, and even between different viruses coinfecting a host cell. However, the putative homologies in the polymerases described by Argos and others (33–38), and other genetic similarities that cross conventional virus families could be used to argue for a more limited, even monophyletic, origin of ancestral viruses. As a cautious expression of this possibility, Strauss and Strauss recently suggested (7) that most RNA viruses could well have had a common ancestor, the present diversity of viral families being the result of evolutionary divergence over time.

Viral taxonomy traditionally uses such characteristics as morphology in the electron microscope and the nucleic acid type (41,42). An extension of molecular analysis is the classification of viruses by superfamilies, analogous to superfamily schemes proposed for eukaryotic genes. Originally developed by Goldbach (43; see the chapter by Goldbach and de Haan), and by the Strausses (7), this is an attractive approach, emphasizing shared properties that may also indicate common ancestry. Most of the work on superfamily classifications of viruses have involved RNA viruses, which are more diverse than DNA viruses. These systems use limited nucleic acid homologies, generally in the RNA replicases, and similarities in genome organization to classify many of the RNA viruses into major "superfamilies." In a sense, Baltimore's seminal 1971 grouping (32) by nucleic acid type (the distinction of positive-stranded and negative-stranded RNA viruses was introduced in this paper and is still used today), which bridged conventional taxonomic groupings for viruses, can be seen as an early superfamily arrangement. Current superfamily systems developed from more detailed genomic comparisons. A considerable advantage of a superfamily arrangement is that it makes order out of the apparent chaos of viral relatedness by emphasizing common features. It thus represents a genuine step forward in evolutionary thinking about viruses, especially when combined with formal phylogenetic analysis. It also makes thinking about possible relationships and origins more apparent, as in Strauss' and Goldbach's conjectures that all the RNA viruses could conceivably have been derived from a small number of common ancestors, or possibly even a single ancestor.

## EVOLVING VIEWS OF VIRAL EVOLUTION

Let us now turn to the other aspects of evolution defined by Futuyma (9). In considering the evolutionary biology of viruses, although some virologists such as Burnet and Andrewes showed enthusiasm for applying evolutionary theory to viruses, this enthusiasm was generally not sustained. Instead, virologists and perhaps even some evolutionary biologists have generally not appeared convinced of the relevance of Darwinian-based evolutionary theory to viruses. Possibly, as a consequence, most expositions of viral evolution have emphasized aspects, such as genomic diversity, while deemphasizing biological interpretation, ecological context, and systematic consideration of natural selection.

This situation is not unique to viruses, the division between these approaches to evolution being reflected in evolutionary theory as a whole. However, I believe that this is both more apparent and more severe with viruses. There are several probable reasons, including inherent difficulties in using viruses as material for traditional methods of evolutionary study (one indication of this feeling is that a reference to the lack of viral fossils can be found in many excellent reviews on viral evolution), as well as certain unresolved questions in evolutionary theory. Many of the traditional approaches to evolution were phenotypic, based on the naturalist tradition of field observation. Virologists (and microbiologists) had become increasingly divided from the naturalists in this sphere. As Bawden explained in 1953 (44), "We are in the post-Darwinian era, but our relevant knowledge of viruses is in the pre-Linnaean stage of knowledge on plants and animals. . . . His [Linnaeus'] groupings have survived largely because he had advantages denied to the virus worker" (i.e., gross morphology and anatomy). The debate over the nature of viruses and whether viruses were living could only have made viruses appear still more remote and less relevant to evolutionists. Later, when the fundamental similarities between viruses and "living" organisms were better appreciated and this debate was resolved, the lack of such potentially important biological characteristics as sexual recombination might have continued to set viruses apart for many evolutionists. Even how to define a viral "species"—and whether the concept was even relevant—was the subject of debate (44,45).

For more biologically oriented thinking in evolution, viral evolution has always presented the problem of what to study. Phenotype is difficult to study with viruses, a problem even greater before tissue culture methods were widely available. The number, range, and resolution of phenotypic characters that could be studied were limited. Because the responsible genetic mechanisms were not clearly understood at the time, the basis of the phenotypic expression was empirical, at best. Despite these limitations, some very interesting work was done. Burnet (46–48), and later Kilbourne (49), used characteristics such as host range, neutralization by certain antisera, and heat stability of certain functions to differentiate strains of influenza and to examine their genetic reassortment. Antigenic relatedness has also been widely used with other viruses, notably with arboviruses (arthropod-borne viruses), and still finds useful application (50). There were a few attempts using morphology, but, among other difficulties, morphology is one of the most stable viral characteristics and is rarely useful as an evolutionary marker. One can also determine selection for virulence; although possible in tissue culture, with which much of this work has subsequently been done, such work is very difficult in vivo. Fenner's myxoma study (51; see the chapters by Garnett and Antia and by Fenner and Kerr), starting 1951, is still the classic example and, even today, remains one of the few in which this approach has been possible under anything approximating natural conditions. Alternatively, one could compare sequential isolates from epidemics of (e.g.) influenza, but how should virulence be assessed, and how does one differentiate sequential changes from geographical variation? Similar questions arise if, instead, an observational, naturalist-like approach is adopted (as was done by the arbovirologists), characterizing isolates in nature. With the advent of techniques for molecular evolution, some of these difficulties can be overcome, as attested by some remarkable work along these lines in recent years, including studies of influenza by Palese and coworkers and others (some cited later), of human immunodeficiency virus (HIV), and of canine parvovirus 2 by Parrish and colleagues (52,53); but the difficulties are not trivial.

In sum, there are few viral characteristics available for evolutionary study that can be studied experimentally in any satisfactory way. Those markers that are available require study by methods distinct from those traditionally used in evolutionary biology. The same problem arose, in a different way, in organismal biology. There, evolutionary studies have long been feasible for morphologically similar organisms, with shared characters that can be evaluated. But how to make comparisons involving morphologically different organisms? This was a problem like the one plaguing viral evolution. Of course, molecular biology provided the general solution to this dilemma. Molecular evolution, the field resulting from the introduction of molecular biology into evolution, was abetted by data and techniques and was also possibly shaped by pressure from the burgeoning field of genetics, which insisted on experimental, rather than observational tests. The major, and probably inevitable, consequence was to shift the emphasis to the genome and, hence, to variation in the genome.

The rise of molecular evolution was the most far-reaching development in evolutionary biology since the 1940s, presenting both breakthrough and challenge. Beginning in the 1960s, observations of protein *polymorphisms* (genetic variants in functionally identical proteins) by Zuckerkandl and Pauling (54) later extended to gene sequences at the nucleotide level, presented a challenge for traditional evolutionary biology, since it was not clear how natural selection could account for this variation. As a seminal paper by Kimura pointed out (55), "calculating the rate of evolution in terms of nucleotide substitutions seems to give a value so high that many of the mutations involved must be neutral ones." The neutral theory developed by Kimura (55–57) as the formal expression of this effect in a population became fundamental to the development of molecular evolutionary theory. The neutral theory gives an important role to genetic drift—random mutations—as a mechanism of evolution.

In the past few years, Temin has extended Kimura's neutral theory to both oncogenes and viruses (58,59). Temin has been particularly interested in the idea of mutation-driven evolution, that mutation can drive the evolution of viruses (59). In Temin's words (60),

> the high rate of virus genetic variation allows mutation-driven evolution. . . . A consequence of the high rate of mutation and recombination in retrovirus replication is that many variant viruses will be present in any retrovirus population. Any variant with a relative advantage will increase, and successful variants can become "frozen" as proviruses. In addition, retroviruses have a very high frequency of phenotypic mixing. This phenomenon allows retroviruses to undergo multiple mutations and recombinations before they are subject to selection (analogous to the effect of recessive mutations in diploid organisms).

Kimura and his colleagues have themselves also been interested in the application of the neutral theory to viruses (61,62).

An especially productive line of thought related to the issue of genetic variability involves molecular evolution of viral nucleic acid sequences. As shown by the contributions to this volume, viruses offer some very appropriate material for evolutionary study and some experimental opportunities not otherwise available, including the possibility of generating phylogenies experimentally (63), as was done in an innovative study (64). Although it has always been difficult to devise ways to study evolution in the laboratory, there have been a few experimental approaches that have appeared promising (63,64). Historical landmarks here include Spiegelman's in vitro replication experiments (65) in the 1960s with the RNA-containing bacteriophage Qβ, one of the earliest attempts to conduct molecular evolution experiments in a test tube; Biebricher and Eigen's development of the "quasispecies" concept (66,67); and Domingo's classic demonstration of variants in the Qβ genome (68). As discussed by Wain-Hobson in his chapter, similar heterogeneous populations have recently been demonstrated in strains of HIV isolated from individual patients (69). The introduction of population thinking in considering viral evolution and variation had long been missing in virology and is an important step. The chapter by Domingo and Holland in this volume reviews this subject, and includes many examples. They note that replication of RNA viruses, especially, can have high error rates, leading to potentially rapid change and yielding, from a single parent, populations of genomes containing many variants centered around a consensus sequence.

## MOLECULAR PHYLOGENETICS AND MECHANISMS OF GENOMIC EVOLUTION

The development of numerical methods, beginning in the late 1960s and early 1970s, for constructing phylogenetic trees with use of molecular sequence comparisons, is one of the most powerful and valuable applications of molecular evolutionary theory (70,71; see chapter by Leigh Brown). Suggestions as to how rates of amino acid or nucleotide substitutions can serve as a molecular clock (54–56) provided an important part of the rationale, although as Felsenstein points out (71), for the purpose of constructing phylogenies "it does not matter whether nucleotide substitutions are neutral or selective." These approaches have been very useful for elucidating relationships among viruses. Successes include valuable insights into the evolution of influenza viruses (72–74) and of retroviruses (24,25,75), including HIV and related viruses (76–80); the latter is analyzed in detail in the chapter by Myers and Korber. The superfamily classification of viruses mentioned earlier can be considered another extension of this approach.

Phylogenetic analysis is based on the comparison of molecular sequences, with the assumption that genes diverge in sequence as they evolve from a common ancestor (see chapter by Leigh Brown). Viral genomes, like other genomes, evolve through such genetic mechanisms as mutation, recombination, and gene duplication (Table 1). The mechanisms for DNA viruses presumably resemble those available to a host cell, and DNA viruses that integrate into host DNA are afforded similar (and possibly fairly frequent) opportunities for recombination with the host genome, allowing the possible capture of new genes from the host or from viral elements previously integrated into the host DNA. For retroviruses, the DNA proviral step in vial replication allows similar opportunities for recombination with host cellular genes and introduction of new sequences into the retrovirus genome (81). With other RNA viruses, these possibilities may seem somewhat more limited, and it was once thought that RNA recombination did not occur. However, RNA recombination (reviewed in 82) has been demonstrated both in the laboratory (83,84) and in vivo (85,86). Recombination by *copy-choice*, in which the RNA polymerase switches templates during replication and incorporates portions of another RNA into the nascent RNA product, has been described in both poliovirus (87,88) and in retroviral RNA (89,90). Such events, as well as recently identified homologies between the esterase protein of influenza C, a negative-stranded RNA virus, and coronaviruses

**TABLE 1.** *Examples of mechanisms in genomic evolution of viruses*

| Mechanism | Example |
| --- | --- |
| Base changes (point mutations) | Genetic drift in influenza |
| Biased hypermutation | Measles virus (U→C) |
| Gene reassortment | Origin of pandemic influenza viruses of 1957, 1968 (surface protein gene(s) from avian virus) |
| Deletions | Precore mutants of hepatitis B (associated with chronic hepatitis infection) |
| Intramolecular recombination | Insertion of gene cassette; WEE (western equine encephalomyelitis) as recombinant of eastern equine (EEE) and Sindbis-like parents |
| Recombination with host gene | Cellular protooncogenes and retroviral oncogenes |
| Recombination between deletion mutants | Regeneration of functional plant virus genome (also reported with retroviruses, ref. 149) |

Adapted, with modification, from Kilbourne, ref. 150.

(positive-stranded) (91) suggest that both gene homologies and genome organization may be more fluid than often believed. This notion of interchangeable functions, swapped among viruses, has been stated more formally as the *modular* scheme of viral evolution (92). With segmented genomes, such as influenza, reassortment of genes when two strains simultaneously infect a single host cell is also an important mechanism for generating new virus strains (74). Some implications of reassortment are as discussed in the chapter by Chao.

## THE NEED FOR SYNTHESIS

With all this salutary explosion of data and thinking concerning viral evolution, are there areas still in need of further development? In my opinion, there is still a great need to integrate the information into a broader theoretical framework. What has been missing is a causal relation: virology still remains outside the paradigm of the *evolutionary synthesis*, or the *synthetic theory* of evolution. It is my conviction that much of the framework exists, in the form of the evolutionary synthesis; integrating natural selection with the consideration of viral variation that is already so well advanced would go a long way toward providing the needed framework. The availability of molecular data makes this an opportune time to build that framework.

Why has evolutionary thinking in virology largely remained outside the evolutionary synthesis? Several reasons for this have already been suggested in this chapter. In essence, the situation with viral evolution is not fundamentally different from the situation within evolutionary biology in general, but the difficulty is compounded for virology by the difficulties of studying viruses in nature and the historical development of virology and microbiology.

Different and diverging traditions of genetics—later to become the cornerstone of today's molecular biology—and evolutionary theory may also have added to this separation. Lewontin (93) noted that variation is central to both genetics and to Darwinism: "For Darwin, evolution was the conversion of the variation among the individuals within an interbreeding group into variation between groups in space and time." He adds, "in striking contrast [to Galton], Mendel placed his emphasis on the *variations* among the offspring, rather than on any average description of them, *and derived his laws from the nature of the variations*. Thus, for Mendelism, as for Darwinism, the fact of variation and its nature was central and essential." But modern founders of these fields did not necessarily see this relation in their views. To Mayr, the split between geneticists and naturalists was so central that he entitled a recent chapter on the evolutionary synthesis "Geneticists and Naturalists Reach a Consensus: The Second Darwinian Revolution" (94). As Mayr notes (95), perhaps somewhat acerbically,

> T.H. Morgan, who showed his lack of understanding of natural selection even in his last book on evolution in 1932 . . . claimed in 1910 that finalism had entered biology through natural selection because "by picking out the new variation . . . purpose enters in as a factor, for selection had an end in view", completely ignoring the randomness of variation and the statistical nature of the selection process.

The tension, and perhaps at times mistrust, between these two approaches, is conveyed in another statement by Mayr (96):

> the emphasis on the role of diversity in evolution was stressed by naturalists from Darwin on, but was almost totally ignored by the Fisherians; the naturalists, for their part, rejected the beanbag genetics of the reductionists and during the post-synthesis period continued their holistic tradition of emphasizing the individual as the target of selection.

This feeling may have been exacerbated by the sort of holistic, historical and comparative arguments used by naturalists, which ran counter to the dicta of well-known geneticists such as Morgan that only experimental results should be considered (97). [I hasten to add that many of Morgan's values are scientific virtues and Morgan, more than anyone, stood for moderation and reason; but reading the literature gives one the impression at times that their different modes of discourse kept these two groups—evolutionary biologists in the "naturalist" tradition and geneticists—quite separate.] As Mayr pointed out (e.g., in the foregoing quote), this continued in many cases even after the synthesis. I would submit that it continues today. Similarly, Mayr, in recently discussing "the evolutionary synthesis as unfinished business," noted (98)

> it must be admitted that, even though the refutation of the major anti-Darwinian theories during the synthesis drastically narrowed down the variation of evolutionary theory, some well defined differences among the Darwinians still existed into the post-synthesis period, and some of these differences are still with us fifty years later,

although he also adds, "by no means are all current intra-Darwinian controversies remnants of the old geneticist-versus-naturalist feud."

Possibly compounding this was that virology began its real development just when the evolutionary synthesis, representing the unification of several branches of evolutionary theory, itself was developing. As a result, evolutionary theory appeared fragmented and able to offer little help at a time when virologists might otherwise have turned to it.

## THE EVOLUTIONARY SYNTHESIS

The recent history of evolutionary theory itself offers interesting parallels with these developments in viral evolution. As Mayr noted (96),

> Some fields enjoy considerable consensus among their active workers, other fields are split into several camps of specialists furiously feuding with each other. This latter description applies well to evolutionary biology between 1859 and about 1935. In 1930 there seemed to be no hope of any consensus; yet a consensus was achieved to a large extent within a dozen years (1936–1947).

The result was the evolutionary synthesis, or the synthetic theory.

The nature of this synthesis, again in the words of Ernst Mayr, one of its chief architects (96), "was not one of great innovations but rather of mutual education.

Naturalists who had not known it before learned from the geneticists that . . . [t]here can be no . . . inheritance of acquired characters. . . . Another finding of genetics, its Mendelian (particulate) character, was also finally universally adopted." Mayr writes (96) that the education of naturalists and geneticists was mutual: "The claim, frequently made, that the evolutionary synthesis was nothing but the application of Mendelian inheritance to evolutionary biology overlooks how much the geneticists had to learn from the naturalists about the importance of population thinking, of the geographical dimension, and of the individual as the unit of selection. . . ." (In another section in the same volume, Mayr suggests that many geneticists assume the gene, erroneously in his view, to be the target of selection.)

Two major points were to accept that small mutations could lead to major changes and, thereby, eventually to macroevolution (perhaps ironically, at the time, Morgan's work was considered some of the strongest evidence in support of this) and to reaffirm, again in Mayr's words (96), "the Darwinian formulation that all adaptive evolutionary change is due to the directing force of natural selection on abundantly available variation."

There have been several challenges to the synthetic theory since then, including Kimura's neutral theory of mutation (discussed earlier) and punctuated equilibrium (99). But it is remarkable that surprisingly few of the former radicals feel today that the synthesis needs to be scrapped totally. In recent conversations several, including Niles Eldredge and Gabriel Dover, emphasized their general agreement with the broad outlines of the "synthesis," especially natural selection. Maynard Smith has stated that, by his own interpretation of Kimura's work, Kimura is not an anti-Darwinist (100). Why have the angry Young Turks become rather more sedate older Ottomans? I can only speculate, but the history of the synthetic theory in the recent past has been a gradual adjustment of the theory to accommodate the successive challenges; each challenge was eventually fit into the synthesis by slightly modifying the emphasis of the orthodoxy.

It was perhaps not always thus. Kimura recently alluded to the chilly reception he felt the neutral theory (or at least the importance of mutational drift) received from Sewall Wright, whom he calls his idol, as well as from other architects of the synthesis (57). Mayr (96) tries to excuse this, in disagreeing with Stephen Jay Gould's similar perception that there had been a hardening of the synthesis after the 1930s:

> It is understandable that in the early stages of the synthesis the universal presence of natural selection should have been emphasized strongly, since a considerable number of Lamarckians still existed among the older evolutionists. However, as soon as this stage had been overcome, one could observe a trend that was exactly the opposite of the one claimed by Gould. More and more authors pointed out the existence of stochastic processes and all sorts of constraints that forever prevent the achievement of "perfection."

Elsewhere (94), Mayr gives an additional reason: "One major reason for the heat of the argument between neutralists and their opponents was that they had a different interpretation of the target of selection. The neutralists are reductionists, and for them the gene—more precisely the base pair—is the target of selection." Maynard Smith (100) appears to interpret Kimura's views similarly.

As discussed in the foregoing, the geneticists were variable in their acceptance of natural selection as a driving force. In the United States, T. H. Morgan, among other, apparently did not accept it, and his influence may have been, at least partly, responsible for the failure of the synthesis to penetrate disciplines then forming around molecular genetics; these disciplines would have included microbiology and virology. As Mayr comments (96), "The classical Mendelians had no use for selection, and Morgan's thinking was very much in that tradition."

As indicated by the earlier quotation from Futuyma, despite an acceptance of natural selection in the evolutionary synthesis, experimental definition of causal factors in evolution has long been problematic. Although natural selection attempts a causal explanation, the traditions of genetics place a far greater emphasis on chance than did the naturalist tradition which most evolutionists represented. Levins and Lewontin (101) note that chance plays two important roles in genetics:

> Ideas of chance play an important role in two aspects of genetics. First, the laws of Mendel and Morgan are couched in terms of probabilities. Given the genotype of the parents, it is not possible to predict the genotype of an offspring exactly, but only to describe the distribution of genotypes in a hypothetical, infinitely large, ensemble of offspring. Some genotypes can be excluded, but in general there is no certainty about which of the possible genotypes an offspring will have. For characters of size, shape, behavior, and so on, this uncertainty is further compounded by the variable relationship between genotype and phenotype. Second, mutation is said to be random, by which is meant that mutagenic agents, like X-rays, do not produce a single kind of mutational change in every treated individual, but rather a variety of possible mutations with different frequencies. The same uncertainty exists with respect to so-called "spontaneous" mutations, which appear unpredictably in individuals and are of many different types.

This emphasizes chance effects of variation, at the expense of more apparently deterministic approaches. Mayr (95) sees the views expressed by Jacques Monod in his book *Chance and Necessity* as a continuation of this view:

> Although Monod is adamantly opposed to determinism, including the invoking of final causes, he totally ignores natural selection as a creative process and ascribes all evolution to pure chance. . . . Except for rare mutations, there is a total invariance of the genotype. This is why for Monod evolution is an "intensely conservative system." Chance determines mutation, favorable mutations are preserved, unfavorable ones are rejected. Since for him evolutionary change is thus entirely due to chance mutation, Monod finds the harmony of nature completely inexplicable.

Mayr also speaks of the typological nature of early geneticists. It may be an overstatement, but one could also see these as two separate typological traditions, and the synthesis as an incomplete reconciling of these traditions. Therefore, when the molecular aspects became dominant both in biomedical science and in the shift of evolutionary studies to molecular evolution, the unresolved differences remained to cause confusion.

## VIRAL ECOLOGY

One tradition of virology that did emphasize interactions between virus and host was the ecological. The ecological line of thought, as expressed in Theobald Smith's pioneering work, *Parasitism and Disease* (102) and Burnet's *Virus as Organism* (1), and as continued by such intellectual kin and descendants as Richard Shope, Frank Fenner, and Sir Christopher Andrewes, tended to encourage viewing viruses as biological entities functioning within an ecological and biological context. One notable early success of the ecological approach was Shope's work on swine influenza. This approach was closer in some respects to the naturalist field tradition that was historically dominant in evolutionary biology. Therefore, the fate of viral ecology as a viable approach to studying viruses would seem to be connected to the emphases chosen in viral evolutionary thinking, and it seems appropriate to digress briefly to ask what became of this tradition.

The ecological tradition might be placed in contrast with the more biochemical approach advocated by Lwoff and others. The view expressed in Lwoff's seminal paper "The Concept of Virus" (20) encouraged viewing viruses as legitimate entities for study in and of themselves, and emphasized the genetic component, concepts essential to our modern understanding of viruses. As knowledge of genetics expanded, and when the nature of the genetic material was finally resolved, this view allowed viruses to be connected with the fruitful new discoveries in genetics. On the other hand, such a picture of viruses, emphasizing their unique biochemical and genetic aspects, tended to relegate the host to a role more as a substrate for viral growth (however essential) than as an active participant. Any dynamic element would be on the part of the virus or phage.

Thus, from being highly influential, the ecological line of thought has almost disappeared since then, although still surviving in field studies, and in an advanced form in the modeling studies of Robert May and Roy Anderson. The disappearance was comparatively rapid: as recently as 1967, Andrewes (14) speaks hopefully of future advances in viral ecology. Several virologists attribute the rapid decline of viral ecology to shifts in funding priorities in the late 1960s and early 1970s, in favor of molecular approaches. Some important work continued (e.g., at the Yale Arbovirus Research Unit), but continuing work using this approach represented a markedly decreased proportion of the total research effort in virology and related sciences. An example often cited is the Rockefeller Foundation Virus Program, a strong impetus for the ecological tradition as well as one of its exemplars until it was phased out in the late 1960s and early 1970s (103,104). The shifting of funds from ecological to molecular approaches resulted in a similar shift of students. With no new generations of ecologically inclined virologists to follow, interest waned.

There were also intellectual causes underlying some of these shifts in emphasis. Historically, the ecological approach and the molecular view were not always compatible. For example, Fenner (105) notes that Burnet "was very reluctant to accept the 'ultimate reductionism' of DNA." Molecular approaches were becoming readily feasible, and demonstrations of their power compelling. The ability to accumulate meaningful experimental data with relative facility was an overwhelming advantage. Thus, some of the shift was inevitable, paralleling similar events in biomedical science as a whole, even in evolutionary biology itself. The comparative difficulty of studying viruses by other means only increased the desirability of this useful approach. By contrast, the ecological approach has been denigrated as "stamp collecting," and the inability to predict which viral isolates had the greatest human disease potential would have made the effort seem amorphous.

## DRIVING FORCES IN VIRAL EVOLUTION: DO VIRUSES OBEY NATURAL SELECTION?

Consideration of viral variation leads us to a central issue. Granted that enormous variation in the genome is the rule with viruses, is this fundamentally different from other organisms; that is, does it lead to fundamentally different consequences? Is such variation, including neutral mutation, necessarily incompatible with natural selection as a directing force, or at least as a key stabilizing force, in viral evolution? Remarkably, in a different context and for rather different reasons, variation was once felt to be a problem for the synthetic theory, but a resolution linking variation and selection apparently proved possible within the theory. As Mayr notes in his chapter,

> natural selection proper is only the second stage of a two-step process. The first step consists of the production of variation in every generation, that is, of suitable genetic or phenotypic variants that can serve as the material of selection, and this will then be exposed to the process of selection. This first step of variation is completely independent of the actual selection process, and yet selection would not be possible without the continuous restoration of variability. Several recent critics have failed to understand the relation of the two steps of the selection process to each other. When an author asks (as several have actually done) "is evolution due to molecular processes or due to selection?" it amounts to asking: "Is evolution a change due to step one or to step two of natural selection?" This question is meaningless since the second step, selection *sensu stricto*, deals with the previously produced variation (*a posteriori*) and is not a process which itself produces variation.

Although Kimura (57) differs in some of the particulars and has recently proposed a somewhat more complex model for evolution (the four-step model), he does not reject a key role for selection. In his model, a key initial step is the relaxation of an existing constraint. Relative to constraints, Kimura (57) observes, "Note that stabilizing selection is the most prevalent type of selection in nature."

Is natural selection operative in viruses, or does the rapid variation of viral genomes preclude this? Finding suitable evidence to answer the question of whether and how selection acts in viruses is complicated by the difficulties of studying viruses under natural conditions and the limitations of demonstrating natural selection in nature. As Felsenstein suggests (71), "The controversies between neutralists and selectionists have continued for 20 years with no clear resolution, primarily due to the low resolving power of the data—natural selection many orders of magnitude weaker than we can detect in the laboratory can be effective in nature." He also notes, "Theories explaining evolutionary change and polymorphism by natural selection have been less well developed [than the neutral theory], partly because there are so many different kinds of selection that it is difficult to choose between them." However, one might look for evidence to decide whether the host influences stability of viral genotype or phenotype. In these terms, I think there is good evidence that natural selection does operate in viruses and would appear to be as important as in more familiar organisms.

A few examples must suffice here; further discussion and examples can be found in the chapter by Kilbourne. The introduction of myxoma virus in Australia, in which a highly lethal strain of virus evolved toward intermediate virulence, is one piece of evidence strongly suggesting that natural selection is stabilizing viral variation. As pointed out by Levin and Pimentel (106) and by Robert May, the level of virulence eventually attained represented a trade-off between virulence and transmissibility; myxoma is readily transmissible, and the level of virulence was relatively high. This has also been noted by Ewald (107). Similarly suggestive is a report of repeated independent isolation from pigs of the identical pig-adapted influenza variant after inoculation of a different variant that was not pig-adapted (108,109). A further indication of strong constraint in influenza, not predictable from gross genetic composition of the virus, comes from work on the ability of various avian influenza A recombinants to replicate in primates, in which several gene combinations that appear as if they ought to replicate adequately in primates are unable to do so (110). In different cases, host range restriction can be caused by various gene combinations that serve in some fashion to attenuate the virus in the new host (110) or occasionally, by the nonfunctioning of an essential gene product in the new host (111).

Despite the great variability and high-mutation rates of virus populations, virus strains in nature can show remarkable relative stability over long periods. One could speculate that this may correspond to a "punctuated equilibrium" proposed by Eldredge and Gould for other organisms (99). In addition to mosquito-borne viruses (112,113), discussed in Chapter 15 by Scott and coauthors, many examples are known. Strains of human T-cell leukemia (T lymphotropic) virus (HTLV), an infective retrovirus, have been characterized extensively in recent years, and several have been described in different populations (114–119). Although there might be some biological differences among these strains, those that have now been characterized appear to be primarily geographic variants, mostly suggestive of past diversification (120–122); they appear relatively static now, so much so that they have been used as population markers to trace migrations (115,116,118). Papillomaviruses are DNA viruses found in many vertebrate species (123). Human papillomaviruses come in numerous varieties (see the chapter by Zhu et al.), but are nevertheless also stable enough to be followed as population markers (124). Several studies show different rates of variation of influenza genes in avian and human hosts (73,74,125,126), with relative stasis in avian hosts and rapid evolution in humans (presumably on introduction). Taken together, such evidence indicates that there are factors strongly stabilizing the viral phenotype and even the viral genome.

The best candidate for this stabilizing force is natural selection. This has been suggested (74) as the explanation for the different rates of variation of influenza genes in avian and human hosts (73,125,126), as well as for the observation that, within a single host species, different viral genes evolve at differing rates (74). One can debate the precise roles of natural selection and neutral mutation, as two groups have recently, with Fitch and associates suggesting that the appearance of influenza A variants is driven by immune selection in a Darwinian manner (127) and Kimura's group favoring random fixation of neutral or near neutral mutations (62). And there may possibly be additional forces stabilizing the viral genome. But these specific uncertainties do not obviate a central role for natural selection.

It might be asked, "What is the phenotype of a virus, and where does selection act?" (Mayr also considers this further in Chapter 2.) The requirement for viability (ability to infect or to replicate) in a host would seem to impose important constraints. One would expect viruses to be subject to selection as phenotype is expressed in progeny produced after each round of replication (with selection acting as those individual progeny compete for new host cells, for example, or new hosts). In animals, viral antigens expressed on infected cells also represent targets for the immune system, and hence would appear subject to selection. Within the cell, gene sequences for products carrying out essential functions also would be constrained by selection, as are any sequences (such as promoters or binding elements) that, even if not expressed, must interact with other molecules or structures. Some of the requirements for function could be overcome by complementation or phenotypic mixing, or during quiescent periods when viral genes are not expressed (e.g., in proviruses), but with obvious limitations. One might expect selection whenever the viral function or product is expressed.

This does not mean that viral variation is not of real or of potential biological importance. As Mayr suggests, this variability also offers a pool of raw material for natural selection. Such data as the increased rate of influenza nucleoprotein gene evolution in humans (compared with avian hosts) and the rapid diversification of HIV (see chapters by Myers and Korber, and by Wain-Hobson) suggest that a virus may undergo rapid variation when introduced into a new host species, perhaps corresponding to the "relaxation of constraints" in Kimura's model. As influenza demonstrates, variants can and do evolve, sometimes with consequences of great biological significance. It also shows that the nature of disease caused by such variants is only rarely very different from that caused by the parental virus. Table 2 lists some of the novel variants that have been identified in nature. The list is not exhaustive, although I have made it as inclusive as I could. I have omitted the poliolike syndromes described in China and South America a few years ago (128), as it is not yet clear whether this is a newly evolved virus or just newly recognized. Most viruses on this list cause diseases typical of their viral families or similar to the parental virus. In both humans and horses, the recombinant western equine encephalomyelitis (WEE), for example, causes a disease similar to eastern equine encephalomyelitis (EEE), but somewhat milder (its apparent evolutionary advan-

**TABLE 2.** *Known or suspected newly evolved viruses*

| Virus | Virus family | Remarks | Disease |
|---|---|---|---|
| Rocio encephalitis (Brazil) | *Flaviviridae* | ?Recombinant | H[a] |
| WEE (western equine encephalomyelitis, US) | *Togaviridae* | Recombinant | H |
| Influenza H5 mutant (chickens, Pennsylvania, 1983) | *Myxoviridae* | New variant | Severe respiratory infection in chickens |
| Influenza H7 (seals, US, 1980) | *Myxoviridae* | | H: Conjunctivitis (upon introduction into humans from seals) |
| Enterovirus 70 | *Picornaviridae* | ?New strain | H: Conjunctivitis |
| Rev-T (strain of avian reticuloendotheliosis virus) | *Retroviridae* | Avian | Fulminant lymphoma in fowl |
| Friend virus, spleen focus-forming strains | *Retroviridae* | Mouse | |
| Canine parvovirus 2 | *Parvoviridae* | | Enteritis, cardiomyopathy in dogs (similar to parvoviruses infecting other species) |

[a]H, Associated with human disease

tage is that WEE has insect and bird hosts different from those of EEE). Lest we become complacent, however, there are less-consoling examples, including an avian influenza strain that became highly virulent in chickens as the result of a single mutation (129). Temin offers the example of an avian retrovirus that acquired greater virulence as a result of several accumulated genetic changes (59).

Several examples of variants with distinct properties arising during natural infection have also been described. Two groups (130,131) recently reported variants in mice infected at birth with lymphocytic choriomeningitis virus (an arenavirus). In this infection, lymphotropic and neurotropic variants of the virus can be identified. Each has a different phenotype, the neurotropic form causing acute, and the lymphotropic chronic, infections. A single amino acid change in the viral glycoprotein (from phenylalanine to leucine at residue 260) was correlated with this altered tropism; both variants could be isolated from many of the same carrier mice that had initially been infected with an acute strain of the virus (131). Thus, some variants with differing tissue tropisms can be internally generated during infection, a situation that may occur with other infections as well. There is some evidence that something similar may happen with HIV (69,132) (see the chapter by Wain-Hobson), for example, although some authors have also suggested strong constraints (132). In what could be an analogous situation, it was recently suggested that chronic fulminant hepatitis B infection was associated with a mutation, generated during infection, in a viral gene for precore protein (133,134). A cause of greater concern, vaccine escape mutants of hepatitis B were also recently described in infected individuals (135). Despite many demonstrations of this phenomenon in vitro, this is one of the few documented vaccine escape mutants isolated from the field. This is a cause for concern, but its relative rarity so far also may indirectly support the argument that, relatively speaking, viral phenotype is not as unstable as usually supposed.

On the other hand, it is reasonable to expect that viral evolution probably can serve to increase adaptedness after introduction of a new virus to a host, as in myxoma. And, rarely, novelties can occur—rare probably for the same reasons as in other organisms (such as constraints), despite the high evolutionary rates of most viruses, which might, however, allow some of these events to occur on shorter time scales (centuries to millennia, rather than millennia to millions of years).

## CONCLUSIONS

Evolutionary biologists have rarely considered viruses and, in turn, virologists have generally not benefited from developments in evolutionary biology. As Ewald notes (136), "The health sciences have devoted relatively little attention to the evolution of virulence presumably because of misunderstandings about the levels at which natural selection acts." In this sense, rather than complementing each other, virology and evolutionary biology appear to have taken separate paths. Other fields of microbiology, until recently, seems to have had a similar fate. In general, as I have discussed, possible reasons for this may include the nature of the historical development of microbiology and virology, especially in its medically oriented aspects, as distinct from other biological disciplines, and its affinities with genetics, all of which served to separate it from the field naturalist tradition of biology.

By contrast, variation in viruses has always been a theme. This was undoubtedly strengthened by the close associations between virology and the early development of molecular biology, including historical associations. (The title of the classic *Phage and the Origins of Molecular Biology*, edited by John Cairns, Gunther S. Stent, and James D. Watson, and published by Cold Spring Harbor Laboratory in 1966, is an example.) Molecular approaches were readily embraced as they became available and, therefore, most recent work on viral evolution has emphasized molecular comparisons. Techniques "imported" into virology from evolutionary biology have generally been molecular tools, for example, phylogenetic *tree building* (analysis of ancestry and relatedness) based on comparisons of nucleotide or amino acid sequences. Kimura's theory of neutral mutation has generally found sympathy among virologists and molecular biologists. An important outgrowth of these approaches is the recognition (most notably by Manfred Eigen, John Holland, Esteban Domingo, and others; see the chapter by Domingo and Holland) that many viral stocks represent a population of genomes with considerable variation centered around a "population mean" represented by a master, or average, set of gene sequences. This seems more in accord with the general emphasis on population thinking in evolutionary biology, although a formal population genetics for viruses, with mathematical treatment of gene distributions, has yet to be developed.

This has all been tremendously exciting and productive. On the other hand, one consequence of the molecular approaches in viral evolution is an emphasis on variation in the genotype, with less clarity about what this variation means. Although this concept is a logical consequence of the types of data available, and not unique to virology or even microbiology, this has led to viruses, and their evolution, being considered in an almost stochastic manner. Other aspects of evolutionary theory, and especially considerations of natural selection and of evolutionary constraints, have not been widely applied to viruses, and an evolutionary framework for virology has long been lacking. This becomes apparent in thinking about the causes of "emerging" viruses (see the last chapter of this volume), which have often been treated on an *ad hoc* basis. Consideration of emerging viruses actually supports the contention that there are strong constraints on the evolution of truly new viruses.

The application of natural selection to viruses has long been neglected. We may feel uncertain about such matters as whether mutations are neutral or selective, the level at which selection operates, or even (for some) whether natural selection is the only or main force driving evolution or whether it operates after the fact. But these questions should not blind us to the evidence for evolutionary constraints operating at the level of the virus–host interaction and the maintenance of the virus in nature. To survive, viruses must be maintained in nature in a living host. There are constraints imposed by the requirements for a means of transmission and the relatively few routes by which a virus can infect a host. Such requirements must impose strong selective pressures on a virus. Therefore, although variants are continually being generated, presumably a stabilizing influence is exerted by natural selection as the virus replicates in its natural hosts. Even if viral mutation is unpredictable at present, genotypic variation and phenotypic change are not equivalent, and there remain evolutionary constraints at the phenotypic level at least.

Until the development of molecular methods for evolutionary comparisons, there had been considerable thought about viral and microbial evolution, but little data. All the data then possible were based on rather gross phenotypic characters, and influence of genotype could not readily be studied. Since the development of molecular evolutionary methods, large amounts of comparative molecular data have been assembled. More recent studies have consequently emphasized genotype, but with little reference to phenotype.

There is perhaps a little historical irony in the way viral evolutionary thinking has developed. Earlier, Burnet and others speculated that viral evolution might be similar to that of other organisms, but too little data and too few experimental approaches were then available. Development of the neutral theory and molecular evolution made it possible to collect more relevant data, but the data collected (from more conventional organisms) also forced some general reexaminations of evolutionary theory. So Burnet's suggestion of similarity has been borne out (perhaps in an unanticipated way, inasmuch as both viruses and organisms show comparatively high mutation rates, albeit higher in viruses), but the framework of evolutionary biology has itself been undergoing adjustment. This debate is still ongoing, although it would appear that a resolution is possible within the overall framework of the evolutionary synthesis. In this sense, viral evolution and evolutionary biology may be closer than they have been in a long time. With the current availability of techniques for studying molecular evolution, and greater quantities of comparative data, it would now seem an opportune time to combine the two at last.

This is also an opportune time to reconcile genotype and phenotype, and to better understand the role of variation in a broader context. Although it is still not generally possible to predict biologic characteristics from genotype, increased understanding is gradually emerging and should be strongly encouraged. Approaches for defining evolutionary constraints in nature, and understanding the molecular bases of host range, virulence, pathogenesis (137,138), and host interactions should be emphasized. Methods for studying these evolutionary events in vivo are just beginning to develop, including comparative approaches using molecular phylogenetic analysis (139,140), and one can hope for increasing depth in understanding ecological interactions and coevolution between viruses and their hosts. These subjects are also just beginning to develop in other fields of evolutionary biology (141,142), and virology could make meaningful contributions here. From such examples as have already come to light, such as myxoma in Australia and the Lake Casitas wild mouse (in which infection by a murine retrovirus selected for mice that were resistant by virtue of containing a truncated integrated retroviral element that prevented infection by the corresponding exogenous virus; 143), there will be some fascinating stories to tell. In a complementary fashion, some recent work, notably by May, Anderson, and colleagues (144,145) and by Ewald (107,146), considers the effects of natural selection on virulence and transmission and offers approaches for modeling such effects (see the chapter by Garnett and Antia).

The development of phylogenetic analyses for viruses from molecular data has shed considerable light and is likely to remain highly productive. Newer techniques of molecular biology, such as the polymerase chain reaction (PCR), have made it even easier to obtain data. From this, a very promising line of molecular epidemiology is developing, comparing strains over time and from different geographic regions (72–74,76,78–80), and tracking viral spread (147). The chapter by Myers and Korber is an exposition by one of the originators and chief proponents of molecular epidemiology. Related to this, further appreciation of geographic variants from the standpoint of evolutionary theory (potentially analogous to geographic variants of animals developing in isolation, as in allopatric speciation for example) has begun (50,115,118,120,121,123,124), and one hopes it will develop further.

Placing viruses in a broader framework of evolutionary biology and ecology, as now seems possible, would benefit both virology and evolutionary biology as a whole. John Maynard Smith stated it succinctly (100): "Of course we need to see further than Darwin, but we shall do so by standing on his shoulders, not by turning our backs on him." H. J. Muller declared (148), "One hundred years without Darwinism are enough." It is time for viruses to join the fold.

## ACKNOWLEDGMENTS

This chapter originated in lectures at the Seventh International School of Biological Sciences, held at Ischia in June 1990 under the auspices of the Stazione Zoologica "Anton Dohrn," Naples. I am most grateful to the organizers, Professors Mirko Grmek, Baruch Blumberg, and Bernardino Fantini, and to my fellow course participants for their stimulating dialogue and questions. An earlier version of this chapter appeared under the title, "Evolving Views of Viral Evolution: Towards an Evolutionary Biology of Viruses," in *History and Philosophy of the Life Sciences* 1992;14:267–300. I thank Ernst Mayr and George Pieczenik for helpful comments, Edwin D. Kilbourne for permission to reprint as Table 1 a slightly modified version of a table from his article "New Viruses and New Disease: Mutation, Evolution and Ecology," *Current Opinion in Immunology* 1991;3:518–524; Malcolm D. Martin, NIH, for the example of Friend virus in Table 2; and Abigail L. Smith, Yale University, for the reference on the Lake Casitas mouse.

I am supported by the National Institutes of Health, US Dept. of Health and Human Services (RR 03121 and RR 01180). Work on emerging viruses was additionally supported by the Division of Microbiology and Infectious Diseases (DMID), National Institute of Allergy and Infectious Diseases, National Institutes of Health. I am grateful to Dr. John R. La Montagne, Director, DMID, and Dr. Ann Schluederberg, Virology Branch Chief, for their support and encouragement.

## REFERENCES

1. Burnet FM.*Virus as organism. Evolutionary and ecological aspects of some human viral diseases* [Dunham Lectures, Harvard University, 1944]. Cambridge, Mass: Harvard University Press; 1945.
2. Gibbs A. Molecular evolution of viruses: "trees," "clocks" and "modules." *J Cell Sci* 1987;Suppl 7:319–337.
3. Reanney DC. Evolutionary virology: a molecular overview. In: Nahmias AJ, Dowdle WR, Schinazi RF, eds. *The human herpesviruses.* New York: Elsevier; 1981:519–536.
4. Smith DB, Inglis SC. The mutation rate and variability of eukaryotic viruses: an analytical review. *J Gen Virol* 1987;68:2729–2740.
5. Steinhauer D, Holland JJ. Rapid evolution of RNA viruses. *Annu Rev Microbiol* 1987;41:409–433.
6. Holland JJ, ed. *Genetic diversity of RNA viruses. Curr Top Microbiol Immunol* 1992;176.
7. Strauss JH, Strauss EG. Evolution of RNA viruses. *Annu Rev Microbiol* 1988;42:657–683.
8. Strauss EG, Strauss JH, Levine AJ. Virus evolution. In: Fields BN, Knipe DM, et al, eds. *Fields' virology,* 2nd ed. New York: Raven Press; 1990:167–190.
9. Futuyma DJ. *Evolutionary biology,* 2nd ed. Sunderland, Mass: Sinauer Associates; 1986:13–14.
10. Green RG. On the nature of filterable viruses. *Science* 1935;82:443–445.
11. Laidlaw PP. *Virus diseases and viruses* [Rode Lecture, University of Cambridge]. Cambridge: Cambridge University Press; 1938.
12. Andrewes CH. *Viruses and evolution* [The Huxley Lectures, 1965–66]. Birmingham: University of Birmingham; 1967.
13. Burnet FM, White DO. *Natural history of infectious disease,* 4th ed. Cambridge: Cambridge University Press; 1972.
14. Andrewes CH. *The natural history of viruses.* New York: WW Norton; 1967.
15. Cech TR. RNA enzymes. *Adv Enzymol* 1989;62:1–36.
16. Gilbert W. The RNA world. *Nature* 1986;319:618.
17. Joyce GF. The rise and fall of the RNA world. *New Biol* 1991;3:399–407.
18. Holland JJ. Replication error, quasispecies populations, and extreme evolution rates of RNA viruses. In: Morse SS, ed. *Emerging viruses.* New York: Oxford University Press; 1993.
19. Waterson AP. The origin and evolution of viruses. I. Virus origins: degenerate bacteria or vagrant genes. *New Sci* 1963;18:200–202.
20. Lwoff A. The concept of virus (The Third Majority Stephenson Memorial Lecture). *J Gen Microbiol* 1957;17:239–253.
21. Temin HM. Malignant transformation of cells by viruses. *Perspect Biol Med* 1970;14:11–26.

22. Temin HM. Origin of retroviruses from cellular moveable genetic elements. *Cell* 1980;21:599–600.
23. Lederberg J. Viruses and humankind: intracellular symbiosis and evolutionary competition. In: Morse SS, ed. *Emerging viruses.* New York: Oxford University Press; 1993.
24. Doolittle RF, Feng D-F, Johnson MS, McClure MA. Origins and evolutionary relationships of retroviruses. *Q Rev Biol* 1989;64:1–30.
25. Xiong Y, Eickbush TH. Origin and evolution of retroelements based upon their reverse transcriptase sequences. *EMBO J* 1990;9:3353–3362.
26. Bangham C, Daenke S, Phillips R, Cruickshank J, Bell J. Enzymatic amplification of exogenous and endogenous retroviral sequences from DNA of patients with tropical spastic paraparesis. *EMBO J* 1988;7:4179–4184.
27. Horwitz MS, Boyce-Jacino MT, Faras AJ. Novel human endogenous sequences related to human immunodeficiency virus type 1. *J Virol* 1992;66:2170–2179.
28. Boeke JD, Corces VG. Transcription and reverse transcription of retrotransposons. *Annu Rev Microbiol* 1989;43:403–434.
29. Temin HM. Retrons in bacteria. *Nature* 1989;339:254–255.
30. Kidwell MG, Kimura K, Black DM. Evolution of hybrid dysgenesis potential following P element contamination in *Drosophila melanogaster. Genetics* 1988;119:815–828.
31. Sakaguchi K. Invertrons, a class of structurally and functionally related genetic elements that includes linear DNA plasmids, transposable elements, and genomes of adeno-type viruses. *Microbiol Rev* 1990;54:66–74.
32. Baltimore D. Expression of animal virus genomes. *Bacteriol Rev* 1971;35:235–241.
33. Kamer G, Argos P. Primary structural comparison of RNA-dependent polymerases from plant, animal and bacterial viruses. *Nucleic Acids Res* 1984;12:7269–7282.
34. Poch O, Sauvaget I, Delarue M, Tordo N. Identification of four conserved motifs among the RNA-dependent polymerase encoding elements. *EMBO J* 1989;8:3867–3874.
35. Poch O, Blumberg BM, Bougueleret L, Tordo N. Sequence comparison of five polymerases (L proteins) of unsegmented negative-strand RNA viruses: theoretical assignment of functional domains. *J Gen Virol* 1990;71:1153–1162.
36. Koonin EV. The phylogeny of RNA-dependent RNA polymerases of positive-strand RNA viruses. *J Gen Virol* 1991;72:2197–2206.
37. Argos P. A sequence motif in many polymerases. *Nucleic Acids Res* 1988;16:9909–9916.
38. Wang TS-F, Wong SW, Korn D. Human DNA polymerase α: predicted functional domains and relationships with viral DNA polymerases. *FASEB J* 1989;3:14–21.
39. McClure MA. Evolution of retroposons by acquisition of deletion of retrovirus-like genes. *Mol Biol Evol* 1992;8:835–856.
40. Baltimore D. Evolution of RNA viruses. *Ann NY Acad Sci* 1980;354:492–497.
41. Murphy FA, Kingsbury DW. Virus taxonomy. In: Fields BN, Knipe DM, et al, eds. *Fields virology*, vol 1, 2nd ed. New York: Raven Press; 1990:9–35.
42. Francki RIB, Fauquet CM, Knudson DL, Brown F, eds. *Classification and nomenclature of viruses. Fifth report of the International Committee on Taxonomy of Viruses. Arch Virol* 1991; suppl 2.
43. Goldbach R. Genomic similarities between plant and animal RNA viruses. *Microb Sci* 1987;4:197–202.
44. Bawden FC. Criticism of binomial nomenclature as applied to plants. *Ann NY Acad Sci* 1953;56:538–544.
45. Mayr E. Concepts of classification and nomenclature in higher organisms and microorganisms. *Ann NY Acad Sci* 1953;56:391–397.
46. Burnet FM. A genetic approach to variation in influenza viruses. 1. The characters of three substrains of influenza virus A (WS). *J Gen Microbiol* 1951;5:46–53.
47. Burnet FM. A genetic approach to variation in influenza viruses. 2. Variation in the strain NWS on allantoic passage. *J Gen Microbiol* 1951;5:54–58.
48. Burnet FM, Lind PE. A genetic approach to variation in influenza viruses. 3. Recombination of characters in influenza virus strains used in mixed infections. *J Gen Microbiol* 1951;5:59–66.
49. Kilbourne ED. The molecular epidemiology of influenza. *J Infect Dis* 1978;127:478–487.
50. Calisher CH. Evolutionary significance of the taxonomic data regarding bunyaviruses of the family Bunyaviridae. *Intervirology* 1988;29:268–276.
51. Fenner F. Biological control, as exemplified by smallpox eradication and myxomatosis [The Florey Lecture, 1983]. *Proc R Soc Lond [B]* 1983;218:259–285.

52. Parrish CR, Have P, Foreyt WJ, Evermann JF, Senda M, Carmichael LE. The global spread and replacement of canine parvovirus strains. *J Gen Virol* 1988;69:1111–1116.

53. Parrish CR, Aquadro CF, Strassheim ML, Evermann JF, Sgro J-Y, Mohammed HO. Rapid antigenic-type replacement and DNA sequence evolution of canine parvovirus. *J Virol* 1991;65:6544–6552.

54. Zuckerkandl E, Pauling L. Molecular disease, evolution and genic heterogeneity. In: Kasha M, Pullman B, eds. *Horizons in biochemistry*. New York: Academic Press; 1962:189–225.

55. Kimura M. Evolutionary rate at the molecular level. *Nature* 1968;217:624–626.

56. Kimura M. *The neutral theory of molecular evolution*. Cambridge: Cambridge University Press; 1983.

57. Kimura M. Recent development of the neutral theory viewed from the Wrightian tradition of theoretical population genetics. *Proc Natl Acad Sci USA* 1991;88:5969–5973.

58. Temin HM. Evolution of cancer genes as a mutation-driven process. *Cancer Res* 1988;48:1697–1701.

59. Temin HM, Is HIV unique or merely different? *J AIDS* 1989;2:1–9.

60. Temin HM. The high rate of retrovirus variation results in rapid evolution. In: Morse SS, ed. *Emerging viruses*. New York: Oxford University Press; 1993:223.

61. Gojobori T, Moriyama EN, Ina Y, Ikeo K, Miura T, Tsujimoto H, Hayami M, Yokoyama S. Evolutionary origin of human and simian immunodeficiency viruses. *Proc Natl Acad Sci USA* 1990;87:4108–4111.

62. Gojobori T, Moriyama EN, Kimura M. Molecular clock of viral evolution, and the neutral theory. *Proc Natl Acad Sci USA* 1990;87:10015–10018.

63. Lenski RE. Evolution, experimental. In: Lederberg J, ed. *Encyclopedia of microbiology*, vol 2. San Diego: Academic Press; 1992:125–140.

64. Hillis DM, Bull JJ, White ME, Badgett MR, Molineux IJ. Experimental phylogenetics: generation of a known phylogeny. *Science* 1992;255:589–592.

65. Mills DR, Peterson RL, Spiegelman S. An extracellular Darwinian experiment with a self-duplicating nucleic acid molecule. *Proc Natl Acad Sci USA* 1967;58:217–224.

66. Biebricher CK. Darwinian selection of self-replicating RNA molecules. *Evol Viol* 1983;16:1–52.

67. Eigen M, Biebricher CK. Sequence space and quasispecies distribution. In: Domingo E, Holland JJ, Ahlquist P, eds. *RNA genetics*, vol 3. Boca Raton: Fla: CRC Press; 1988:211–245.

68. Domingo E, Sabo D, Taniguchi T, Weissmann C. Nucleotide sequence heterogeneity of an RNA phage population. *Cell* 1978;13:735–744.

69. Goodenough M, Huet T, Saurin W, Kwok S, Sninsky J, Wain-Hobson S. HIV-1 isolates are rapidly evolving quasispecies: evidence for viral mixtures and preferred nucleotide substitutions. *J AIDS* 1989;2:344–352.

70. Fitch WM, Margoliash E. Construction of phylogenetic trees. *Science* 1967;155:279–284.

71. Felsenstein J. Phylogenies from molecular sequences: inference and reliability. *Annu Rev Genet* 1988;22:521–565.

72. Buonagurio DA, Nakada S, Parvin JD, Krystal M, Palese P, Fitch WM. Evolution of human influenza A viruses over 50 years: rapid, uniform rate of change in *NS* gene. *Science* 1986;232:980–982.

73. Gammelin M, Altmüller A, Reinhardt U, Mandler J, Harley VR, Hudson PJ, Fitch WM, Scholtissek C. Phylogenetic analysis of nucleoproteins suggests that human influenza A viruses emerged from a 19th-century avian ancestor. *Mol Biol Evol* 1990;7:194–200.

74. Webster RG, Bean WJ, Gorman OT, Chambers TM, Kawaoka Y. Evolution and ecology of influenza A viruses. *Microbiol Rev* 1992;56:152–179.

75. McClure MA, Johnson MS, Feng D-F, Doolittle RF. Sequence comparisons of retroviral proteins: relative rates of change and general phylogeny. *Proc Natl Acad Sci USA* 1988;85:2469–2473.

76. Smith TF, Srinivasan A, Schochetman G, Marcus M, Myers G. The phylogenetic history of immunodeficiency viruses. *Nature* 1988;333:573–575.

77. Penny D. Origins of the AIDS virus. *Nature* 1988;333:494–495.

78. Myers G, MacInnes K, Korber B. The emergence of simian/human immunodeficiency viruses. *AIDS Res Hum Retroviruses* 1992;8:373–386.

79. Allan JS, Short M, Taylor ME, et al. Species-specific diversity among simian immunodeficiency viruses from African green monkeys. *J Virol* 1991;65:2816–2828.

80. Gao F, Yue L, White AT, et al. Human infection by genetically diverse $SIV_{SM}$-related HIV-2 in West Africa. *Nature* 1992;358:495–499.

81. Zhang Y, Temin HM. 3' junctions of oncogene–virus sequences and the mechanisms for formation of highly oncogenic retroviruses [Minireview]. *J Virol* 1993;67:1747–1751.
82. Lai MMC. RNA recombination in animal and plant viruses. *Microbiol Rev* 1992;56:61–79.
83. Weiss BG, Schlesinger S. Recombination between Sindbis virus RNAs. *J Virol* 1991;65:4017–4025.
84. Liao C-L, Lai MMC. RNA recombination in a coronavirus: recombination between viral genomic RNA and transfected RNA fragments. *J Virol* 1992;66:6117–6124.
85. Hahn CS, Lustig S, Strauss EG, Strauss JH. Western equine encephalitis virus is a recombinant virus. *Proc Natl Acad Sci USA* 1988;85:5997–6001.
86. Keck JG, Matsushima GA, Makino S, Fleming JO, Vannier DM, Stohlman SA, Lai MMC. In vivo RNA–RNA recombination of coronavirus in mouse brain. *J Virol* 1988;62:1810–1813.
87. Kirkegaard K, Baltimore D. The mechanism of RNA recombination in poliovirus. *Cell* 1986;47:433–443.
88. Jarvis TC, Kirkegaard K. Poliovirus RNA recombination: mechanistic studies in the absence of selection. *EMBO J* 1992;11:3135–3145.
89. Hu WS, Temin HM. Retroviral recombination and reverse transcription. *Science* 1990;250:1227–1233.
90. Zhang J, Temin HM. Rate and mechanism of nonhomologous recombination during a single cycle of retroviral replication. *Science* 1993;259:234–235.
91. Vlasak R, Luytjes W, Spaan W, Palese P. Human and bovine coronaviruses recognize sialic acid containing receptors similar to those of influenza C viruses. *Proc Natl Acad Sci USA* 1988;85:4526–4529.
92. Botstein D. A modular theory of virus evolution. In: Fields BN, Jaenisch R, Fox CF, eds. *Animal virus genetics*. New York: Academic Press; 1980:11–20. ICN-UCLA Symposia on Molecular and Cellular Biology, vol 18.
93. Lewontin RC. *The genetic basis of evolutionary change*. New York: Columbia University Press; 1974.
94. Mayr E. Geneticists and naturalists reach a consensus: the second Darwinian revolution. In: Mayr E. *One long argument. Charles Darwin and the genesis of modern evolutionary thought*. Cambridge, Mass: Harvard University Press; 1991:132–140.
95. Mayr E. The concept of finality in Darwin and after Darwin. In: Mayr E. *Toward a new philosophy of biology: observations of an evolutionist*. Cambridge, Mass: Harvard University Press; 1988:233–257.
96. Mayr E. On the evolutionary synthesis and after. In: Mayr E, *Toward a new philosophy of biology: observations of an evolutionist*. Cambridge, Mass: Harvard University Press; 1988:525–554.
97. Mayr E. *The growth of biological thought*. Cambridge, Mass: Belknap Press of Harvard University Press; 1982:30–31.
98. Mayr E. *One long argument. Charles Darwin and the genesis of modern evolutionary thought*. Cambridge, Mass: Harvard University Press; 1991.
99. Eldredge N, Gould SJ. Punctuated equilibria: an alternative to phyletic gradualism. In: Schopf TJM, ed. *Models in paleobiology*. San Francisco: Freeman, Cooper, 1972;82–115.
100. Maynard Smith J. Do we need a new evolutionary paradigm? *New Sci* 1985;1447:38–39. [Reprinted in Maynard Smith J. *Did Darwin get it right?* New York: Chapman and Hall; 1989;157–161].
101. Levins R, Lewontin RC. *The dialectical biologist*. Cambridge, Mass: Harvard University Press; 1985.
102. Smith T. *Parasitism and disease*. Princeton: Princeton University Press; 1934.
103. Downs WG The Rockefeller Foundation Virus Program: 1951–1971 with update to 1981. *Annu Rev Med* 1982;33:1–29.
104. Theiler M, Downs WG. *The arthropod-borne viruses of vertebrates. An account of the Rockefeller Foundation Virus Program, 1951–1970*. New Haven: Yale University Press; 1973.
105. Fenner FJ. Frank Macfarlane Burnet, 3 September 1899–31 August 1985. *Biogr Mem Fellows R Soc* 1987;33:101–162.
106. Levin SA, Pimentel D. Selection for intermediate rates of increase in parasite–host systems. *Am Nat* 1981;117:308–315.
107. Ewald PW. The evolution of virulence. *Sci Am* 1993;268:86–93.
108. Both GW, Shi CH, Kilbourne ED. Hemagglutinin of swine influenza virus: a single amino acid change pleiotropically affects viral antigenicity and replication. *Proc Natl Acad Sci USA* 1983;80:6996–7000.

109. Kilbourne ED, Easterday BC, McGregor S. Evolution to predominance of swine influenza virus hemagglutinin mutants of predictable phenotype during single infections of the natural host. *Proc Natl Acad Sci USA* 1988;85:8098–8101.
110. Treanor J, Murphy B. Genes involved in the restriction of replication of avian influenza A viruses in primates. In: Kurstak E, Marusyk RG, Murphy FA, van Regenmortel MHV, eds. *Applied virology research*, vol 2. New York: Plenum Press; 1990:150–176.
111. Subbarao EK, London W, Murphy BR. A single amino acid in the *PB2* gene of influenza A virus is a determinant of host range. *J Virol* 1993;67:1761–1764.
112. Rico-Hesse R. Molecular evolution and distribution of dengue viruses type 1 and 2 in nature. *Virology* 1990;174:479–493.
113. Weaver SC, Scott TW, Rico-Hesse R. Molecular evolution of eastern equine encephalomyelitis virus in North America. *Virology* 1991;182:774–781.
114. Hall WW, Takahashi H, Liu C, et al. Multiple isolates and characteristics of human T-cell leukemia virus type II. *J Virol* 1992;66:2456–2463.
115. Gessain A, Gallo RC, Franchini G. Low degree of human T-cell leukemia/lymphoma virus type I genetic drift in vivo as a means of monitoring viral transmission and movement of ancient human populations. *J Virol* 1992;66:2288–2295.
116. Gessain A, Boeri E, Yanagihara R, Gallo RC, Franchini G. Complete nucleotide sequence of a highly divergent human T-cell leukemia (lymphotropic) virus type I (HTLV-I) variant from Melanesia: genetic and phylogenetic relationship to HTLV-I strains from other geographical regions. *J Virol* 1993;67:1015–1023.
117. Sherman MP, Saksena NK, Dube DK, Yanagihara R, Poiesz BJ. Evolutionary insights on the origin of human T-cell lymphoma/leukemia virus type I (HTLV-I) derived from sequence analysis of a new HTLV-I variant from Papua New Guinea. *J Virol* 1992;66:2556–2563.
118. Dube DK, Sherman MP, Saksena NK, et al. Genetic heterogeneity in human T-cell leukemia/lymphoma virus type II. *J Virol* 1993;67:1175–1184.
119. Bastian I, Gardner J, Webb D, Gardner I. Isolation of a human T-lymphotropic virus type I strain from Australian aboriginals. *J Virol* 1993;67:843–851.
120. Komourian F, Pelloquin F, de Thé G. In vivo genomic variability of human T-cell leukemia virus type I depends more upon geography than upon pathogenesis. *J Virol* 1991;65:3770–3778.
121. Ratner L, Philpott T, Trowbridge DB. Nucleotide sequence analysis of isolates of human T-lymphotropic virus type 1 of diverse geographical origins. *AIDS Res Hum Retroviruses* 1991;7:923–941.
122. Schultz TF, Calabro ML, Hoad JG, Carrington CVF, Matutes E. Catovsky D, Weiss RA. HTLV-I envelope sequences from Brazil, the Caribbean, and Romania: clustering of sequences according to geographic origin and variability in an antibody epitope. *Virology* 1991;184:483–491.
123. Chan S-Y, Bernard H-U, Ong C-K, Chan S-P, Hofmann B, Delius H. Phylogenetic analysis of 48 papillomavirus types and 28 subtypes and variants: a showcase for the molecular evolution of DNA viruses. *J Virol* 1992;66:5714–5725.
124. Chan S-Y, Ho L, Ong C-K, et al. Molecular variants of human papillomavirus type 16 from four continents suggest ancient pandemic spread of the virus and its coevolution with humankind. *J Virol* 1992;66:2057–2066.
125. Gorman OT, Bean WJ, Kawaoka Y, Webster RG. Evolution of the nucleoprotein gene of influenza A virus. *J Virol* 1990;64:1487–1497.
126. Gorman OT, Bean WJ, Kawaoka Y, Donatelli I, Yuanji G, Webster RG. Evolution of influenza A virus nucleoprotein genes: implications for the origins of H1N1 human and classical swine viruses. *J Virol* 1992;65:3704–3714.
127. Fitch WM, Leiter JME, Li X, Palese P. Positive Darwinian evolution in human influenza A viruses. *Proc Natl Acad Sci USA* 1991;88:4270–4274.
128. Gibbons A. New "China syndrome" puzzle [News]. *Science* 1991;253:26.
129. Kawaoka Y, Webster RG. Molecular mechanism of acquisition of virulence in influenza virus in nature. *Microb Pathogen* 1988;5:311–318.
130. Salvato M, Borrow P, Shimomaye E, Oldstone MBA. Molecular basis of viral persistence: a single amino acid change in the glycoprotein of lymphocytic choriomeningitis virus is associated with suppression of the antiviral cytotoxic T-lymphocyte response and establishment of persistence. *J Virol* 1991;65:1863–1869.
131. Ahmed R, Hahn CS, Somasundaram T, Villarete L, Matloubian M, Strauss JH. Molecular basis of organ-specific selection of viral variants during chronic infection. *J Virol* 1991;65:4242–4247.
132. Kasper P, Kaiser R, Kleim J-P, Oldenburg J, Brackmann HH, Rockstroh J, Schneweis KE. Diver-

sification of HIV-1 strains after infection from a unique source. *AIDS Res Human Retroviruses* 1993;9:153–157.

133. Omata M, Ehata T, Yokosuka O, Hosoda K, Ohto M. Mutations in the precore region of hepatitis B virus DNA in patients with fulminant and severe hepatitis. *N Engl J Med* 1991;324:1699–1704.

134. Liang TJ, Hasegawa K, Rimon N, Wands JR, Ben-Porath E. A hepatitis B virus mutant associated with an epidemic of fulminant hepatitis. *N Engl J Med* 1991;324:1705–1709.

135. Carman WF, Zanetti AR, Karayiannis P, Waters J, Manzillo G, Tanzi E, Zuckerman AJ, Thomas HC. Vaccine-induced escape mutant of hepatitis B virus. *Lancet* 1990;336:325–329.

136. Ewald PW. Waterborne transmission and the evolution of virulence among gastrointestinal bacteria. *Epidemiol Infect* 1991;106:83–119.

137. Almond JW. The attenuation of poliovirus neurovirulence. *Annu Rev Microbiol* 1987;41:153–180.

138. Cohen JI. Toward understanding the molecular basis for attenuation of picornaviruses. *Adv Virus Res* 1989;36:153–180.

139. Brooks DR, McLennan DA. *Phylogeny, ecology, and behavior. A research program in comparative biology.* Chicago: University of Chicago Press; 1991.

140. Harvey PH, Pagel MD. *The comparative method in evolutionary biology.* Oxford: Oxford University Press; 1991.

141. Berry RJ, Crawford TJ, Hewitt GM, eds. *Genes in ecology.* Oxford: Blackwell Scientific; 1992.

142. Dover GA. The tangled bank revisited. *Nature* 1993;362:672–673.

143. Gardner MB, Kozak CA, O'Brien SJ. The Lake Casitas wild mouse: evolving genetic resistance to retroviral disease. *Trends Genet* 1991;7:22–27.

144. May RM, Anderson RM. Epidemiology and genetics in the coevolution of parasites and hosts. *Proc R Soc Lond [B]* 1983;219:281–313.

145. May RM, Anderson RM. *Infectious diseases of humans. Dynamics and control.* Oxford: Oxford University Press; 1991.

146. Ewald PW. Transmission modes and the evolution of virulence. With special reference to cholera, influenza, and AIDS. *Hum Nature* 1991;2:1–30.

147. Ou C-Y, Ciesielski CA, Myers G, et al. Molecular epidemiology of HIV transmission in a dental practice. *Science* 1992;256:1165–1171.

148. Muller HJ. One hundred years without Darwinism are enough. *School Sci Math* 1959;59:304–316.

149. Li Y, Hui H, Burgess CJ, Price RW, Sharp PM, Hahn BM, Shaw GM. Complete nucleotide sequence, genome organization, and biological properties of human immunodeficiency virus type 1 in vivo: evidence for limited defectiveness and complementation. *J Virol* 1992;66:6587–6600.

150. Kilbourne ED. New viruses and new disease: mutation, evolution and ecology. *Curr Opin Immunol* 1991;3:518–524.

*The Evolutionary Biology of Viruses,*
edited by Stephen S. Morse.
Raven Press, Ltd., New York © 1994.

# 2

# Driving Forces in Evolution

## An Analysis of Natural Selection

### Ernst Mayr

*Museum of Comparative Zoology, Harvard University, Cambridge, Massachusetts 02138*

Among all the new scientific concepts, perhaps none has been as revolutionary in its effect on our thinking as Darwin's theory of natural selection. In its unprecedented departure from the time-honored traditional thinking of philosophy and in its anti-biblical implications, as well as for various other reasons, it encountered at once the most determined opposition. Furthermore, since it was not the discovery of a concrete object or process that one could examine physically, the concept of natural selection was interpreted in various ways, even among its followers. It was as true for this concept as for all new concepts that everyone had to incorporate it into his own conceptual framework and adjust it to that which he already believed. The result was that everyone looked at natural selection in a somewhat different way, and much of the ensuing argument about the validity of natural selection suffered from misunderstandings and semantic difficulties.

Among biologists, there are still opposing opinions concerning the target of selection (gene, individual, group, species), the role of sexual selection, "neutral evolution," the role of development, deterministic versus probabilistic aspects of selection, and the nature and power of constraints, to mention only a few. It is these largely biological questions that I will address here.

For Darwin, the concept was simplicity itself. He said, "The preservation of favorable variations and the rejection of injurious variations, I call Natural Selection" (1). His opponents, however, raised the most diverse objections to this simple formulation. These objections, in part, were based on ideology, religion, or philosophy, and in part on the claim that Darwin had failed to apply the proper scientific method. I shall not cover this long-lasting controversy in this essay, since I have done so elsewhere (2).

But it must be admitted that Darwin's opponents raised some valid points. First of all, his argument was largely deductive, and he was not able to produce anything that his contemporaries were willing to accept as proof. The term *selection* was a particularly formidable obstacle in the way of acceptance, because it implied some-thing forward-looking, some teleological process, some selecting agency. Yet, was there such an agency? And if it existed, why were its activities so haphazard and unpredictable? This teleological implication was transferred to adaptation, the prod-uct of selection; even today an evolutionist may say that a species of desert plants had adapted itself to its arid environment. A long time passed before even evolu-tionary biologists fully realized that selection is, if I may use this seemingly contra-dictory terminology, an *a posteriori* process. Selection is simply that in every gen-eration a few individuals among the hundreds, thousands, or millions of offspring of a set of parents survive and are able to reproduce because these individuals hap-pened to have a combination of characteristics that favored them under the constel-lation of environmental conditions they encountered during their lifetime.

I have avoided the somewhat misleading, because seemingly circular, expression "survival of the fittest," which has so often led to the claim that the whole concept of natural selection is based on tautology. That this claim is not valid has been shown by many recent authors (3–10). In fact, it is possible to formulate the theory in a clearly nontautological manner (9,11). Actually, this had already been done by Darwin. He pointed out that in every generation there is a great overproduction of individuals, only a small percentage of whom can survive and reproduce. Second, all these individuals differ in their endowment and, therefore, differ, at least in principle, in their adaptedness to their common environment. And third, the causes of the differences in adaptedness are in part heritable. It follows by simple logic that those with the highest adaptedness have the greatest chance to survive and repro-duce.

So much evidence for the reality of the process of natural selection has been produced since the publication in 1859 of the *Origin* that systematic opposition has more or less died down, at least within science. There is, first of all, abundant experimental evidence for the efficacy of selection under laboratory conditions, as summarized in recent textbooks. But there is also—and perhaps this is more impor-tant—abundant evidence for the occurrence of natural selection in nature (12). The theory of natural selection has an enormous heuristic value and permits so-called predictions under specified environmental conditions. For instance, if on a certain island there are two species of finches, a small-billed one, specializing on small seeds, and a large-billed one, specializing on large seeds (with a certain amount of overlap of the food niche), there should be a higher mortality of the small-billed species during a drought period that hits the plants that produce small seeds partic-ularly hard. And this, indeed, is what is found (13). I do not think I exaggerate if I say that every year some 20 to 30 papers (perhaps far more) are published in which confirmation is provided for such "predictions" derived from the selection theory (14,15).

It has been argued that the theory of natural selection is not a scientific theory at all, because it cannot be refuted: "Natural selection explains nothing, because it explains everything" (11). Indeed, it must be made clear what the nature of "the" theory of natural selection is. The principle of natural selection is more comprehensive than a specific theory; therefore, it has been referred to as a generic theory or basic general principle (9,16–18), which as such can neither be refuted nor does it have predictive powers. It becomes a genuine theory, called by Tuomi (16,17) a theoretical model, only when enriched with specific ancillary assumptions, for instance, with particulate inheritance, the postulate of small mutations, and the occurrence of genetic recombination. The principle can be dissected into individual theories, and such specific theories can be tested and refuted.

There are various ways in which definitions used in selection theory can be made more precise—for instance, by expressing fitness, not in terms of the actual contribution to the gene pool of the next generation, but rather, as the propensity to make such a contribution. If two identical twin brothers go out on a walk and one is killed by lightning, it will drastically affect his contribution to the gene pool of the next generation, but not his original propensity for making such a contribution. "The fitness of an organism is its propensity to survive and reproduce in a particular specified environment and population" (4,19). Some philosophers have questioned the validity of the propensity concept.

Actually, for every situation in nature, we can imagine innumerable solutions that could *not* be explained by natural selection. Returning to our earlier example of the finches, this time consider a slightly different situation, two islands with one species of finch on each. On island A only plants with large heavy-shelled seeds occur, no plants with small seeds, and hardly any insects. On island B, by contrast, only plants with small seeds occur, and an abundance of small insects. Frankly, I would not know how to explain it by natural selection if the finch on island A were thinbilled and the one on island B a grosbeak with a heavy bill. The fact of the matter is that we take adaptation in nature for granted to such an extent that we would consider situations such as the fictitious scenario of the two finches as totally absurd. Actually, the normally observed "harmony of living nature" is spectacular evidence for the power of natural selection. Maynard Smith has pointed this out also, illustrating it with a different example (20).

## A TWO-STEP PROCESS

One of the favored arguments among Darwin's opponents has always been, "How can natural selection produce favorable variations?" Or the even more absurd comparison, made from Jonathan Swift and J. F. Herschel to contemporary critics, of natural selection to a troupe of monkeys. Natural selection, they claimed, has no more chance to produce an adaptation than a troupe of monkeys to compose *Hamlet* by aimlessly throwing the letters of the alphabet together. These critics overlooked that natural selection proper is only the second stage of a two-step process. The first step consists of the production of variation in every generation; that is, of suitable genetic or phenotypic variants that can serve as the material of selection, and this will then be exposed to the process of selection. This first step of variation is completely independent of the actual selection process, and yet selection would not be possible without the continuous restoration of variability. Several recent critics have failed to understand the relation of the two steps of the selection process to each other. When an author asks (as several have actually done): "Is evolution due to molecular processes or due to selection?" it amounts to asking: "Is evolution a change due to step one or to step two of natural selection?" This question is meaningless, since the second step, selection *sensu stricto*, deals with the previously produced variation (*a posteriori*) and is not a process that itself produces variation. Darwin saw clearly that a fixed type, a constant Platonic essence, cannot evolve. And yet the struggle for existence, the mechanism responsible for change during the second step, has no influence whatsoever on the first step, that is, on the amount and nature of the variation produced before a bout of reproduction.

As aware as Darwin was of the importance of variation [he devoted to it three of the first five chapters of the *Origin* and later (1868) an entire two-volume work (21)], neither he nor anyone else solved the riddle of the source of variation and its inheritance until the rediscovery of Mendel's principles in 1900.

The two points Darwin stressed particularly were that a seemingly inexhaustible supply of variation was available at all times and that the production of any new variation was independent of the needs of the organism. In contrast with the evolutionary theories of some of his opponents, variation for Darwin was nondirected. This is what the evolutionist means when he says new mutations are random. This terminology has often been misunderstood. It does not in the least mean that any variation can occur anywhere, any time. On the contrary, mutations, in a given species, are highly "constrained," which means that only a very restricted range of changes of the phenotype is possible. An eye of *Drosophila* can vary from white to pink, red, or brown, but never to blue or green. A tetrapod can vary in the modifications of its two pairs of extremities, but the occurrence of one extra pair (as in insects) or two extra pairs (as in spiders) is not part of the possible normal variation of a tetrapod. When it is said that mutation or variation is random, the statement simply means that there is no correlation between the production of new genotypes and the adaptational needs of an organism in the given environment. Owing to numerous constraints, the statement does not mean that every conceivable variation is possible (22). Even in the viruses, in spite of their high mutation rates and opportunities to acquire new genes from host cells, we take for granted the existence of characteristics stable enough (e.g., morphology) to be used for defining viral taxa (see the chapters by Morse and by Kilbourne).

After 1900 it became customary to see the origin of new variants as the result of mutation. This was the thinking not only of the early Mendelians, but also of much of the Morgan school, and one could read in biology textbooks until the 1960s or even 1970s that evolution is due to "mutation and selection." Throughout much of this period, genetic recombination, as produced during meiosis and by other mechanisms, was rather neglected. Although crossing-over had been known since the 1880s (Weismann), it was rarely invoked in evolutionary discussions until C. D. Darlington's "equating of chiasmata and cross-overs led to the recognition of the universality of intrachromosomal recombination in natural species" (23). Recombination is the process responsible for the fact that each individual in sexually reproducing species is different from all others; and even if at a much lower frequency, recombination also occurs in viruses in a variety of ways, between host and viral genes (or between an infectious virus and endogenous viral elements) and between two viruses coinfecting a host cell (24,25).

Evolution is not a smooth, continuous process, but consists, in sexually reproducing organisms, of the formation of a brand-new gene pool in every generation. Furthermore, there is a steady alternation of the first step, which consists of the meiotic production of new gametes and their fate before fertilization, with the second step, the "struggle" of the new zygotes to reach the reproductive stage and to reproduce successfully. Similar or analogous processes generating variation in progeny also occur in viruses, presumably serving similar purposes. Reassortment in influenza viruses has been suggested as a close analogy of sex (see the chapter by Chao). Other types of recombination events have already been mentioned. Finally, the relatively high mutation rate of many viruses has also been noted (discussed in the chapter by Domingo and Holland). The individual progeny will later be subject to selection as they compete to infect new host cells. Thus, this analysis of natural selection as a two-step process, the first step generating variation and the second being the competition to survive and reproduce, seems equally applicable to viruses. Selection is not a forward-looking process but simply a name for the survival of those few individuals that have successfully outlasted the "struggle for existence." Each individual is, so to speak, a new experiment that is tested for its fitness in the struggle for existence. No other process is known in nature that could achieve this in such an extraordinarily successful manner as the combination of the first and second steps of natural selection.

## THE TARGET OF SELECTION

In Darwin's time and, indeed, in the first years of Mendelism ("unit characters"), the difference between genotype and phenotype was not understood. The organism as a whole, that is, its phenotype, was considered the target of selection; and something representing the phenotype was handed from generation to generation. Darwin's theory of pangenesis was very much in this tradition. However, it was also realized that the genotype was not simply a condensed projection of the adult phenotype, as is evident from the theory of recapitulation. But it was not until after 1910 that the question of the relation of genotype and phenotype during development became a problem. At that time V. Haecker, Woltereck, Goldschmidt, and other German geneticists worried considerably over it. In that literature there was a tendency to consider both genotype and phenotype, rather holistically, with a stress on genetic interactions. As a result the individual, as a whole, was treated as the target of selection, as it had been by Darwin. By contrast, a rather strong reductionist trend prevailed in the English-language scientific literature, and genes were studied more or less in isolation from each other and from the phenotype. Ever since the rise of mathematical population genetics and, as a matter of fact, already in the thinking of some members of the Morgan school (Sturtevant, Muller), the gene was considered the currency of evolution. Consequently, in the mathematical approaches of R. A. Fisher and J. B. S. Haldane, individual genes were accepted as the units on which calculations were based. Evolution was now defined as a change in gene frequencies, and relatively simple mathematical formulae were believed to permit an adequate representation of what happens during evolution. Accepting the change of gene frequencies as the earmark of evolution, some authors unfortunately began to ignore whether such change was due to genetic drift or to selection. And in the single-minded concentration on genes, it was often forgotten how important individuals, populations, and species are in determining the course of evolution.

It is rather unfortunate that the customary question to ask was "What is the 'unit of selection'?" The term *unit* was never properly defined, and this led to all sorts of misunderstandings (discussed in 26). As a result, a protracted controversy developed over whether the gene, the genotype, the individual (phenotype), the group, or the species is the "unit of selection," or all of them. The misunderstandings go back to the early post-Darwinian period, when certain essentialists wrote that selection was "for the good of the species." A long controversy ensued, one which saw the geneticists, on the whole, on the side of championing the gene, and the naturalists favoring the individual or the group (27). I, myself, have favored the individual since the 1950s, and an increasing number of recent authors have come to the same conclusion.

The insistence that the individual, as a whole, rather than each separate gene, is the target of selection would be unimportant if the genotype were an aggregate of separate, independent genes. However, those who favor the individual consider the genotype to be a well-integrated system, analogous to an organism with structure and organs. This view was developed by the Russian evolutionary geneticists, particularly Schmalhausen, and by Dobzhansky and his school. It was forcefully presented in Lerner's 1954 book *Genetic Homeostasis* (28), and in my own writings (22,29–31). The controversy between this viewpoint and the reductionist one (both within the Darwinian framework) has not yet been resolved. There is, however, a great deal of recent evidence that the system of interactions among components of the genotype is far more complex than conceived of in classic population genetics, and the system nature of the genotype is strongly corroborated by the recent recognition of the great functional diversity of kinds of DNA (31; and see following). Much of the recent criticism of neo-Darwinism is directed against a reductionist–atomistic conception of the genotype. These critics ignore that other representatives of the synthetic theory, such as Schmalhausen, Waddington, Rensch, Lerner, members of Dobzhansky's school, and myself, have for many decades adhered to a far more holistic concept of the genotype.

The clarification of this problem has been helped by Sober's analysis (32). He pointed out that much of the confusion came about because people did not distinguish between asking "selection of" and "selection for." In the controversy over whether the gene, individual, or group is the unit of selection, the question to be asked is "selection of." And this makes it quite clear why the individual, and not the gene, must be considered the target of selection. There are many ways of documenting this primary importance of the individual. First, it is the individual as a whole that either does or does not have reproductive success. Second, the selective value of a particular gene may vary greatly, depending on the genotypic background on which it is placed. Third, since different individuals of the same population differ at many loci, it would be exceedingly difficult to calculate the contributions of each of these loci to the fitness of a given individual.

When we say that the individual is the target of selection, we mean the individual at all stages of its life cycle. Consideration of the individual as a whole, rather than single genes, to be the target of selection clarifies several evolutionary puzzles. After the enormous variability of enzyme genes had been discovered in the 1960s through the use of electrophoresis, Kimura and others proposed the theory that most changes in gene frequencies are of no selective significance, but are neutral. Not that Kimura (33) and those who followed him completely denied the existence of natural selection, but they certainly reduced its importance as a factor in the evolutionary change of the genotype.

A great deal of recent research has indeed established that some changes at the molecular level seem to be either entirely neutral or of very small selective significance. Nevertheless, serious objections can be raised against the neutrality concept of molecular evolution. First, many of the supposedly neutral mutations have subsequently been found to be subject to selection (34). Also, Kimura's models have been challenged, and it has been asserted that the number of neutral mutations appears to be far smaller if a different method of calculation is employed (35,36). Second, the selective value of a given allele is not necessarily constant and may change in a different environment or in a different constellation of the genotype. Finally, and most importantly, if the individual, as a whole, is the target of selection, and if certain individuals are favored by selection because of other characteristics, then numerous neutral genes can be carried along as "hitchhikers." Hence, the neutrality of certain mutations is not in the slightest conflict with the theory of selection, which rewards only the integrated selective value of the individual, *as a whole*. The DNA sequences believed to be functionless, such as pseudogenes and certain introns, behave as if selectively neutral and, thus, may be subject to rapid change, owing to genetic drift and to their being immune to stabilizing selection.

It has been questioned whether it is legitimate to refer to changes in "neutral" genes as evolution. Since, by definition, these molecular changes have no effect on the fitness of the individual nor on the evolutionary changes of the phenotype, they are considered by some authors as nothing but evolutionary "noise." Whether this reasoning, based on the situation in higher organisms, is equally applicable to viruses, with their very much shorter pathway from gene to phenotype, has not yet been determined. There now appears to be no good reason for recognizing neutral evolution as a separate process from Darwinian evolution and in conflict with it.

The normal process of mutation is the replacement of an allele by a new allele, or, in molecular terms, the replacement of an amino acid residue, or of a base pair in the DNA. However, genetics and molecular biology have discovered numerous other processes that may lead to a change in the genotype and to increased variability. Some of these deserve to be mentioned because they seem to contradict the concept of natural selection. This includes genes for the distortion of segregation (meiotic drive), various modes of production of "selfish DNA," and so-called molecular drive. These molecular processes are not random and might be referred to as *biased variation*. It is a mistake, in my opinion, to claim that the phenomena of biased variation prove that genes can serve as the target of selection. This is not true, because it is always the resulting zygote (for multicellular organisms, or resulting progeny in viruses and microorganisms) that is the actual target. Normal selection begins to operate as soon as the biased variation affects the fitness of the phenotype.

There are several problems of definition relative to all those genes that are responsible for biased variation. But, as far as I am concerned, their activity, without exception, affects the first step of natural selection, that of the generation of variability, whereas the second step, exposing individuals to the forces of selection, is not affected. There is a continuing selection for unlinked modifiers for the reduction of the intensity of the meiotic drive. To be sure, strongly biased variation can be more potent than selection, as in the segregation–distorter genes, but this does not affect the conceptual issues. I believe one is thus justified in stating categorically that the answer to the question, what is the target of selection (*selection of*), is the individual (in the viruses, the infecting viral particles or their progeny).

Although I discuss species selection elsewhere (37), I want to mention at least one frequently cited instance of species selection. If two species of flour beetles (*Tribolium*) are introduced into a container with flour, experience shows that only one of the two species will remain after several generations. This has been considered a case of species selection. Actually, there is no typological competition between two species, but rather, each individual of either species is competing with all the coexisting individuals (of both species). Accordingly, this is again a typical case of selection of individuals. Again, usually, an analogous situation will be true of viruses, as each individual progeny virus particle competes for new host cells.

## THE OBJECTIVES OF SELECTION

At this point we have to remember once more that one must distinguish between selection *of* and selection *for*. Having discussed selection of under the heading of "The Target of Selection," we are now discussing selection for; we are discussing what kind of traits might be favored by selection.

The success of selection in a given generation can be expressed in terms of the numerical contribution to the gene pool of the next generation. However, successful reproduction—and Darwin saw this much more clearly than any of his contemporaries—might be due to two entirely different causes. Darwin designated as *natural selection* anything that favored survival, whether this was an increase or decrease of body size, a broadening of the niche or its more efficient utilization, better protection against the environment or an increased tolerance of environmental extremes, a superior ability to cope with diseases or to escape enemies. Natural selection would favor anything that would accomplish greater ecological–physiological efficiency or save energy. Any individual favored by such selection would contribute genotypes to the gene pool that were likely to spread in future generations and, thus, enhance the adaptedness of the population as a whole.

However, Darwin saw also—and, again, much more clearly than any of his contemporaries—that not all selection leads to improved adaptation, as this term is currently understood. An individual might contribute more genes to the next generation not through physiological efficiency or any other component of viability, but simply through being more successful in reproduction. This kind of selection Darwin called *sexual selection*. He further realized that there was a potential conflict between these two kinds of selection, the natural one for improved fitness (in the vernacular meaning of this word) and the sexual one for mere reproductive success of individuals. A purely egotistical selection for reproductive success might favor the evolution of traits that do not add to fitness (as traditionally understood), but might actually make the species more vulnerable. It leads to the gorgeous plumes of the birds of paradise, the extraordinary tail of the peacock, and the gigantic size of the elephant seal bulls. Darwin devoted almost two thirds of the text of his *The Descent of Man* to the discussion of sexual selection (38). Yet this important process was largely ignored during the ensuing 100 years, in part, owing to the reductionist definition of the target of selection by the mathematical population geneticists. By making the gene the unit of selection and defining anything as selectively superior that leads to an increased representation in the gene pool of the next generation, the entire difference between survival selection and selection for reproductive success was eliminated. Unfortunately, this concealed some of the most interesting aspects of selection.

Even though I called attention in 1963 (22) to the importance of "selection for reproduction success," and so did Ghiselin (39), this phenomenon continued to be neglected until the commemoration of the centennial (1971) of Darwin's *Descent of Man*. Then, within a few years, literally hundreds of cases of behaviors and devices were discovered that tended to favor the reproductive success of an individual, without contributing to its survival qualities. This new interest in selection for reproductive success is undoubtedly one of the few major developments in post-synthesis evolutionary biology (40).

The existence of selfish selection for reproductive success poses a dilemma for the evolutionary biologist. Classic natural selection ordinarily resulted in genotypes that were adaptively superior. This was so obvious that it was even suggested to define an *adaptation* as anything that was a "product of selection." However, this definition does not fit sexual selection at all. Is the gigantic size of the elephant seal bulls of adaptive value for this species? Is the survival of birds of paradise enhanced by the brilliant plumes of the adult males? Surely not. On the contrary, it is quite possible that an excessive development of certain characters favored by sexual selection contributed to the extinction of some species.

## DOES SELECTION LEAD TO PERFECTION?

The opponents of natural selection have often used the following argument: Since selection by necessity leads to perfection, they said, and since evidence for insufficient perfection and maladaptiveness abound in nature, then obviously natural selection does not operate. But which modern evolutionist has ever claimed that natural selection would lead to perfection? Darwin already had emphasized "natural selection tends only to make each organic being as perfect as, or slightly more perfect than, the other inhabitants of the same country with which it has to struggle for existence" (1). He cites as proof the rapid extinction of fauna and flora in New Zealand when European competitors were brought to that island.

To be sure, selection is an optimization process, but one of a very special kind. It is neither teleologically programmed nor controlled by any law, but is entirely opportunistic. As many recent authors have pointed out, the process of selection is subject to so many constraints that it could not possibly achieve perfection. The most striking refutation of the frequently made claim that "selection can do anything" is provided by the frequency of extinction. More than 99.9% of all species that ever existed on earth have become extinct. This includes even such temporarily so-flourishing groups as the trilobites, ammonites, and dinosaurs. Examples of viruses or diseases that have briefly flourished and suddenly disappeared suggest that this has also happened with viruses. Here, it does not matter what factor was responsible for the extinction: competition, a pathogen, a climatic catastrophe, or an asteroid's impact. In each instance, selection had been unable to find, among the available variants, an appropriate answer to the new situation.

## CONSTRAINTS

Curiously, the existence of constraints was already well known before the establishment of Darwinian evolution. Thus, Cuvier said prior to 1820 that no carnivore would ever have horns. The whole concept of the archetypes of the idealistic morphologists was based on the silent acceptance of developmental constraints. The naturalists of the 19th century knew very well that each taxon had severe limits to its potential variation. Weismann, who was a panselectionist, if there ever was one, frequently stressed the constraints encountered by selection (41).

The importance of such constraints was, however, unfortunately neglected after 1900, when the geneticists thought of evolution as a matter of genes, rather than of whole organisms. Consequently, it has been wholesome that authors such as Gould and Lewontin (42) have again called attention to the power of the constraints on selection (but see also ref. 43).

There are many different kinds of constraints that prevent an optimal response to selection forces. They affect all aspects of the selection process, from the production of variation to the interactions between genotype and environment. I shall discuss some of them. Additional ones are recorded in the literature (8,44–48), and lists of constraints were previously given by a number of authors (42,49,50).

## A Capacity for Nongenetic Modification

The more pliable the phenotype (owing to developmental flexibility), the more this reduces the force of an adverse selection pressure. Plants, and particularly microorganisms, have a far greater capacity for phenotypic adaptation than higher animals. In microorganisms and viruses, this can often occur through regulation of gene expression, which can be finely tuned in viruses (51) and is sensitive to host factors and the state of the host cell (52,53; reviewed in 54). The existence of this capacity is, however, present even in humans, as shown, for instance, by a person's capacity for physiological adjustment when he or she goes from the lowlands to high altitudes. In the course of days and weeks this person can become reasonably well adapted to the lowered atmospheric pressure, but even under these circumstances, natural selection is not entirely neutralized. First of all, the very capacity for nongenetic adaptation is under strict genetic control; but also when a population shifts to a new, specialized environment, genes will be favored by selection during the ensuing generations that reenforce and eventually largely replace the capacity for nongenetic adaptation.

## The Individual as Target

To be favored by natural selection, it is sufficient that an individual is competitively superior to most other individuals of its population (species). This superiority may be achieved by particular features, indeed sometimes, by a single gene. In that event, natural selection "tolerates" the remainder of the genotype, even when some of its components are more or less neutral or even slightly inferior (see foregoing). Genetic linkage, therefore, can also exert considerable constraints.

## Availability of Variation

There is a limit to the kind and amount of genetic variation that can be contained in a population; therefore, the kind of genes needed for an appropriate response to new selection pressure may not be present in the gene pool. One can see the great variability and high mutation rates of viruses as a strategy for providing raw material to increase the chances of the needed genes or functions being present if selection pressures change.

Darwin apparently took it for granted that there was always sufficient variability in natural populations to satisfy the demands of natural selection. Many geneticists tended to disagree. De Vries, Bateson, and other Mendelians did not actually deny the existence of variation, but they thought that ordinary continuous variation was evolutionarily irrelevant and that evolution proceeded by occasional more or less drastic mutations. This was contradicted by the work of Chetverikov (55) and his school and by the ecological geneticists (56,57), who showed that abundant natural variation was always available. But, said Muller, and this opinion was widely adopted, this variation simply consists of deleterious recessives that are eliminated as quickly as they become homozygous. Finally, in 1866 Darwin seemed to be vindicated when Lewontin and Harris demonstrated the enormous variability of enzyme genes as revealed by electrophoresis. But that by no means ended the argument, because Kimura and others presented evidence that much of this variation was neutral, that is, not suitable as material for natural selection.

More importantly, the high frequency of extinction, as well as the high frequency of instances when artificial selection was unable to achieve its objective, indicate that more is needed than abundant variability. It must be variability of the needed portions of the genotype. Success in evolution may depend on highly specific genes or combinations of genes, and these are sometimes not present when needed.

## Multiple Pathways

Several alternate responses are usually possible for any environmental challenge. This is well illustrated by pelagic swimmers that have developed either from sessile or benthic (crawling) or actively swimming ancestors belonging to many different animal phyla and that have become adapted to the pelagic mode of life through entirely different adaptations. Each solution is a very different compromise between the new need and the previously existing structure. This illustrates splendidly Jacob's principle of opportunistic "tinkering" (58). Whatever solution is chosen depends on a constellation of circumstances. The adoption of a particular solution may greatly constrain the possibilities of future evolution, for instance the acquisition either of an internal or external skeleton. The size and symmetry of a viral capsid may limit the size of the genome or impose other constraints, such as the need to retain recognition sequences. The route of infection and the need to maintain an ability to infect may impose significant constraints (59; see also chapter in this volume by Kilbourne). Similarly, it has also been noted by several authors that many pathogens evolve in new host populations, not to avirulence (near complete loss of virulence), but toward a balance between transmissibility and virulence (see chapter by Garnett and Antia). I have elsewhere discussed, with reference to the well-studied myxoma virus, why I believe this results from individual selection and is not an example of group selection (60).

## Interference by Evolutionary "Noise"

Interference by evolutionary noise refers to all the chance factors that interfere with the superficially deterministic process of selection. They include not only the rich array of chance factors during the first step of selection, but also the breaking apart of favorable gene constellations through crossing-over as well as the numerous accidents to which a zygote is exposed during its maturation and reproduction. With viruses, similar events can occur during replication, as mentioned earlier for reassortment and recombination, for example. Hence, the propensity for fitness is rarely fully expressed in the realized fitness. (For a further discussion of this noise factor see also the later section, "Chance.")

## Cohesion of the Genotype

As I have shown in a number of recent analyses (31), and as is not at all unknown to geneticists, genes are frequently so tightly functionally interconnected with each other that any change is deleterious. In animals, presumably this is usually due to epistatic interactions during ontogeny. New information turns up all the time, supplying further evidence for the resistance of the genotype to any but the most superficial changes. More drastic changes may lead to the production of deleterious phenotypes.

Evolutionary rates, as well, are highly dissimilar at the species level. Large, populous species, with wide distribution, show great evolutionary inertia, not so much, as was once said, because it takes so long for an allele to disperse through the entire species range, but rather, because the coadapted gene complex is highly resistant to the incorporation of new genes. On the other hand, founder populations may undergo a genetic evolution within an extremely short time and may possibly replace alleles at 30% to 50% of the loci within 5,000 to 10,000 years for some animal species. One can imagine that similar events might occur as a virus is introduced into a new geographic area, although perhaps on a shorter time scale.

Considering these drastic differences at the species level it is difficult to advance generalizations concerning phyletic evolution. There is little doubt, however, that those who enter biology from the physical sciences make rather unrealistic assumptions about the evolutionary rate of changes in macromolecules. From the fact that some molecules, such as the histones, evolve very slowly, whereas others have intermediate or very rapid evolution, some authors have drawn the conclusion that such molecules have a built-in, so to speak orthogenetic, rate of evolution. Depending on the particular macromolecule, they assume that there will be an amino acid replacement every 2 million, 5 million, or 10 million years. All the known facts contradict this naive assumption.

It seems to me that it leads to a far more realistic interpretation of evolutionary rates if we assume that the evolutionary change of a given type of macromolecule is not due to a built-in rate, but rather, is due to the need for coadaptation with other molecules with which it has to interact. As the other molecules change in the course of time in response to ad hoc selection pressures, every molecule occasionally has to adjust to its changed molecular environment; it must adjust to the cohesion of the genotype. This, undoubtedly, is the reason for the continuous evolutionary change of molecules, even of those in which the active site has not changed since the days of the most primitive eukaryotes or even prokaryotes.

I might add that there is perhaps now more interest in evolution among molecular biologists than among workers in any other nonstrictly evolutionary branch of biology. Molecular biologists have discovered that every molecule has an evolutionary history; that one can establish phylogenies of molecules; and that the study of such phylogenies must use methods analogous to those of comparative anatomy. Hence, the rise of molecular biology has not, in the long run, prevented the unification of biology. In fact, it has strengthened it.

Let me summarize these considerations concerning the cohesion of the genotype:

The genes are not the units of evolution nor are they, as such, the targets of natural selection. Rather, genes are tied together into balanced adaptive complexes, the integrity of which is favored by natural selection. One can think of the viral genome as such a complex.

The study of the mechanisms by which the cohesion of the genotype is achieved is a promising area of evolutionary research.

It is important to understand this cohesion of the genotype, because it permits the explanation of many previously puzzling phenomena of speciation and macroevolution.

## Developmental Channeling

Natural selection does not piece together each organism from scratch. All it can do is to attempt to modify slightly an already existing highly complex structure. Consequently, evolutionary pathways are largely channeled. Some developments are possible, others are not. The Hawaiian finches (Drepanididae) show that there is little channeling relative to the shape of the bill. The birds of paradise of New Guinea document that in that family there is a broad leeway in the elaboration of plumes. Yet, in most higher taxa, there is an extraordinary basic similarity of all members (consider the stability of viral morphology mentioned earlier). This is why a good zoologist can usually tell for any animal not only to what phylum, but often also to what class, order, or even family it belongs, and a viral morphologist can often tell to what family an unknown virus belongs from examining an electron micrograph. Curiously, it is rarely emphasized what a convincing demonstration of developmental constraint this taxonomic similarity is.

The existing genotype, thus, prescribes definite channels for future evolution. Is this fact in any conflict with the universal statement that natural selection is the only direction-giving factor in evolution? The answer is, yes, and, no. Directional changes in evolution are caused by natural selection, but constrained by the potential of the existing genotype. Indeed, there is much evidence that the cohesion of the genotype is so tight that there is usually very little change of the phenotype over millions of years (see foregoing section, "Cohesion of the Genotype"). It needs some special events or processes to destabilize the genotype. This has often happened during domestication (61), and this is what seems to happen sometimes during speciation in founder populations (62). Perhaps the introduction of a virus into a new host population can be considered analogous.

## THE GENETICS OF DEVELOPMENTAL CONSTRAINTS AND PHYLOGENETIC CHANNELING

One of the great new insights of the animal systematists of the early 19th century was that animals cannot be arranged in a smooth, highly continuous series from the simplest to the most perfect, as the proponents of the *scala naturae* had claimed. Instead, a limited number of discrete types could, be recognized. This new insight was developed by Cuvier, von Baer, Oken, Owen, Agassiz, and their followers. Even after everybody had adopted Darwin's theory of common descent, most major taxa remained discrete entities, each with its own *Bauplan* (ground plan), or morphological type.

Atomistic genetics has been incapable of coming up with an explanation for the stability of the Bauplan. The fact that all tetrapods have basically one pair of anterior and one pair of posterior extremities can perhaps be functionally explained. Yet why all insects should have three pairs of extremities, and all spiders four pairs, can be explained only by the conservatism of the developmental system built into their genotypes. The same argument is true for the five-rayed foot (or hand) of terrestrial vertebrates. Even where it is reduced, as in the foot of the horse or in the wings of birds, or supplemented by additional rays, as in certain marine vertebrates, the structure is always laid down in ontogeny with five rays. One could go on and on about such conservative aspects of the Bauplan, either extended into the adult phenotype or visible only during ontogeny, but this does not add to the solution of the problem. There are presumably internal structures of the genotype that reductionist genetics has not even begun to explain (31). One must ask: What are these structures, and what happens to them during the short evolutionary periods when rather drastic ("saltational") evolutionary innovations are initiated?

No answers can yet be given to these questions. However, the recent advances in molecular genetics justify the hope that we are not too far from a solution. It is now clear that there are many different kinds of DNA, including middle and highly repetitive DNA, mobile genetic elements, and inducible mutational systems. All of them may, and probably do, have regulatory functions, but classic genetics was unable to work with this heterogeneity of DNA. The classic genes are structural genes and, for the most part, seem to be involved in the *production* of substrate, rather than in its regulation. However, the aspect of gene function that is crucial for evolution is how this substrate is used and when. For some 20 years everyone has been proclaiming that regulatory genes are the crucial element in evolution, but so far extremely little is known about which parts of the DNA function as regulatory genes and how these genes, in turn, are being controlled. The only thing that is clear, at least to me, is that the classic beanbag-type genetics is altogether insufficient to provide satisfying evolutionary explanations.

## CHANCE

Selection is often described as a deterministic process, indeed sometimes even as a teleologic process, because it seems to result in long-term evolutionary trends; however, these designations are quite misleading. A close analysis of long-term evolutionary trends has shown almost invariably that they are actually quite irregular and, often, even terminated by reversals.

The importance of chance during evolution has been stressed by certain authors for more than 100 years (22). As early as 1871, Gulick insisted that the differences among snail populations on Oahu Island in the Hawaiian Islands were due to random variation and not due to selection. Since that time no one has stressed the role of chance factors more emphatically than Sewall Wright. Chance operates at every level in the process of reproduction, from crossing-over to the survival of newly formed zygotes (63). This includes the locus at which mutations occur, the location of chiasmata involved in crossing-over, and, for sexually reproducing eukaryotes, the segregation of chromosomes during the reduction division, the survival of the millions or billions of gametes, the meeting of two gametes of opposite sex before fertilization, and, finally, the untold interactions of a zygote with its environment (in the widest sense of the word). There is also genetic drift in all its forms, particularly significant in small populations, and all the effects of linkage. It is thus evident that a considerable percentage of differential survival and reproduction is not the result of ad hoc selection, but rather, of chance (64,65). Several of these factors are applicable to viruses, as well as the important role of chance in other aspects of the viral life cycle; for example, in finding a host. Chance is also introduced by the phenomenon of pleiotropy. If a gene has multiple expressions, it will be selected for the most important of these, and other expressions of the gene will be carried along incidentally.

The large number of stochastic processes in populations of finite size, as well as the constraints that operate during selection, prevent selection from ever being a deterministic process. Rather, it must always be remembered that selection is probabilistic. This is true even for the success of the zygote. Each individual encounters in its environment numerous unpredictable adverse forces, such as catastrophes, epidemics, and unexpected encounters with enemies in which the outcome is largely probabilistic.

Finally, survival may depend on aspects of population structure. A certain genotype may have a high survival probability in a small founder population, whereas it would be clearly inferior in a very large, widespread population.

It should be evident from this discussion how misleading is the picture of natural selection that some authors have. A careful reading of Darwin's version shows that his concept of natural selection was far more mature than that of most of his opponents. Nevertheless, even he did not fully appreciate the power of the constraints and of chance.

## WHAT HAPPENED IN THE EVOLUTIONARY SYNTHESIS?

I have referred in several places to the evolutionary synthesis, and I would like to close by briefly summarizing what the synthesis achieved in my opinion and what remained unresolved. During the evolutionary synthesis of the 1930s and 1940s, selection was universally accepted by evolutionary biologists as the only cause of inherited adaptation and as the only direction-giving factor in evolutionary change. Nevertheless, it became quite clear during the ensuing 50 years how many misconceptions and uncertainties were still attached to this concept. It is particularly gratifying that various modern philosophers, instead of trying to refute the principle, have attempted to sharpen definitions and eliminate equivocations and confusions. This includes analyses by Brandon (66), Mills and Beatty (4), and Tuomi (17), the important book by Elliott Sober (32), and the introductory discussions in Brandon and Burian (27).

Two questions have been asked repeatedly within the last score of years. First, were some of the theories adopted during the evolutionary synthesis erroneous? And second, to what extent was the synthesis incomplete? As far as the first question is concerned, the theory of neutral evolution, the punctuated equilibrium theory, and the attack on the adaptationist program are most frequently mentioned. As the subsequent discussions established, none of these three theories was actually in conflict with the synthesis. The neutral theory is clearly a product of the erroneous assumption of some geneticists that the gene is the target of selection. As soon as one returns to Darwin's assumption that the individual as a whole is the target, the occurrence of neutral gene replacements is no longer in conflict with Darwinism. Neither is the punctuated equilibrium theory, because Darwin already had realized how unequal rates of evolution can be, and he had been aware, particularly in his early work, of the evolutionary significance of isolated founder populations, and this has been confirmed by perceptive authors ever since. Finally, the occurrence of stochastic processes during the action of natural selection had been recognized long before the synthesis. If the individual, as a whole, is the target of selection, then there is no need to explain an adaptive significance of every aspect of the phenotype. The adaptationist program, which, for that matter, preceded Darwin by several centuries, is legitimate and highly heuristic, provided it is properly applied (43).

But there has also been an incompleteness of the synthesis in the integration of the basic concepts of the experimentalists and the naturalists. Even today there are some geneticists who still act as if the gene was the only target of selection, and this is the reason for their difficulties with the bottleneck effect of founder populations and with so-called neutral evolution.

Historians, perhaps even Mayr and Provine (67), have overemphasized the unity achieved by the synthesis. To be sure, the anti-Darwinians were so decisively defeated that none of their theories retained any credibility. Furthermore, the differences between the experimentalists and naturalists were greatly reduced, but honesty compels us to admit that conspicuous differences still remained. To begin with, for the geneticists, evolution continued to be a change of gene frequencies, with the gene considered the target of selection, whereas for the naturalists, evolution was a series of processes resulting in adaptedness and diversity, with the individual being the target of selection. Population thinking, stressed to the utmost by the representatives of the new systematics, was still unable to make much headway in paleontology and, in a multipopulational sense, even in genetics. Geneticists almost invariably dealt with the processes taking place in a single, closed gene pool. Being interested in the individual as a whole, the naturalists were particularly concerned with the genotype as a whole, including all of its epistatic interactions and all phenomena that indicate a certain "cohesion" of the genotype, whereas the experimental geneticists concentrated on single genes and on additive inheritance, thereby being unable to do justice to the bottleneck effects taking place in founder populations. As a result, the two camps also differed drastically in the evaluation of neutral evolution, that is, the high frequency of base pair replacements that do not change the fitness of the phenotype. For the experimentalists, such neutral evolution is a very important evolutionary phenomenon, whereas for the naturalists, it is little more than evolutionary noise.

## CONCLUSIONS

No other concept in evolutionary biology is as important as that of natural selection; and, owing to its novelty, no other concept has encountered as much resistance. It was in conflict not only with some of the tenets of Christianity but also with major axioms of the accepted philosophy of the mid-19th century (2). With few exceptions, natural selection continued to be rejected by most philosophers for the next 100 years or more. And when philosophers finally began to accept it as a legitimate scientific theory, their interpretations encountered unexpected complexities. Is selection a teleologic concept? Is the simple principle of natural selection a theory or is it a theory only when applied to concrete cases? What is the target of selection? How deterministic and predictive is selection? What is the material of selection? Is selection a one-step or two-step phenomenon? Can selection produce perfection? These are among the questions I have tried to address in this chapter and to show that natural selection is part of a two-step process, with variation the necessary first step, providing the raw material on which selection can act.

## ACKNOWLEDGMENT

This essay is adapted from: An analysis of the concept of natural selection. In: Mayr E. *Toward a new philosophy of biology*. Cambridge, Mass: Harvard University Press; 1988, with modification and the addition of new material. Some material has also been added from other essays in that volume, and from Mayr E. "What was the evolutionary synthesis?" Trends Ecol Evol 1993;8(1):31–34.

# REFERENCES

1. Darwin C. *On the origin of species*. London: John Murray; 1859. [Facsimile ed. Cambridge, Mass: Harvard University Press; 1964.]
2. Mayr E. *The growth of biological thought*. Cambridge, Mass: Harvard University Press; 1982.
3. Caplan AL. Tautology, circularity, and biological theory. *Am Nat* 1977;111:390–393.
4. Mills S, Beatty J. The propensity interpretation of fitness. *Philos Sci* 1979;46:263–288.
5. Ruse M. *Philosophy of biology*. London: Hutchinson; 1973.
6. Stebbins GL. In defense of evolution: tautology or theory? *Am Nat* 1977;111:386–390.
7. Williams MB. The logical status of the theory of natural selection and other evolutionary controversies. In: Bunge M, ed. *The methodological unity of science*. Dordrecht: Reidel; 1973:84–102.
8. Riddiford A, Penny D. The scientific status of modern evolutionary theory. In: Pollard JW, ed. *Evolutionary theory: paths into the future*. London: John Wiley & Sons; 1984.
9. Brandon RN. A structural description of evolutionary theory. *PSA 1980* [Philosophy of Science Association] 1981;2:427–439.
10. Hodge J. The development of Darwin's general biological theorizing. In: Bendall DS, ed. *Evolution from molecules to men*. Cambridge: Cambridge University Press; 1983:43–62.
11. Lewontin RC. Testing the theory of natural selection. *Nature* 1972;236:181–182.
12. Endler JA. *Natural selection in the wild*. Princeton: Princeton University Press; 1986.
13. Grant PR. *Ecology and evolution of Darwin's finches*. Princeton: Princeton University Press; 1986.
14. Naylor BG, Handford P. In defense of Darwin's theory. *Bioscience* 1985;35:478–484.
15. Ferguson A. Can evolutionary theory predict? *Am Nat* 1976;110:1101–1104.
16. Tuomi J, Hankioja E. Predictability of the theory of natural selection: an analysis of the structure of the Darwinian theory. *Savonia* 1979;3:1–8.
17. Tuomi J. Structure and dynamics of Darwinian evolutionary theory. *Syst Zool* 1981;30:22–31.
18. Beatty J. What's wrong with the received view of evolutionary theory. *PSA 1980* [Philosophy of Science Association] 1981;2:397–426.
19. Brandon RN. Adaptation and evolutionary theory. *Stud Hist Philos Sci* 1978;9:181–206.
20. Maynard Smith J. *On evolution*. Edinburgh: Edinburgh University Press; 1972.
21. Darwin C. *The variation of animals and plants under domestication*, 2 vol. London: John Murray; 1868.
22. Mayr E. *Animal species and evolution*. Cambridge, Mass: Harvard University Press; 1963.
23. White MJD. Tales of long ago. *Paleobiology* 1981;7:287–291.
24. Doolittle RF, Feng D-F, Johnson MS, McClure MA. Origins and evolutionary relationships of retroviruses. *Q Rev Biol* 1989;64:1–30.
25. Lai MMC. RNA recombination in animal and plant viruses. *Microbiol Rev* 1992;56:61–79.
26. Mayr E. Philosophical aspects of natural selection. In: Mayr E. *Toward a new philosophy of biology. Observations of an evolutionist*. Cambridge, Mass: Belknap Press of Harvard University Press; 1988:116–125.
27. Brandon RN, Burian RM. *Genes, organisms, populations*. Cambridge, Mass: MIT Press; 1984.
28. Lerner M. *Genetic homeostasis*. Edinburgh: Oliver and Boyd; 1954.
29. Mayr E. *Populations, species and evolution*. Cambridge, Mass: Harvard University Press; 1970: Chap. 10.
30. Mayr E. *Evolution and the diversity of life*. Cambridge, Mass: Harvard University Press; 1976.
31. Mayr E. The unity of the genotype. In: Mayr E. *Toward a new philosophy of biology. Observations of an evolutionist*. Cambridge, Mass: Belknap Press of Harvard University Press; 1988:423–438.
32. Sober E. *The nature of selection: evolutionary theory in philosophical focus*. Cambridge, Mass: MIT Press; 1984.
33. Kimura M. *The neutral theory of molecular evolution*. Cambridge: Cambridge University Press; 1983.
34. Nevo E. Adaptive significance of protein variation. In: Oxford GS, Rollinson D, eds. *Protein polymorphism: adaptive and taxonomic significance* [Systematics Association Special Volume 24], New York: Academic Press; 1983:239–282.
35. Gillespie JH. Rates of molecular evolution. *Annu Rev Ecol Syst* 1986;17:637–655.
36. Gillespie JH. *The causes of molecular evolution*. New York: Oxford University Press; 1991.
37. Mayr E. Speciational evolution through punctuated equilibria. In: Mayr E. *Toward a new philosophy of biology. Observations of an evolutionist*. Cambridge, Mass: Belknap Press of Harvard University Press; 1988:457–488.

38. Darwin C. *The descent of man, and selection in relation to sex*, 2 vol. London: John Murray; 1871.
39. Ghiselin M. *The triumph of the Darwinian method*. Berkeley: University of California Press; 1969.
40. Trivers R. *Social evolution*. Menlo Park, Calif: Benjamin/Cummings; 1985.
41. Mayr E. On Weismann's growth as an evolutionist. In: Mayr E. *Toward a new philosophy of biology. Observations of an evolutionist*. Cambridge, Mass: Belknap Press of Harvard University Press; 1988:491–524.
42. Gould SJ, Lewontin RC. The spandrels of San Marco and the Panglossian paradigm. *Proc R Soc Lond [B]* 1979;205:581–598.
43. Mayr E. How to carry out the adaptationist program. *Am Nat* 1983;124:324–334.
44. Peters DS. Mechanical constraints canalizing the evolutionary transformation of tetrapod limbs. *Acta Biotheor* 1985;34:157–164.
45. Cole BJ. Size and behavior in ants: constraints on complexity. *Proc Natl Acad Sci USA* 1985; 82:8548–8551.
46. Cheverud JM, Dow MM, Leutenegger W. The quantitative assessment of phylogenetic constraints in comparative analyses: sexual dimorphism in body weight among primates. *Evolution* 1985; 39:1335–1351.
47. Gould SJ. The evolutionary biology of constraint. *Daedalus* 1980;109(2):39–52.
48. Dawkins R. *The extended phenotype*. San Francisco: WH Freeman; 1982.
49. Mayo O. *Natural selection and its constraints*. New York: Academic Press; 1983.
50. Reif W-E, Thomas RDK, Fischer MF. Constructive morphology: the analysis of constraints in evolution. *Acta Biotheor* 1985;34:233–248.
51. Cullen BR. Human immunodeficiency virus as a prototypic complex retrovirus. *J Virol* 1991; 65:1053–1056.
52. Wasylyk B, Imler JL, Chatton B, Schatz C, Wasylyk C. Negative and positive factors determine the activity of the polyoma virus enhancer alpha domain in undifferentiated and differentiated cell types. *Proc Natl Acad Sci USA* 1988;75:7952–7956.
53. Engel DA, Muller U, Gedrich RW, Eubanks JS, Shenk T. Induction of c-fos mRNA and AP-1 DNA-binding activity by cAMP in cooperation with either the adenovirus 243- or the adenovirus 289-amino acid E1A protein. *Proc Natl Acad Sci USA* 1991;88:3957–3961.
54. Shenk TE. Virus and cell: determinants of tissue tropism. In: Morse SS, ed. *Emerging viruses*. New York: Oxford University Press; 1993:79–90.
55. Chetverikov SS. On certain aspects of the evolutionary process. [Russian; 1926.] Eng. trans: *Proc Am Philos Soc* 1961;105:167–195.
56. Dobzhansky T. *Genetics and the origin of species*. New York: Columbia University Press; 1937.
57. Ford EB. *Ecological genetics*. London: Methuen; 1964.
58. Jacob F. Evolution and tinkering. *Science* 1977;196:1161–1166.
59. Fields B. Pathogenesis of viral infections. In: Morse SS, ed. *Emerging viruses*. New York: Oxford University Press, 1993:69–78.
60. Mayr E. Myxoma and group selection. *Biol Zentralbl* 1990;109:453–457.
61. Belyaev DK. Destabilizing selection as a factor in domestication. *J Heredity* 1979;70:301–308.
62. Mayr E. Speciation and macroevolution. In: Mayr E. *Toward a new philosophy of biology. Observations of an evolutionist*. Cambridge, Mass: Belknap Press of Harvard University Press; 1988:439–456.
63. Mayr E. Accident or design: the paradox of evolution. In: *The evolution of living organisms* [Symp R Soc Victoria Melbourne 1959]. Victoria: Melbourne University Press; 1962:1–14.
64. Beatty J. Chance and natural selection. *Philos Sci* 1984;51:183–211.
65. Beatty J. Natural selection and the null hypothesis. In: Dupré J, ed. *The latest on the best: essays on evolution and optimality*. Cambridge, Mass: MIT Press; 1987.
66. Brandon RN. The levels of selection. *PSA 1982* [Philosophy of Science Association] 1982;1:315–324.
67. Mayr E, Provine W, eds. *The evolutionary synthesis*. Cambridge, Mass: Harvard University Press; 1980.

# PART I

Defining Viral Evolution

The Evolutionary Biology of Viruses,
edited by Stephen S. Morse.
Raven Press, Ltd., New York © 1994.

# 3

# Population Biology of Virus–Host Interactions

## G. P. Garnett and R. Antia

*Parasite Epidemiology Research Group, Imperial College, London SW7 2BB, United Kingdom*

The term *virus* describes a wide range of simple organisms that owe their simplicity to their reliance for reproduction on the exploitation of host cells. Such obligate parasitism means that their population biology (the factors regulating numbers of virus) can be studied only in the light of the interaction between virus and host. A mathematical framework representing the population biology and epidemiology of parasites allows the interacting factors to be clearly described and the qualitative consequences of certain evolutionary routes to be explored (1). In this chapter we provide a general introduction to this epidemiological framework both at an inter- and intrahost level. We then illustrate how insights about the evolution of virulence and mechanisms for persistence in host populations are generated. There is an understandable concern about the morbidity and mortality caused by viruses in human populations, which means these infections have been studied extensively. To illustrate this discussion we will concentrate on viruses that infect humans.

There are two levels from which the population biology of virus–host interactions may be approached: (a) the within-host dynamics of virus and (b) the between-host dynamics of virus. The host population can be thought of as a patchy environment, where a virus population grows within a host under constraints imposed by the defenses of that host. In response either to the host's defenses successfully neutralizing a virus or to the death of the host, the virus must move to new hosts. Although there have been advances in our knowledge of the processes of viral growth and the host's immune response, the qualitative description of these phenomena is still rudimentary and is a fertile area for research (2). In contrast with this level of study, contemporary epidemiology provides us with detailed empirical and theoretical studies of how viruses spread through populations of hosts. However, the two levels of within- and between-host population biology of viruses are not independent. For a virus to be transmitted between hosts, it must reach high densities within individual hosts or reach sites from which it can be disseminated. Viral properties required for the survival and growth of viruses within individual hosts will, in general, differ from those required for the transmission of viruses between hosts.

In the next section we describe simple mathematical models representing the population dynamics of viruses, first between hosts and then within hosts. Some of the implications of these models are outlined in the subsequent two sections.

## A MATHEMATICAL FRAMEWORK

### Modeling the Density of Infected Hosts

In a simple representation of viral infections, the population of hosts can be divided into compartments representing the density of individuals who are susceptible to infection (*X*), individuals who are infected and infectious (*Y*), and individuals who are immune to infection (*Z*) (if such immunity arises). The transitions between these compartments, which are illustrated in a flow diagram in Fig. 1, are described

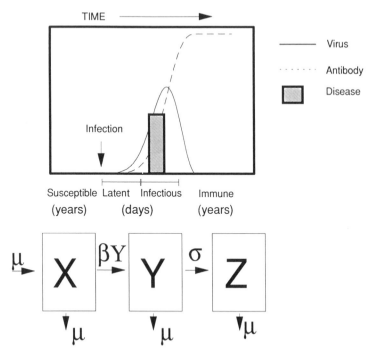

**FIG. 1.** The progress of a simple viral infection, such as measles, through a host. The growth of the virus population, the immune response to the virus, and the timing of acute disease are illustrated. *Below* these curves a flow diagram of the transmission between infection categories represented in a simple model is illustrated. In a constant population the birth and death rate per capita are equal ($\mu$). People are born susceptible *X*, are infected at a rate $\beta Y$ and recover from the infectious category *Y*, at a rate $\sigma$.

by ordinary differential equations relative to time. If, in directly transmitted viruses, it is assumed that hosts mix randomly, then a mass action term determines the number of new infections occurring over a unit of time. The *mass action* term is the product of the density of susceptible individuals ($X$), the density of those infectious ($Y$), and a constant ($\beta$) that is a composite of the probability of contacts being made that would allow the transmission of infection and the probability that infection is transmitted in such a contact. Those infected are assumed to recover from infection at a constant rate ($\sigma$), which is the reciprocal of the average duration of an infection. On recovery, individuals could either enter an implicit permanently immune class, enter a temporarily immune class, return immediately to the susceptible class, or die, depending on the biology of the particular host–virus interaction. In the former case, if a constant population is assumed, with an equal per capita birth and death rate, then the system is described by the following two ordinary differential equations:

$$\frac{dX}{dt} = \mu N - (\mu + \beta Y)X \qquad [1]$$

$$\frac{dY}{dt} = \beta XY - (\mu + \sigma)Y \qquad [2]$$

In this constant population of size $N$ the recovered class $Z$ is simply described as

$$Z = N - (X + Y)$$

This system is often referred to as the SIR model, which refers to the compartments susceptible, infectious, and recovered, and can be altered to take account of, and provide insights about, specific complexities. For example, the model can be altered to include other stages of infection, such as those incubating infection, or those with maternally derived antibodies. The host population could be stratified into age, behavioral, or spatial groups that differ in their exposure to infections.

The framework can also be altered to take account of other transmission routes. The range of possibilities and a far from exhaustive list of examples is shown in Fig. 2. It is noticeable that there are many more examples of human infections transmitted in aerosols, such as measles, mumps, rubella, chickenpox, influenza, or smallpox. It could be that these viruses are more noticeable, as their route for release from the host requires greater virulence; however, it could also be that this route is a particularly suitable one and there are many more infections that use it. Arboviruses are described by the inclusion of compartments describing the dynamics of the vector species, which mediate transmission (3), enteroviruses would be described with an environmental component that delayed transmission. Viruses with distinct host populations, such as rinderpest which infects both domestic and wild ungulates, could be included by a transmission term that may be either constant or may vary seasonally. Sexually transmitted viruses, such as human immunodeficiency virus (HIV), can be modeled by altering the transmission term. The contact required

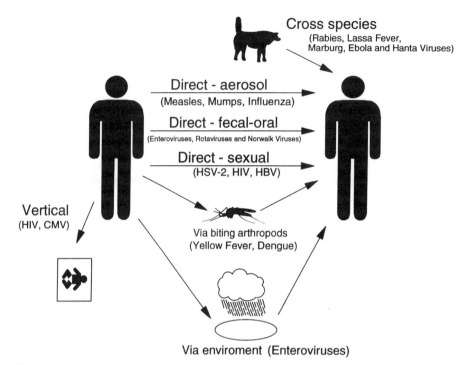

**FIG. 2.** A schematic representation of the transmission routes of viral infections in humans with some examples. Direct transmission can include sexual contact or fecal–oral contact. The sexually transmitted viruses, human immunodeficiency (*HIV*) and hepatitis B virus (*HBV*), may also be transmitted by other processes during which blood is exchanged. Some viruses that are transmitted by fecal–oral contact may also be released into the environment, which provides a source of infection. The cross-species virus transmission can be from rodents (the Hanta viruses and the arenaviruses such as Lassa fever), from monkeys (African hemorrhagic fever—the Marburg and Ebola viruses), from domestic animals infected by wild animals (rabies). There is evidence of some transmission between humans for these infections, but the extent of this interperson transmission is limited.

for transmission is much more restricted and, hence, less frequent than that required for most directly transmitted viruses (4). Some of the implications of differences between transmission routes are considered in a following section "Persistence of Virus within Host Populations."

The framework described is deterministic and, therefore, predicts the average trends in large populations. However, chance events can play a role in the dynamics of a viral population, particularly if we are determining when a virus will become extinct (5). Stochastic methods describing the probability of transitions between compartments in the model can also be used to gain insights when this is the case (6).

## The Reproductive Rate

The basic reproductive *rate* of a virus ($R_0$) can be derived from the mathematical model. This is defined as the average number of new cases of infection generated by one infectious individual in an *entirely susceptible* population. The value is central to viral evolution for two reasons. First, when normally only a single strain of virus infects a host at one time, then, if there are different viral genotypes in the host population, the genotype with the greatest reproductive rate will be preponderant. Second, if the reproductive rate is greater than 1, then an emergent virus will invade a host population. However, if the reproductive rate falls below 1 for an extant virus, then the virus will become extinct. For equations [1] and [2] the value is given by

$$R_0 = \frac{\beta N}{\sigma + \mu} \qquad [3]$$

The value is dependent on the duration of infectiousness $[1/(\sigma + \mu)]$, the number of potential hosts $N$, and the probability that these are contacted and infected $\beta$. If this value is below 1 then a virus will not enter a population and will become extinct.

The initial reproductive rate applies only when the population is entirely susceptible; when persons are infected or have already experienced infection, making them insusceptible, then the *effective* reproductive rate $R_e$ describes the number of new cases a single infection causes:

$$R_e = \frac{\beta X}{\sigma + \mu} \qquad [4]$$

where $X$ is the density of susceptibles. An endemic state within a host population is achieved when this effective reproductive rate equals 1.

The model framework described so far looks at the infected individual as a unit and not at individual viruses. The interaction between a host and its virus population is also important in determining the interaction between transmission and duration of infections, and natural selection will act on individual viruses. In models of virus populations within the host the important elements are the manner in which the small number of viruses in the initial infection increase to sufficiently high densities for transmission between hosts, the number of virus released to allow this transmission, and the regulation of virus by the hosts defenses. Defense mechanisms, such as the host's immune system, impose important constraints on viral evolution. In the following discussion we will describe briefly models of virus populations in hosts when there is no host immunity and, then, when host immunity is present.

## The Growth of a Virus Population in the Absence of Host Immune Defenses

The total transmission of virus from an infected host equals the integral of the rate of transmission of the virus over the duration of infection. Both the rate of transmission of virus and the virus-induced mortality is typically thought to increase with increasing virus density within the host. We might then expect some form of trade-off between the increased rate of transmission of faster-growing viruses (as they reach a higher density within the host), and the shorter duration of infection arising because of increased host mortality (or the rapid evolution of host immunological responses) at high viral densities. Mathematical models of the within-host density of the virus have been used to explore this trade-off between current transmission and host longevity. The treatment we will describe was proposed by Sasaki and Iwasa (7), and the reader is referred to the original paper for details.

The increase in the density $V(t)$ of virus within an individual host will, at a very simple level, depend on the current density of virus and the per capita rate of reproduction of the virus $r$ such that the rate of growth of the population is given by

$$\frac{dV(t)}{dt} = r(t)V(t) \tag{5}$$

In the absence of host immunity, we can assume that the rate of mortality of the host $\mu(V)$ is a function of the density of infection within the host $V(t)$. Here, the probability of survival of a host to time $t$ is given by

$$S(t) = e^{-\int_0^t \mu(V)ds} \tag{6}$$

Sasaki and Iwasa have assumed that the rate of mortality increases linearly with increasing virus density as $u + \alpha V(t)$ where $u$ is the hosts natural mortality and $\alpha V(t)$ the disease-associated mortality. The total transmission of virus from an infected host $Q$ will equal the product of a transmission rate, which depends on the density of virus $f(V(t))$, and the survival probability $S(t)$ over the full course of infection.

$$Q = \int_0^\infty f(V(t))S(t)dt \tag{7}$$

## The Effect of Host Immunity on the Growth of the Virus Population Within the Host

In the virus–host systems discussed here, the density of virus within a host is subject to regulation by the host's immune system. This can be modeled by an inclusion of a population of specific immune cells $I$ along with the virus population. Here, the rate of growth of the population is reduced by a mass action term for the interaction of virus and immune cells, such that

$$\frac{dV}{dt} = rV - kVI \tag{8}$$

where $k$ is the rate constant for the elimination of virus by the specific immune response. The variable $I$ represents the density of immune cells (cytotoxic T cells and antibody-producing B cells) specific to the virus and, hence, the magnitude of the immune response. The immune cells are assumed to be generated by the expansion of a few T and B cells into a large population at a rate proportional to the density of the virus in the formula

$$\frac{dI}{dt} = \frac{\rho I V}{V + \psi}$$ [9]

where $\rho$ is the maximum rate of growth of the immune response, and $\psi$ is the within-host virus population density at which the growth rate of the immune response is half its maximum. For simplicity, the model assumes that the nonspecific defenses and the other causes of parasite mortality are time- and density-independent and implicit in the value of $r$, and that there is a threshold density of virus, $D$ beyond which the host and consequently the virus die. Hence, Eq. 8 holds only for values of $V$ less than $D$. To determine the total number of virus released by an infected host, it is necessary to know the relation between within-host virus density and transmissibility. If the rate of transmission is linearly dependent on the within-host density, then the total number of virus released from an infected host $Q$ equals the integral over the duration of infection ($t$ days) of the rate of release of transmissible virus per unit time

$$Q = \int_0^\infty \phi V(t) dt$$ [10]

where $\phi$ is the constant of proportionality relating the within-host density of the virus to its rate of transmission from the host.

We next proceed to discuss some of the insights about virus evolution that the theoretical frameworks described can provide.

## THE EVOLUTION OF VIRULENCE

### Maximizing Transmission Between Hosts

The morbidity and mortality caused by a virus is naturally of great concern and has often meant that discussions of viral evolution have focused on virulence. An early consensus was that parasites should evolve to a state of avirulence, and that virulence is a product of recent or unnatural relations between parasites and their hosts. This view was based on the notion that the group of viruses that managed to inhabit their host for the longest duration would be the most successful. The death of the host, which also terminates the existence of the virus population, should in this hypothesis be avoided. However, this concentrates on only a part of the equation, the duration of infectiousness. The number of new infections a virus population in one host causes is the effective reproductive rate. Mortality associated with viral infection can be incorporated in the model (see Eqs. 1,2) by the addition of a rate $\alpha$ (in our simple model this will cause decline in the population unless the birth rate is increased). The new effective reproductive rate is

$$R_e = \frac{\beta X}{\sigma + \alpha + \mu}$$ [11]

A virus population with a higher value for $R_e$ than another will have more of its progeny in future generations; hence, the value of $R_e$ should be maximized. A reduction in the death rate does increase the reproductive rate of the virus, as does a reduction in the recovery rate. However, it is most unlikely that the mortality associated with an infection is independent of the other variables in this equation. It is likely that there will be a trade-off between the rate of mortality and the probability of transmission of the virus. The mechanism of growth of the virus population in the host and transmission of virus will mediate this trade-off, but, in its crudest form, a slow-replicating virus is least likely to kill its host; however, at the same time, is least likely to be transmitted. A balance is likely to occur between the two parameters: the mortality rate and the transmission probability. At the same time, an evolutionary arms race will occur between the host, which will develop defenses against disease, and the fast-replicating virus, which will try to optimize its transmission. This conclusion allows one to think of infectious disease, not as an unnatural relation between virus and host, but as the natural consequence of their coevolution.

The view that viruses evolve to avirulence was supported with data showing that infections were less virulent in their normal hosts. For example, Asian cattle are more resistant to rinderpest than African cattle and wildebeest. However, in a host population experiencing an infection, those resistant to the infection that do not die will dominate future generations. Leading to an evolutionary arms race between host defenses against infection and the viruses' ability to reproduce before these defenses come into play or to resist the defenses (8). The observed resistance of populations that have been exposed to infections over a long period could be the result of two factors. First, the selection for avirulence in the virus itself or, alternatively, the selection of resistance in the host population. Often, as described, the resistance has been cited as evidence for the former. However, if an infection is virulent in a previously inexperienced population, this suggests that the avirulence in normal hosts is a product of selection of hosts for resistance. An example of this effect is provided by myxoma virus, which is discussed in a following section on the virulence of myxomatosis.

## Maximizing Transmission From Individual Hosts

Recent studies have examined this issue at the level of the virus population within the host using the models described in Eqs. 5–10. In their model of a host without a specific immune response Sasaki and Iwasa (7) assume that both the transmission function, $f(V(t))$, and the likelihood of host mortality, $\mu(V)$, increase with greater viral densities, $V(t)$. For these biologically realistic assumptions, they derive the function for the rate of growth $r(t)$, which maximizes the transmissible virus released $Q$. If the transmission rate increases faster than linearly with the density of virus, then the optimum growth schedule of the virus is the fastest possible proliferation. However, if the transmission rate has an asymptotic saturation, one would then expect the optimum growth schedule to be composed of three consecutive stages: (a) an initial rapid proliferation at the fastest possible rate, (b) remaining present in the host at a constant density $\hat{V}$, and (c) a restarting of proliferation just before any increase in host mortality that might be triggered by the host's senescence or superinfection of the host by other parasites. If the host mortality $\mu(V(t))$ equals $u + \alpha V(t)$ where $u$ is the virus-independent host mortality and $\alpha V(t)$ the parasite-dependent host mortality, then the steady-state density $\hat{V}$ can be obtained from the formula:

$$\frac{df(\hat{V})}{d\hat{V}} = \frac{f(\hat{V})}{u/\alpha + \hat{V}} \qquad [12]$$

They show that the stationary level $\hat{V}$ increases as natural host mortality increases and decreases as the virulence of the virus increases. An increase in mortality because of host aging or superinfection by other parasites has the effect of shifting the optimal virus density upward, which requires a reactivation of proliferation.

In the model with the incorporation of specific immunity, for which the replication rate $r$ is assumed to be constant, Antia and Levin (9) demonstrate that, at low rates of replication, the virus is cleared by the immune system well before it reaches a lethal density $D$. At intermediate rates of replication the virus is cleared by the immune system just as its density reaches $D$, and at the fastest rate of replication the virus population density soon exceeds $D$, in which event the host dies before the virus is cleared. Thus, in this model the virulence of the virus is directly related to its rate of replication $r$ in the host. In Fig. 3 we plot the total number of viruses released from an infected host $Q$, as a function of the rate of replication of the virus $r$. For these calculations, we integrated Eq. 10 for the entire period of the infection, using the within-host virus densities determined from Eqs. 8 and 9. Slowly growing viruses are cleared by the immune response before they reach high within-host densities and, therefore, are released at lower rates than faster-growing virus. In contrast, very fast growing virus kill the host before releasing as much transmissible virus as virus populations with a lower growth rate that persist longer in the living host. Virus with intermediate rates of replication release the greatest number of free virus, and this transmission is maximized at that value of $r$ at which the immune response is just able to control the virus population before it reaches its lethal density $D$. Thus, it is predicted that virus populations within the host with intermediate rates of replication maximize the number of new infections they generate.

What does this result tell us about the evolution of parasitic virulence? The term *virulence* has been variously associated with the ability of virus to induce morbidity and mortality in infected hosts. A quantitative measure of the virulence of a given virus–host pair can be expressed in terms of the *lethal dose* (LD): the size of a virus inoculum that will kill the host in a specified period. If genetic and environmental variability are introduced into the virus and the host populations, the LD corresponds to the standard $LD_{50}$ (the dose that will kill 50% of infected hosts) used in dose–response curves in toxicological studies (10). In the model described in the foregoing the log (LD) decreases approximately linearly with increasing $r$. There-

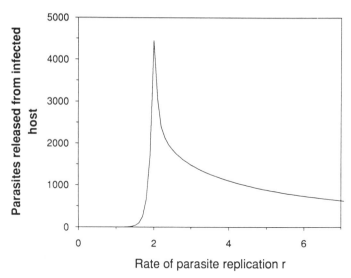

**FIG. 3.** The numbers of parasites released from an individual host (arbitrarily scaled) for a range of population growth rates in a model of the growth of the parasite for which an immune response eliminates virus (9). The peak shows that there is a distinct optimum rate of replication during which the number of released parasites is maximized.

fore, by predicting that virus with intermediate growth rates are favored, the model also predicts that selection will favor virus with intermediate levels of virulence.

### Competition Between Viral Strains

The argument just advanced suggests that viruses with intermediate rates of growth will release the most transmissible forms (i.e., have the highest $R_0$). In this section we examine the epidemiological consequences of competition among multiple strains of virus following their introduction into a population of hosts.

When a single virus strain infects a population of hosts, we can use the simple epidemiological model presented in the earlier section and Eqs. 2 and 3 to examine the dynamics of spread of the virus (here taken to be proportional to the density of infected hosts) as well as its density when the system reaches equilibrium. We can show that the density of infected hosts at equilibrium is given by

$$Y^* = \frac{\mu N}{\mu + \sigma} - \frac{\mu}{\beta}$$  [13]

We can see that the viruses with the highest $R_0$ not only transmit the most when present in a population of susceptible hosts, but also attain the highest steady-state density.

This result, however, does not address the question of the competition between different virus strains. To examine this question, we can extend the single-strain model of Eqs. 1 and 2 to include a second strain. For simplicity, we first assume that hosts infected with one strain are resistant to infection with the other strain (i.e., there is no superinfection of hosts by two strains). Here, the rates of change in the densities of susceptible hosts $X$ and hosts infected by strains 1 and 2 (i.e., $Y_1$ and $Y_2$) and immune hosts $Z$ will be given by

$$\frac{dX}{dt} = \mu N - (\mu + \beta_1 Y_1 + \beta_2 Y_2)X \qquad [14]$$

$$\frac{dY_1}{dt} = \beta_1 X Y_1 - (\mu + \sigma_1)Y_1 \qquad [15]$$

$$\frac{dY_2}{dt} = \beta_2 X Y_2 - (\mu + \sigma_2)Y_2 \qquad [16]$$

$$\frac{dZ}{dt} = \sigma_1 Y_1 + \sigma_2 Y_2 - \mu Z \qquad [17]$$

The outcome of competition between two strains can be shown to depend on the relative magnitudes of their reproductive rates (11). If the strains are chosen so that the reproductive rate of strain 1 is greater than that of strain 2, we can show that only the former will persist at equilibrium at a density given by Eq. 13.

Since the other strain has been driven to extinction, the persisting strain reaches the same steady-state density that it would have if it alone infected the population of hosts. The details of the "competitive exclusion" of different parasite strains has been considered by Bremerman and Thieme (12), who show that, provided the host density is bounded, in the absence of superinfection the parasite strain with the highest $R_0$ will persist, driving all other strains to extinction.

When individual hosts can be concurrently infected by more than one strain of parasite, the situation becomes more complex. Consider, for example, when super-infection benefits the strain with the lower $R_0$, by allowing it to superinfect hosts infected with the other strain. Levin and Pimentel (13) show that different outcomes are possible, depending on a trade-off in the advantage of being able to monopolize hosts through faster replication and the advantage of producing more free virus in slower-growing strains. When the differences in mortality and immune response to the two strains of virus are small compared with the advantage derived from super-infection, the more virulent strain will drive the other to extinction. When the advantage derived by the virulent strain from its ability to superinfect hosts infected by the less virulent strain is about the same as the disadvantage in having a lower reproductive rate in singly infected hosts, both strains can coexist. And when the advantage derived from superinfection is small compared with the differences in reproductive rate between the strains, the less virulent parasite can drive the more virulent one to extinction.

### Competing Virus From the Same Infection

An interesting hypothetical possibility is that the diversities of virus genotypes evolve during the infection of an individual after the transmission of a limited range of types. Here, those virus types replicating fastest are most likely to be transmitted, which will favor an increase in the average rate of replication of the virus population. However, some slower-growing viruses may also be transmitted. If subsequent infection from a slower-growing virus strain can last for a longer period, then it may cause more new infections. One would then expect a population with a range of viral strains, the replication rates of which would depend on the relative probability of faster-growing virus strain being transmitted early and slower-growing populations producing more new infections. The situation is analogous to the trade-off described by Levin and Pimental (13) during which coexistence can be achieved by different viral genotypes. The argument is illustrated schematically for an infection with an average reproductive rate of ⁸/₇ (Fig. 4). Here, the evolution of different

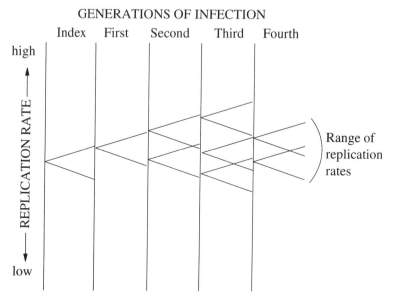

**FIG. 4.** A schematic illustration of the chain of diversity in replication rates expressed by virus in a chain of infections, assuming that the virus evolves new strains in the course of an individual infection. The vertical lines represent transmission to novel hosts. Initially a single-replication rate is assumed, but as infection progresses, the number of strains increases. The faster-replicating viruses within a host are more likely to be transmitted. However, virus populations with a range of less rapidly replicating virus are likely to infect more susceptible hosts. Therefore, it is likely, assuming that other influences such as cell tropisms, antigenicity, and structural viability, are not acting, that, for rapidly changing viral genotypes, a distribution of types will persist in the host population.

replication rates in the hosts is shown. It is clear from this hypothetical example that stochastic effects allowing different infections to generate new infections (which will depend on the range of types transmitted) will influence the outcome, producing the observed distribution of viral types within a host population. This illustrates how the rates of replication and, hence, the virulence of parasites could reach a steady intermediate level. However, the argument advanced here is an obvious simplification, one would expect the distribution of replication rates of the virus within a population to be determined by the mechanisms of pathogenesis, transmission, and virus reproduction. Clearly, this subject requires much more careful and rigorous analysis in the future.

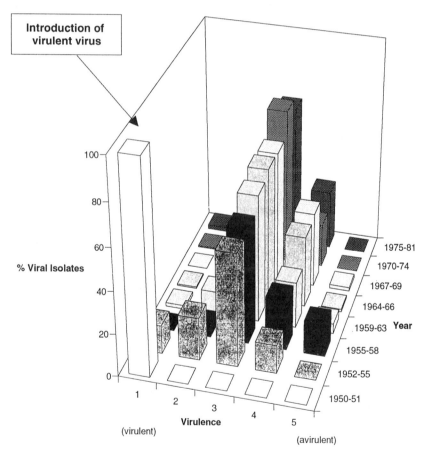

**FIG. 5.** Changes in the virulence of myxoma over time are shown. The proportions of myxoma virus with different degrees of virulence observed in Australia since the release of a virulent strain. The strains shown are: *1*>99% mortality in infected rabbits; *2*, 95–99% mortality; *3*, 70–95%; *4*, 50–70%; and *5*<50%. A rapid evolution to between 50% and 90% mortality was observed. (From Fenner, ref. 15.)

## The Virulence of Myxomatosis

The theoretical observations derived in the foregoing require testing through empirical observations. The interaction of myxoma virus released into a rabbit population remains the classic natural "experiment" on this subject, which has an extensive set of observations of the process of virus–host evolution, as well as a direct, albeit inadvertent and incomplete, test of the theory described. Following the introduction of highly virulent myxoma virus to control rabbit populations in Australia and Europe, viruses with intermediate, rather than low or high levels of virulence attained the maximum density (14). The proportion of infections belong to strains with different virulences (ranging from more than 99% mortality in strain 1 to less than 50% mortality in strain 5) at different times since the release of the virus are presented in Fig. 5. These observations are consistent with the main prediction of our analysis in that an intermediate, rather than a high or low, level of myxoma virulence evolved and was maintained. The reasons postulated for this result (13,16,17) are consistent with the predictions of our analysis. Very virulent myxoma strains kill their hosts early and, consequently, are transmitted between hosts to a lesser extent than are strains of lower virulence. Benign strains have a transmission disadvantage because they are cleared before they are transmitted.

## THE PERSISTENCE OF VIRUS WITHIN HOST POPULATIONS

A virus is inextricably dependent on a host; consequently, if individual hosts can clear viral infections and produce lifelong resistance to subsequent reinfection, then the success of a virus will be limited by the size of the host population, or by its ability to overcome these obstacles.

### The Threshold Population Size

For a virus to persist in a host population, the mean effective reproductive rate $R_e$ must be equal to 1. This value for a directly transmitted virus is given in Eq. 4 in a simple model where lifelong host immunity is produced by infection. For values of the transmission coefficient and the rate of recovery that will be particular to each viral infection there will be a threshold host population size that allows persistence of the virus $N_T$. This threshold will depend on the rate at which new susceptible hosts enter the population, which in this model is denoted by the death rate $\mu$, which equals the birth rate. Hence

$$N_T = \frac{\sigma + \mu}{\beta} \qquad [18]$$

If the population size is not constant, then the death rate is no longer the same as the birth rate and should be replaced by it. This simple derivation of the size of a host population that can maintain a virus population has been widely applied.

By using similar methods, but also including the effects of stochastic events that may allow a virus population to become extinct in a host population greater than $N_T$, as just derived, Bartlett (18) predicted that measles would be unable to persist in a population of fewer than approximately 250,000. This prediction was investigated by observing the number of times measles infections were absent from reports within cities in both the United Kingdom and the United States (18,19). Measles was much more likely to "fade-out" in cities with populations of about 300,000 or fewer. For example, in Dallas, which had 295,000 inhabitants in 1940, there were 18 fade-outs between the years 1921 and 1940, whereas there were none in San Francisco with a population of 635,000 (19). To reduce the influence of reintroductions masking fade-outs and to control for population density, Black (20) studied discontinuities in measles incidence in island populations. In all islands with fewer than 500,000 inhabitants, measles was not endemic, and it was observed that the more dispersed the population, the longer epidemics lasted. The evolutionary implication of such studies is that measles must have emerged since the formation of human civilizations with population groups large enough to support the infection.

One factor that determines the threshold population size below which an infection cannot persist is the birth rate, which controls the supply of new susceptibles ($\mu$ in Eq. 18). In small populations with high turnovers, such as mice colonies, it has been observed in experimental situations that mousepox can persist in populations of about 300 hosts (21,22). When host populations have a slower generation time, directly transmitted infections are then more likely to be found in animals that live in large colonies or herds. However, the example of rinderpest in the large herds of wildebeest also illustrates how seasonal changes in transmission can influence viral persistence in host populations. Before the vaccination of domestic cattle, it was believed that rinderpest would persist in the vast wildebeest herds of the Serengeti (23). However, this was not true (24): rinderpest has a very high transmission probability, and wildebeest have a short-breeding season (25), which means that susceptible numbers are high for only a short time. This seasonal variation in susceptible numbers appears to have caused a break in the spread of the infection, which is no longer endemic in the wildebeest population.

## Mechanisms that Facilitate the Persistence of Virus Populations

Other viral infections have mechanisms that allow their populations to remain viable while there is a buildup in the number of susceptible hosts after the fade-out of acute infections. The herpes viruses remain latent within hosts for long periods before reactivating to cause infectious disease, which allows them to persist within very small populations, such as those found on the Shetland Islands (26). Another example is hepatitis B virus, which has a chronic carrier state in some hosts. Such mechanisms can be captured within the expression for the reproductive rate. For example infection with varicella–zoster virus (VZV) causes an acute phase (chickenpox), with a transmission parameter $\beta_1$ and a recovery rate $\sigma_1$. There is also the possibility of reactivation to cause shingles, which is transmissible. Thus, there remains a much lower transmission probability $\beta_2$ that remains for the rest of the host's life. The reproductive rate, which must be greater than 1 for persistence is then

$$R_0 = N\left(\frac{\beta_1}{\mu + \sigma_1} + \frac{\beta_2}{\mu}\right) \qquad [19]$$

The transmission parameter $\beta_2$ may be very small, but it acts for a much longer period than that produced by chickenpox.

The enteroviruses, although being transmitted mainly by direct contact, have environmental reservoirs for infection, such as water supplies, where the virus will persist. Other viruses infect more than one host. An example of this is provided by rinderpest, which, as described earlier, appears to have been maintained in wild ungulates by a reservoir of infection in domestic animals. In most instances of cross-species infection, the virus population is maintained in a main host, such as rodents for the arenaviruses, or foxes and raccoons for rabies, and occasionally enters the other species such as humans. The division of hosts into separate populations with a degree of interaction could also increase the initial reproductive rate, because persistence will depend on the sum of the within-group reproductive rate and the between-group reproductive rate (17).

Sexually transmitted viruses are not affected by the threshold population size. The transmission of these infections is not dependent on population density, but only on the formation of sexual partnerships. This means that sexually transmitted diseases (STDs) can persist in small populations. In a population of identical individuals the reproductive rate of HIV is given by

$$R_0 = \beta c D \qquad [20]$$

where $D$ is the mean duration of infectiousness and $c$ is the rate of acquisition of new sexual partners. The value of $\beta$ has been estimated as about 0.15 per sexual partnership (the value appears to be approximately 0.1 from women to men and 0.2 from men to women) and $D$ as 8 years (27). For HIV to invade a homogeneous population, the rate of sexual partner acquisition would have to be over 0.8 per person per year. However, there is behavioral heterogeneity within the population, which can allow the persistence of STDs in what have been called core groups (28). In our simple model, these can be allowed for by using an *effective* rate of partner acquisition, which is the mean rate plus its variance divided by its mean (29). This reliance on small subsets of the population, called core groups, for much of their spread limits the prevalence of most STDs. The rapid spread of HIV to many people is more a consequence of its long asymptomatic infectious period than of the sexual route of transmission.

## The Emergence of New Viral Infections

The emergence of HIV, with its attendant mortality, has concentrated attention on the ways in which new viruses arise and on how their success is determined. Once a new virus emerges through (a) radical genetic change in another virus; (b) a change that allows the virus to cross a species barrier, such as the new morbillivirus infecting seals; (c) sympatric or allopatric speciation of old virus types; or (d) rogue oncogenes from a host, then the ability of a new virus to invade its host will depend on the initial reproductive rate. It has been suggested that social and economic developments have made the emergence of new viruses more likely. The emergence of new viruses will be hastened by the existence of more hosts and, hence, more viruses overall in which mutations may generate new species. The major influence of larger human population sizes and densities and more frequent contacts between population groups will be to provide more susceptible individuals and higher transmission probabilities for a new virus, making its invasion more likely. There are several examples of viruses that occasionally pass from animals to humans, such as the Marburg and Ebola viruses from monkeys and the arenaviruses from rodents. These will emerge as serious problems only if their transmission between humans becomes sufficient for their persistence in the absence of their original hosts, as happened with HIV.

## ANTIGENIC VARIATION

The persistence of a viral infection, such as measles, depends on the replacement of individuals who have no previous experience of infection, because of the long-lasting immunity developed during an infection. This same immunity is produced by infections with influenza. However, protection is only partial against new antigenic variants of influenza. The importance of antigenic change to the success of influenza virus is often referred to, but is complicated because of the two types of antigenic change involved (30). Influenza undergoes shifts in the genotype coding for the two surface glycoproteins, hemagglutinin (HA) and neuraminidase (NA), which are caused by reassortment of different genotypes infecting the same cell. A *shift* in the HA type has been defined as involving a change in the amino acid composition of all four antigenic regions identified on its globular tip (31). Antigenic *drift*, on the other hand, is the gradual mutation of the gene sequences for the glycoproteins. This drift occurs constantly at a slow rate. For example, in a recent study of hemagglutinin from influenza cases in Finland between 1984 and 1990, 12 amino acid changes from 345 residues were observed (32). Because these changes are in the antigenic components of the viruses, there may be some loss of protective immunity in previously infected hosts as drift increases, and it has been suggested that the antigenic drift is "the answer of the virus to the progressive build up of serum antibodies during recurrent waves of the same virus subtype" (33). However, the transmission dynamics of influenza suggest that no such answer is required. In a serological survey for influenza antibodies, a mean age of infection of 5.5 to 7 years was estimated for the population of South Yorkshire (34). This allows a rough calculation of the initial reproductive rate of infection, which is approximately the mean life expectancy of the host population divided by the mean age of infection (1). If we estimate the mean life expectancy of population of South Yorkshire as 75 years, then the threshold population size for influenza's persistence is approximately 70,000. The population of South Yorkshire is roughly 1 million. Even if this estimate of the threshold population size is far too low, it is likely that influenza would currently persist without antigenic drift. If antigenic drift did have a function in the maintenance of the virus, then it must have been within the smaller populations of the past.

The influence of the immune system on the process of antigenic drift is readily apparent in the phylogenetic tree for influenza (Fig. 6a). There is little branching in this tree because the rate of evolution of influenza is slow in comparison with an infection such as HIV which evolves within the host. Additionally, strains do not seem to remain within the population; as new strains emerge, old strains die out (35). This suggests that the immune system is effective against individual viral types and, when combined with cross-reactive neutralization from newly emerged types, will prevent the previous types from persisting. In other words, because the immune system applies a substantial selection pressure on influenza, there is competitive exclusion of the viral types as antigenic drift progresses. In the same way, because there is less previously acquired specific protection against new antigenic types, they will predominate. This selective advantage should exist as soon as any proportion of the population is immune to a previous type, which raises the interesting question as to why antigenic drift progresses at the rate observed.

The interaction of the immunity of hosts with antigenic drift is markedly different

**FIG. 6.** Phylogenetic trees derived for influenza and HIV. The influenza tree (**a**) shows the relationships between Hong Kong subtypes for the years 1968 to 1980 (35). The gene coding for the H3 hemagglutinin was sequenced for virus in different populations at different times. The difference between the 1968 NT68 strain and the 1979 BK179 strains is 73 bases. A single type was derived for each patient. In the HIV tree (**b**), a range of types were sequenced from each patient. In the tree the two extreme types are represented with an *x* and a *y*. The phylogenetic tree was derived by sequencing the HIV-1 *env* V3 region for a single dentist and his HIV-infected patients (patients A to F). Additionally two local control HIV-positive persons were included (LC2 and LC3) (36). In both trees the horizontal differences represent the percentage of bases that are different between the sequences; the vertical distances are for convenience. It is clear from the radically different time scales that, whereas influenza changes gradually over the years between different hosts within a population, HIV evolves very rapidly within the individual hosts and on transmission. The comparison between the two viruses may be influenced by a difference in the number of amino acid substitutions that are inviable. However, it does appear that the HIV tree is much more branched than that for influenza.

(a)

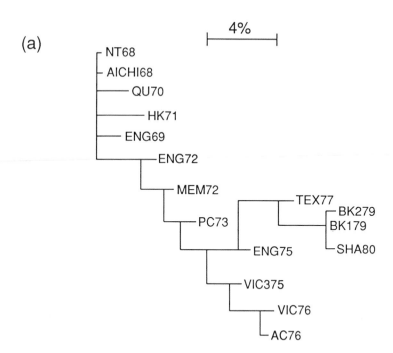

(b)

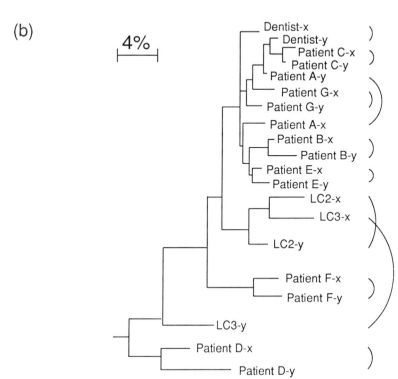

from that with shifts. Here, there is little, if any, cross-immunity, and major influenza pandemics occur, such as those seen with Spanish, Asian, and Hong-Kong influenza (30). The "success" of such emergent types has been cited as an evolutionary advantage of sex (37). However, the long gap between occurrences of shift and the presence of antigenic drift, which would maintain infection in small populations, suggests that this is not so, and that the recombination of virus from different host species is a chance event. It has been suggested that serotypes cycle over a 60- to 70-year period to allow the host population to become totally susceptible to the virus (38). This long wait would be unnecessary; it would take much less than 60 years for the birth of sufficient susceptibles to allow a new pandemic.

The human immunodeficiency virus (HIV) shows a fundamentally different rapid antigenic change, as is clear from its phylogenetic tree (see Fig. 6b). The tree describing influenza's evolution branches over a period of years, whereas HIVs branches between individual hosts, and there is a wide range of antigenic types within each host. The comparison shown is between surface antigens of the two viruses, but there may be functional constraints on the possible amino acid changes. However, the shape of the trees suggest that the immune system is much less influential in controlling the evolution of HIV. The immune response to HIV is not effective in clearing infection, and new antigenically distinct strains (quasispecies) emerge within the host. The difference in the rate of change is also obvious; the change in influenza observed over a period of a few years can occur for HIV during the passage through a few individuals, although the duration of infection is of a different time scale.

The main viremia occurs at two stages in HIV infection, both early and late in the course of infection. This is when transmission is believed to be most likely. Limited evidence suggests that only certain strains of HIV are transmissible (39). It is interesting to speculate whether these strains are those that replicate fastest and are preponderant in the population, or whether they have particular cell tropisms that facilitate transmission. In this instance, the long duration of infection means that it may not be the quasispecies that flourish earliest that are the most likely to infect new hosts. There is clearly rapid evolution of HIV, enabling it to invade a wide range of new cell types following its initial entry in a host. However, many virus types will be selected out by the transmission event. Because we have no definite idea of the role hypervariability plays in the persistence and extent of an HIV infection, we can say little about the selective pressures acting on HIV within the host.

Evidence that there are some selection pressures constraining the divergence of retrovirus strains is present when the genotypes of different genera of retrovirus are compared (40). The relationship between genera can be traced from shared amino acid sequences in several proteins. The rapid rate of evolution of retroviruses observed would not have allowed such similarities if much of the divergence was not selected for. One important selection pressure is the use of antiviral drugs. The rapid change in HIV genotypes means that there can be a rapid response to the selection pressures induced by the use of antiviral drugs. Virus with tropisms for cells where levels of the drugs are low or virus where reproduction is least interfered with may be selected for. Such drug resistance is observed in HIV (41), and it appears that resistance commonly evolves during an infection (42), which may help explain why treatment with some antivirals is unable to effect a cure.

It is interesting to speculate how prominent a role antigenic variation plays in other viral infections, particularly those in which the virus remains in the host for long periods. Other RNA viruses, such as poliovirus, undergo rapid evolution, but the significance of this is unknown (43). Hepatitis B virus, which produces chronic infections, is particularly interesting, as it appears to have an unusual replication mechanism that makes it very like retroviruses in terms of an altering genotype change (44). It could be that rapid antigenic change plays a part in the persistence of hepatitis B infections. This does not appear to be true with the herpesviruses, such as varicella–zoster virus. This member of a family of DNA viruses does exhibit some genomic variation (45). However, it appears that latency occurs in the dorsal root ganglia where little of the viral genome is expressed on the surface of infected cells (46).

## CONCLUSIONS

In this chapter we have described simple mathematical models that provide a precise framework with which to discuss and explore the population dynamics of viral infections and how they relate to the evolution of viruses. A large part of the chapter is devoted to the question of virulence. This is the topic that has tended to dominate the subject, partly because, as humans, we suffer from viral-induced morbidity and mortality and partly because the intuitively attractive view that parasites will evolve to benignity was held for so long. In the chapter we have also explored some of the other areas for which population dynamics can provide insights into viral evolution. In particular, the constraints on the ability of viruses to successfully invade and persist in host populations.

Much of the work described is based on theoretical insights. However, the hypotheses generated by such methods must be tested by observation and experiment. Here we have described some observations from viral epidemics that have been recorded in the interests of public health, which provide some support for the hypotheses presented, but virus–host systems with rapid generations that can be readily manipulated are necessary for experiments. One such system is provided by phage infections of bacteria and valuable work using this system is described by Lenski and Levin (47).

In recent years the advances in molecular biology have provided many insights into molecular epidemiology. These we have illustrated by a comparison of the molecular epidemiology of influenza virus and HIV. The former changes on a long time scale between hosts and is selected for by host immunity preventing new infections, the other evolves rapidly within hosts and is likely to undergo selection on transmission, depending on the mechanism of transmission, rather than on its ability to avoid herd immunity. Our understanding of such processes and the modeling of viral populations within hosts is at an early stage, but with the rapid developments in molecular biology, our understanding of virus population dynamics is likely to improve rapidly in the coming years.

## REFERENCES

1. Anderson RM, May RM, *Infectious diseases of humans: dynamics and control*. Oxford: Oxford University Press; 1991.
2. Novak MA, Anderson RM, McLean AR, Wolfs TFW, Goudsmit J, May RM. Antigenic diversity thresholds and the development of AIDS. *Science* 1991;254:963–969.
3. Dietz K. Transmission and control of arbovirus diseases. In: Ludwig D, Cooke KL, eds. *Epidemiology*. Proceedings of the Society for Industrial and Applied Mathematics, Philadelphia; 1975:104–121.
4. May RM, Anderson RM. The transmission dynamics of human immunodeficiency virus (HIV). *Philos Trans R Soc Lond Ser B* 1988;321:565–607.
5. Bartlett MS. *Stochastic population models in ecology and epidemiology*. London: Methuen; 1960.
6. Bailey NTJ. *The mathematical theory of infectious diseases and its application*. London: Griffin; 1975.
7. Sasaki A, Iwasa Y. Optimal growth schedule of pathogens within a host: switching between lytic and latent cycles. *Theor Popul Biol* 1991;39:201–239.
8. Dawkins R, Krebs JR. Arms races between and within species. *Proc R Soc Lond Ser B* 1979; 205:489–511.
9. Antia R, Levin BR. Within-host population dynamics and the evolution and maintenance of microparasite virulence. *Am Nat* 1992; in press.
10. Davis BD, Dubelco R, Eisen HN, Ginsburg HS. *Microbiology*. New York: Harper & Row; 1980.
11. Anderson RM, May RM. The invasion, persistence and spread of infectious diseases within animal and plant communities. *Philos Trans R Soc Lond [B]* 1986;314:533–570.
12. Bremermann HJ, Thieme HR. A competitive-exclusion principle for pathogen virulence. *J Math Biol* 1989;27:179–190.
13. Levin S, Pimentel D. Selection of intermediate rates of increase in parasite–host systems. *Am Nat* 1981;117:308–315.
14. Fenner F, Ratcliffe FN. *Myxomatosis*. Cambridge: Cambridge University Press; 1965.
15. Fenner F. Biological control, as exemplified by smallpox eradication and myxomatosis. *Proc R Soc Lond [B,]* 1983;218:259–285.
16. Anderson RM, May RM. Coevolution of hosts and parasites. *Parasitology* 1982;1985:411–426.
17. May RM, Anderson RM. Parasite–host coevolution. In: Futuyama DJ, Slatkin M, ed. *Coevolution*. Sunderland, Mass: Sinauer Associates; 1983.
18. Bartlett MS. Measles periodicity and community size. *J R Stat Soc Ser A* 1957;120:48–70.
19. Bartlett MS. The critical community size for measles in the United States. *J R Stat Soc Ser A* 1960; 123:37–44.
20. Black FL. Measles endemicity in insular populations: critical community size and its evolutionary implication. *J Theor Biol* 1966;11:207–211.
21. Greenwood M, Bradford-Hill A, Topley WWC, Wilson J. *Experimental epidemiology*. Medical Research Council Special Report No. 209. London: HMSO; 1936.
22. Fenner F. The epizootic behaviour of mousepox (infectious ectromelia of mice). II. the course of events in long continued epidemics. *J Hyg* 1948;46:383–393.
23. Mack R. The great African cattle plague epidemic of the 1890's. *Trop Anim Health Prod* 1970;2: 210–219.
24. Plowright W. The effects of rinderpest and rinderpest control on wildlife in Africa. *Symp Zool Soc Lond* 1982;50:1–28.
25. Watson RM. Reproduction of wildebeest *Connochaetes taurinus albjubastes* and *thomas* in the Serengeti and its significance to conservation. *J Reprod Fert* 1969; 6 (suppl):287–310.
26. Hope-Simpson RE. Studies on shingles—is the virus ordinary chickenpox virus. *Lancet* 1954; 2:1299.
27. Anderson RM, May RM, Boily MC, Garnett GP, Rowley JT. The spread of HIV-1 in Africa: sexual contact patterns and the demographic impact of AIDS. *Nature* 1991;352:581–589.

28. Yorke JA, Hethcote HW, Nold A. Dynamics and control of the transmission of gonorrhoea. *Sex Transm Dis* 1978;5:51–56.
29. Anderson RM, Medley GF, May RM, Johnson AM. A preliminary study of the transmission dynamics of the human immunodeficiency virus (HIV), the causitive agent of AIDS. *IMA J Math Appl Med Biol* 1986;3:229–263.
30. Cliff AD, Haggett P. Influenza: a changing virus with unique epidemic waves. In: *Atlas of disease distributions: analytical approaches to epidemiological data.* Oxford: Blackwell; 1988:235–244.
31. Glezen WP, Couch RB. Influenza viruses. In: Evans AS, ed. *Viral infections of humans: epidemiology and control,* 3rd ed. New York: Plenum Medical; 1989:419–449.
32. Kinnunen L, Ikonen N, Pöyry T, Pyhälä R. Evolution of influenza B/Victoria/2/87-like viruses: occurrence of a genetically conserved virus under conditions of low epidemic activity. *J Gen Virol* 1992;73:733–736.
33. Stuart-Harris SH, Schild GC, Oxford JS. *Influenza: the virus and the disease,* 2nd ed. Baltimore: Edward Arnold; 1985.
34. Edmonds PM. *Models and seroepidemiology of co-existing virus strains.* [PhD thesis]. London: Imperial College, 1991.
35. Both GW, Sleigh MJ, Cox NJ, Kendal AP. Antigenic drift in influenza virus H3 hemagglutinin from 1968 to 1980: multiple evolutionary pathways and sequential amino acid changes at key antigenic sites. *J Virol* 1983;48:52–60.
36. Ou C-Y, Ciesielski CA, Myers G, Bandea CI, Luo C-C, Korber BTM, Mullins JI, Schochetman G, Berkelman RL, Economou AN, Witte JJ, Furman LJ, Satten GA, MacInnes KA, Curran JW, Jaffe HW, Laboratory Investigation Group, and Epidemiologic Investigation Group. Molecular epidemiology of HIV transmission in a dental practice. *Science* 1992;256:1165–1171.
37. Hamilton WD. Recurrent viruses and theories of sex [Letter]. *Trends Ecol Evol* 1992;7:277–278.
38. Dimmock NJ, Primrose SB. The evolution of viruses. In: *Introduction to modern virology,* 3rd ed. Oxford: Blackwell Scientific; 1987:275–286.
39. Holmes EC, Zhang LQ, Simmonds P, Ludlam CA, Leigh Brown AJ. Convergent and divergent sequence evolution in the surface envelope glycoprotein of human immunodeficiency virus type 1 within a single infected patient. *Proc Nat Acad Sci USA* 1992;89:4835–4839.
40. Coffin JM. Genetic diversity and evolution of retroviruses. *Curr Top Microbiol Immunol* 1992; 176:143–164.
41. Richman DD. HIV drug resistance. *AIDS Res Hum Retroviruses* 1992;8:1065–1071.
42. Land S, McGavin C, Lucas R, Birch C. Incidence of zidovudine-resistant human immunodeficiency virus isolated from patients before, during, and after therapy. *J Infect Dis* 1992;166:1139–1142.
43. Kinnunen L, Pöyry T, Hovi T. Genetic diversity and rapid evolution of poliovirus in human hosts. *Curr Top Microbiol Immunol* 1992;176:49–61.
44. White DO, Fenner FJ. Hepadnaviruses. In: *Medical virology,* 3rd ed. Orlando: Academic Press; 1986:365–386.
45. Weller TH. Varicella–herpes zoster virus. In: Evans AS, ed. *Viral infections of humans: epidemiology and control,* 3rd ed. New York: Plenum Medical; 1989:659–683.
46. Cohrs R, Mahalingam R, Dueland AN, Wolf W, Wellish M, Gilden DH. Restricted transcription of varicella–zoster virus in latently infected human trigeminal and thoracic ganglia. *J Infect Dis* 1992; 166(suppl 1):s24–s29.
47. Lenski RE, Levin BR. Constraints on the coevolution of bacteria and virulent phage: a model, some experiments, and predictions for natural communities. *Am Nat* 1985;125:585–602.

*The Evolutionary Biology of Viruses,*
edited by Stephen S. Morse.
Raven Press, Ltd., New York © 1994.

# 4

# Methods of Evolutionary Analysis of Viral Sequences

## Andrew J. Leigh Brown

*Centre for HIV Research, Division of Biological Sciences, University of Edinburgh, Edinburgh EH9 3JN, Scotland*

There are many reasons why a clear understanding of the genetic relation among different strains of a virus and of their evolution is desirable. Such knowledge can provide information on the origins and geographic distribution of particular strains, on their routes of transmission, and for the development of vaccines. With the increasingly widespread use of rapid nucleotide-sequencing methods and, particularly, since the advent of the polymerase chain reaction, extensive genetic data have been obtained on many viruses. Whereas, previously, serological methods were mainly used to distinguish strains, detecting limited numbers of variable epitopes, now a much more complete description of the genetic relatedness among strains is possible. This has been particularly prominent in the analysis of the evolution and origins of influenza viruses (1–7) and of human immunodeficiency virus (HIV)-1, for which the determination of relatedness by sequence analysis has preceded a successful serological classification (8–14). Recently, there have been significant developments in the phylogenetic analysis of other viruses as well, notably papillomaviruses, feline immunodeficiency virus, and hepatitis C virus (15–17).

The principles involved in the analysis of genetic relation between viral strains, or phylogenies, are similar, whether the data consist of oligonucleotide fragments, restriction fragments, or nucleotide sequences. Evolution is conventionally described in terms of a bifurcating tree. Although not necessarily always formally correct (as strong selection can cause significant deviation from a continuously divergent tree), for the purposes of this discussion, it will be assumed. However, in all phylogenetic comparisons, it is necessary to be confident that the sequences being compared between strains did, at some point, derive from a common ancestral gene (i.e., they are truly *homologous*). By virtue of this definition, this is an all-or-none state. It is appropriate to refer to *percentage identity* and, if amino acids are grouped on the basis of physicochemical characteristics, *percentage similarity* can be used to refer to more distantly related proteins.

Phylogenetic trees can be *rooted*, which implies that the direction of evolution is known a priori, or *unrooted*, when it is not. In practice it is often most appropriate to adopt a procedure that gives an unrooted tree, but to impose a root by including an "outgroup"—perhaps a strain that is clearly the most distant relative of those of which relationships are under investigation. The major practical problems associated with the inference of phylogenies are the following:

1. The number of possible trees, from which the correct tree has to be chosen.
2. The branches do not diverge continually, often because strains vary in their rate of change.
3. There has been incomplete sampling of the tree.

The last is an issue in data collection, but the first two are problems of analysis and will be discussed in more detail.

## THE NUMBER OF EVOLUTIONARY TREES

It is often not fully appreciated how many possible trees exist that could describe the relation between even small numbers of strains. The number of unrooted trees is given by: $\pi(2i - 1)$ where $i$ takes values from 1 to $n - 2$ ($\pi$ is the symbol for multiplication, as $\epsilon$ is the symbol for summation). In Table 1 these figures are given for 1 to 10 strains (18). The number of possible unrooted trees for $n$ strains is the same as the number of rooted trees for $n - 1$ strains. For 20 strains the number of possible different unrooted trees will be approximately $3 \times 10^{23}$. With rapid-sequencing methods becoming widely used, data sets substantially larger than 20 are being generated. Clearly, any complete evaluation of that number of topologies is impossible and, in general terms, strategies must be used that can find the "correct" tree by searching only a subset of all possible trees.

## DIVERGENT AND CONVERGENT EVOLUTION

In the simplest evolutionary models, it is assumed that all changes are unique and no reversion or back-mutation occurs at any site in the entire phylogeny. The

**TABLE 1.** *Tabulation for Number of Unrooted Trees*

| Number of strains | Number of unrooted trees |
|---|---|
| 3 | 1 |
| 4 | 3 |
| 5 | 15 |
| 6 | 105 |
| 7 | 945 |
| 8 | 10,395 |
| 9 | $1.4 \times 10^5$ |
| 10 | $2.0 \times 10^6$ |

branches of the tree, therefore, will continually diverge from each other and all data sets obtained from the genomes being studied (for example, from different genes) will give the same topology. This view is completely appropriate only when individual characters are so complex that they are likely to arise only once. For molecular data, of which characters are increasingly nucleotides, amino acids, or restriction sites, this is not true. For example, a particular restriction site could be lost in parallel on different branches of the tree, but it is less likely to be reacquired. With sequence data, and especially for nucleotide sequences, it is quite likely that a given site could undergo parallel changes in different branches or take different paths before reaching the same state (convergence). Three different approaches to phylogenetic analysis have been developed that attempt, in different ways, to deal with the many possible trees and the problem of convergence.

## MEASURES OF GENETIC DISTANCE

One set of methods has been developed based on estimates of the pairwise genetic similarity (or distance) between strains. These originated in work on the classification of bacteria (19), but more recently metrics of similarity for restriction site (20) and nucleotide and amino acid sequence data (21,22) have been developed. Calculation of the metric for all pairwise comparisons of strains reduces the data to a single figure for each comparison. This generates a *distance matrix* that then forms the basis for estimation of the evolutionary relationship.

### Estimation of Nucleotide Distance

To an increasing extent the data becoming available for analysis of relationships among virus strains consist of nucleotide sequences. Although they certainly provide the most complete and direct information about genetic relations, nucleotide sequences suffer from an important limitation at the analysis stage. Because of the few possible character states—the four nucleotides—evolutionary convergence onto the same state by two unrelated sequences will occur relatively frequently. This will result in a significant *underestimation* of genetic distances, a bias that will increase when distantly related strains, or rapidly evolving genes, are being considered. Therefore, it is not sufficient to simply adopt a measure based on the observed similarity between strains; rather, it is essential to estimate the real divergence from the data by using one of several estimators that make some form of correction for "multiple hits," (i.e., multiple substitutions) at the same residue (for recent review see ref. 23). Those in most common use are the one-parameter model of Jukes and Cantor (21) and the two-parameter model of Kimura (22). In the former, it is assumed that any nucleotide is equally likely to be substituted by any other. Here, the estimated number of nucleotide substitutions ($K$) that have occurred since two strains diverged is given by

$$K = -3/4 \ln(1 - 4/3 \lambda) \qquad [1]$$

where $\lambda$ is the observed fraction of sites at which the two sequences differ. In the two-parameter model, transitions (purine–purine and pyrimidine–pyrimidine substitutions) and transversions (purine–pyrimidine substitutions) are considered separately. The estimate of the mean number of nucleotide substitutions is then

$$K = -1/2 \ln[(1 - 2P - Q)\sqrt{(1 - 2Q)}] \qquad [2]$$

where $P$ is the fraction of sites at which the two sequences differ by a transition substitution and $Q$ is the fraction at which they differ by a transversion; $P + Q = \lambda$. When transitions and transversions differ in frequency, then an improvement in the estimate can be obtained by using the Kimura expression.

However sophisticated the estimator, there are biologically realistic situations for which no correction will work well: when the probability of substitution varies according to position, all will give an underestimate, because they average across the entire sequence. This effect can be quite serious, for example, in the V4-C3-V5 regions of the *env* gene of HIV-1 analyzed by Balfe et al. (12), in which two rapidly evolving regions flank a conserved sequence, it would be appropriate to subdivide the sequence for analysis into fast and slowly evolving segments.

In protein-coding sequences, in which the amino acid sequence is relatively conserved, *synonymous substitutions* (those nucleotide substitutions that do not change in amino acid) and *nonsynonymous substitutions* (those that lead to an amino acid change) should be considered separately (9,24). Synonymous substitutions are usually then used in the estimation of evolutionary distance, because they are less affected by variation in the rate of change. A good example is the investigation of sequence variability in Newcastle disease virus by Sakaguchi et al (25).

## Phylogenies From Distance Matrices

Once a distance matrix has been constructed, the phylogeny can be inferred by applying one of a variety of clustering procedures. Several simple cluster analysis methods are available that give a representation of these relations. One of the simplest is the *unweighted pair–group mean* (UPGM) method (19). Developed to represent the phenotypic distance between organisms, this often used method is limited in its value in evolutionary analyses, as it assumes equal rates of change down different branches of the phylogenetic tree.

From a matrix of distances, such as that given in Table 2, the two strains with the shortest distance (*d*) are grouped. Here, strains A and B are grouped together: as the

**TABLE 2.** *Distance Matrix*

| Strain | A | B | C |
|--------|------|------|------|
| B | 4 | | |
| C | 9 | 7 | |
| D | 20 | 18 | 17 |

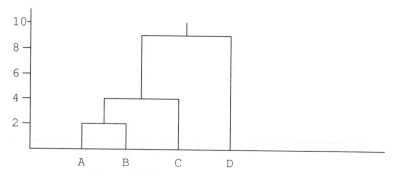

**FIG. 1.** UPGM tree based on data from Table 2.

rate of evolution is assumed to be constant, the branch length to each from their common ancestor, $d_{AB}/2$, is 2 units. A new matrix is now constructed treating A and B as a single composite group. The distances from the composite group (AB) to the other strains is taken as the arithmetic mean of their individual distances from A and B; that is,

$$d_{(AB)C} = \frac{(d_{AC} + d_{AB})}{2}$$

Here, the new matrix is therefore:

| Strain | (AB) | C |
|--------|------|---|
| C      | 8    |   |
| D      | 19   | 17 |

Again, the two closest units, (AB) and C, are grouped. The distance from each to their common ancestor is determined: $(d_{AC} + d_{BC})/2 \times 1/2 = 4$. The mean distance of the new composite group [(AB)C] to D, $(d_{(AB)D} + d_{CD})/2 = 18$, and so the branch length to D is 9 units. By convention, the tree is rooted at the midpoint of the longest branch. Therefore, the tree looks like that shown in Fig. 1.

A substantial amount of information can be lost when the UPGM method is used if rates of change often differ between branches. This can readily be demonstrated by clustering the same strains by using a different method devised by Fitch and Margoliash, which allows rates of change to differ within the tree, rather than assuming them to be equal.

In this procedure, the branch lengths are estimated from a series of simple simultaneous equations. These are set up using the data on the observed distances for each nearest-neighbor pair and the next closest sequence and permit us to assign individual lengths below each node. Following this procedure for the data in Table 2, we obtain the tree shown in Fig. 2, which even on this simple data set is rather different from that given in Fig. 1. Although this procedure is clearly better, it still

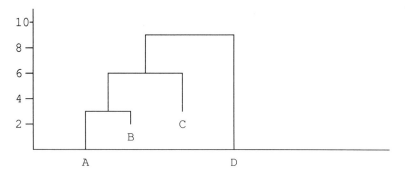

**FIG. 2.** Fitch–Margoliash tree based on data from Table 2.

assumes additivity in the data, and substantial departures from this will result in negative branch lengths.

A more recent procedure, which is extremely fast and highly efficient at finding the correct tree, is the neighbor-joining method of Saitou and Nei (26), which sequentially finds the nearest pairs of neighboring sequences that give the shortest overall length of the tree. For a comparison of these methods and an excellent introduction to molecular phylogenetics, the reader is referred to Li and Graur (27). A more detailed review of phylogenetic methods suitable for molecular data can be found in Felsenstein (28).

## CHARACTER-BASED METHODS

### Maximum Parsimony Trees

The most popular current approach to phylogeny reconstruction is that based on the principle of *maximum parsimony*. Unlike distance methods, parsimony methods are character-based and examine each site (character) in a sequence separately. Under the parsimony principle, it is assumed that the "best" tree is that which has the shortest *overall* length; therefore, the aim is to minimize the number of convergent substitutions. The principle behind the parsimony approach is as follows (18): The data are reduced first by elimination of all invariant sites and second by removal of those in which the variant residues (nucleotides or amino acids) are found only in a single sequence. Sites at which two or more residues are each represented more than once in the data, are termed *phylogenetically informative*. The tree topology that minimizes the total number of substitutions at these sites is selected. Quite often, there is not one single most parsimonious tree, in which event some additional criterion might be used to distinguish the alternatives.

In Table 3, four 7-nucleotide sequences are shown of which site 7 has no variation, and sites 3, 4, and 6 distinguish only a single sequence. Three sites (1, 2, and 5) are phylogenetically informative. At these three sites, strains b and c shared

**TABLE 3.** *Informative Nucleotide Sequences*

| Strain | Sequence | | | | | | |
|--------|---|---|---|---|---|---|---|
| a | T | A | C | T | C | G | G |
| b | T | T | A | C | A | C | G |
| c | G | T | A | C | A | C | G |
| d | G | A | A | C | T | C | G |
| Site | 1 | 2 | 3 | 4 | 5 | 6 | 7 |

nucleotides at both positions 2 and 5, whereas strains a and b and c and d share nucleotides at position 1 only. The tree with the fewest changes overall is then adopted. In this simple example, the parsimony tree takes the structure shown in Fig. 3.

The number of changes in each branch between a pair can be deduced by comparison of the sequences involved, together with those of the nearest neighbor. Thus, b and c are grouped together, but to determine on which branch from node α the T/G change at site 1 occurred, they are compared with sequence d. As there is a G at this site, this is deduced to be the ancestral state, and the change is assigned to branch b, written as "1,1," indicating one change at site 1, with the ancestral sequence being GTACACG. Use of sequence a in the same way allows the common ancestral sequence at node β to be partly reconstructed. At site 2, sequences b and c both have T, and d has an A. As sequence a also has A, this is deduced to be the ancestral state, and the change at this site is assigned to the internode between β and α. However, no decision concerning site 5 can be reached from these data, hence, the ancestral sequence at β is written GAACA/TCG.

In data sets for which the evolutionary pattern is generally divergent, parsimony methods may perform well and give not only a good estimate of the correct tree, but also of the branch lengths. Under such circumstances, the programs in the PAUP package distributed by David Swofford (see the concluding section for details) will be particularly useful because of their speed of execution and ease of use, and their use was demonstrated in the analysis of sequence diversity in type 3 reoviruses by Dermody et al. (29).

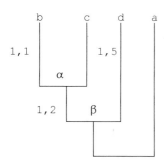

**FIG. 3.** Maximum parsimony tree based on the data given in Table 3 (not to scale).

## Limitations to Parsimony

The circumstances for which parsimony methods will not perform well are those in which there are many convergent substitutions in the data. This is particularly relevant when evolution is rapid and the rate is variable (30). In influenza A and HIV-1, regions of certain proteins where rapid change is observed have been implicated as being involved in escape from immune selection (1,31,32). If there are also constraints on the direction of sequence change, then convergent substitution will occur. Parsimony methods then are likely to give the wrong tree, because they are intended to minimize the number of such events. Recent results suggests that the V3 hypervariable region of gp120 in HIV-1 is an instance in which such effects occur (14).

# STATISTICAL SIGNIFICANCE OF TREE TOPOLOGIES

## Bootstrapping

The topology obtained by any of these procedures should be viewed as only an estimate of the phylogenetic relation among a group of strains or sequences. In the interpretation of the tree, it is important to know whether two groups on the tree are really distinct, or whether their apparent separation falls within the error associated with estimating the tree. Most methods for estimating tree topologies do not give any information on the statistical significance of their structure. This limitation has usually been dealt with by the use of a nonparametric statistical procedure, known as bootstrapping, to evaluate the significance of a particular topology (33).

Bootstrapping involves resampling the data on which the tree was based to generate a distribution of data sets, from each of which a new tree is determined. The frequency at which particular branches are observed in the resampled data sets then allows probability statements to be attached to them in the original tree. This procedure is highly suitable for use in conjunction with maximum parsimony and neighbor-joining procedures and, as a matter of routine, should be incorporated into all studies that are based on them.

## Maximum Likelihood Trees

An alternative approach to phylogenetic inference that is based on maximum likelihood estimation has the very important advantage in that it immediately provides an estimate of the probability associated with a given branch or internode distance. In addition, it can be used to evaluate the relative likelihood of two different topologies.

Maximum likelihood procedures involve the estimation of the likelihood of a particular tree given a certain model of nucleotide substitution (34,35). Their advantages are threefold. First, they are based on an explicit model of sequence evolution, which, in principle, can be tailored to the particular problem under investigation. Second, likelihood procedures associate a probability statement with each internodal segment, as indicated in the foregoing. Third, in contrast with distance matrix methods, the complete nucleotide sequence information is used in the determination of the phylogeny. However, the individual computations of likelihood are very time-consuming, and the use of the likelihood approach has generally been restricted to small data sets. With the increasing availability of fast work stations that can be dedicated to such problems, the DNAML likelihood program in the PHYLIP 3.4 package should become applicable to much larger data sets.

In conclusion, it should always be remembered that any tree, whether it is "most parsimonious" or has the "maximum likelihood" is no more than a working hypothesis for the relations among a set of strains or species. All methods make some assumptions about the nature of the evolutionary process; hence, to increase confidence in the result, it is advisable to use more than one method for all major analyses. Concordance between different methods then adds significantly to one's confidence in the result, whereas differences may point up some interesting features of the data, such as convergence arising from selection.

## SOFTWARE FOR PHYLOGENETIC ANALYSES

There is a large choice of methods available for phylogenetic analyses, many of which incorporate the parsimony principle. Programs for executing such methods, which can often be run within a reasonable time on a personal computer or desk top work station, can be selected from software packages distributed at little if any cost by Drs. J. Felsenstein ["PHYLIP" (PHYLogeny Inference Package), version 3.4], and D. Swofford ["PAUP" (phylogenetic analysis using parsimony), version 3.0] (Addresses in Bibliographic Note). Other programs that might be considered include "MacClade," written for Apple Macintosh computers by Wayne Maddison at the University of Arizona (suitable for character state data only); "CLUSTAL" (36), a multiple alignment program developed by Dr. Desmond Higgins, now at EMBL, which estimates phylogenies while aligning sequences and now incorporates the neighbor-joining procedure with a bootstrap analysis; and "TreeAlign," similar to CLUSTAL, which was developed by Dr. Jotun Hein and is also now available from EMBL. It should be noted that confidence intervals cannot be placed on the phylogenies produced by simultaneous alignment/tree-building programs.

## REFERENCES

1. Buonagurio DA, et al. *Science* 1986; 232:980–982.
2. Both GW, et al. *J Virol* 1983; 48:52–60.
3. Daniels RS, et al. *J Gen Virol* 1985; 66:457–464.
4. Hayashida H, et al. *Mol Biol Evol* 1985; 2:289–303.
5. Yamashita M, et al. *Virology* 1988; 163:112–122.
6. Cox NJ, et al. *J Gen Virol* 1989; 70:299–313.

7. Donis RO, et al. *Virology* 1989; 169:408–417.
8. Alizon M, et al. *Cell* 1986; 46:63–74.
9. Li W-H, et al. *Mol Biol Evol* 1988; 5:313–330.
10. Yokoyama S, et al. *Mol Biol Evol* 1988; 5:237–251.
11. Smith TF, et al. *Nature* 1988; 333:573–575.
12. Balfe P, et al. *J Virol* 1990; 64:6221–6233.
13. Myers G, et al. *Human retroviruses and AIDS 1992*. Los Alamos, NM: Los Alamos National Laboratory, 1992.
14. Holmes EC, et al. *Proc Natl Acad Sci USA* 1992; 89:4835–4839.
15. Chan S-Y, Bernard H-U, Ong C-K, Chan S-P, Hofman B, Delius H. *J Virol* 1992; 66:5714–5725.
16. Rigby M, Holmes EC, Pistello M, Mackay N, Leigh Brown AJ, Neil J. *J Gen Virol* 1993; 74:425–436.
17. Simmonds P, McOmish F, Yap PL, Chan S-Y, Lin CK, Dusheiko G, Saeed AA, Holmes EC. *J Gen Virol* 1993; in press.
18. Fitch WM. *Am Nat* 1977; 111:223–257.
19. Sokal RR, Sneath PHA. *Principles of numerical taxonomy*. San Francisco: WH Freeman, 1963.
20. Nei M, Li W-H. *Proc Natl Acad Sci USA* 1979; 76:5269–5273.
21. Jukes TH, Cantor CR. In: Munro HN, ed. *Mammalian protein metabolism III*. New York: Academic Press, 1969.
22. Kimura M. *J Mol Evol* 1980; 16:111–120.
23. Gojobori T, et al. *Methods Enzymol* 1990; 183:531–550.
24. Leigh Brown A, Monaghan P. *AIDS Res Hum Retroviruses* 1988; 4:313–330.
25. Sakaguchi T, et al, *Virology* 1989; 169:260–272.
26. Saitou N, Nei M. *Mol Biol Evol* 1987; 4:406–425.
27. Li W-H, Graur D. *Fundamentals of molecular evolution*. Sunderland, Mass: Sinauer Associates, 1991.
28. Felsenstein J. *Annu Rev Genet* 1988; 22:521–565.
29. Dermody TS, et al. *J Virol* 1990; 64:4842–4850.
30. Felsenstein J. *Syst Zool* 1978; 27:401–410.
31. Wolfs TFW. *Virology* 1991; 185:195.
32. Wolfs TFW, et al. *Virology* 1992; 189:103.
33. Felsenstein J. *Evolution*. 1985;39:783–791.
34. Felsenstein J. *J Mol Evol* 1981; 17:368–376.
35. Bishop MJ, Friday AE. *Proc R Soc Ser B* 1985; 226:271–302.
36. Higgins DG, Sharp PM. *Gene* 1988; 73:237–244.

# PART II

## Viral Phylogeny: Origins and Relatedness

*The Evolutionary Biology of Viruses,*
edited by Stephen S. Morse.
Raven Press, Ltd., New York © 1994.

# 5

# Evolution of Viruses as Recorded by Their Polymerase Sequences

Jaap Heringa and Patrick Argos

*European Molecular Biology Laboratory, Postfach 10.2209,*
*D-69012 Heidelberg, Germany*

Viral polymerase sequences have proved themselves scientifically significant. Among all the primary structures on the viral genomes, they are the most closely related, especially as juxtaposed to the most distant, often unrecognizable, correlated capsid proteins. Viral molecular taxonomy and evolutionary phylogenetic trees became realities with the polymerase structures and suggested surprising relationships across various viral species, hosts, and symptomatic expressions. Conserved subsequence motifs critical for polymerase structural stability and functional attributes were delineated and were also found in many host polymerase sequences, establishing categorically the intimate ancestral relations between parasite and the infected. Some of these motifs have also been associated with the primary sequence of *Escherichia coli* DNA polymerase I, the tertiary fold of which is known from x-ray crystallographic experiments. This connection has enabled development of models for other polymerases and has led to a barrage of experiments that have used site-directed mutagenesis and have attempted polymerase engineering.

There are four basic types of polymerases, which are classified according to their template and product dependencies among DNA and RNA. The RNA-dependent RNA polymerases bind RNA, which they depend on for a template, and catalyze the production of an RNA polymer. The RNA-dependent DNA polymerases are reverse transcriptases, as they associate with RNA, but yield in a reverse sense a DNA product. The remaining two classes comprise DNA-dependent polymerases, each with respective products of DNA and RNA. The term *directed* is sometimes used in lieu of *dependent*. In this review, the following short-hand notation will be used to identify the polymerase types: D for DNA, R for RNA, the first-named letter referring to the polymerase dependency, and the following letter indicating the product type, and pol for polymerase. For example, DR pols are simply DNA-directed RNA-yielding polymerases.

The RR and RD pols and their motifs and evolutionary relationships have been extensively studied, and many recent reviews abound. In this chapter only the results for RNA-pols (RD and RR pols) will be summarized and some of the major achievements made salient; knowledge of the bulk of the research efforts can be had (directly or indirectly) by the reader from the references given. The situation is somewhat different for the DNA-pols (DR and DD pols) and, thus, some details will be related in the present exposition. The present work will concentrate on the identification of conserved 3'-5' exonuclease subsequence motifs in some DD pols, as well as related motifs in polymerase domains of some DNA and RNA pols. Phylogenetic trees constructed from the motifs will be discussed. Experiments justifying or supporting the critical importance of the subsequence regions to DNA-pol function will be given. Finally, a suggestion for a possible three-dimensional fold for many of the DD and DR pols is presented.

## RNA POLYMERASES: A SUMMARY

Kamer and Argos (1) were the first to identify sequence relationships between the known poliovirus RR polymerase and those from several positive-stranded RNA viruses, which included picornaviruses, Sindbis virus, and cowpea, tobacco, alfalfa, and brome mosaic viruses. Haseloff et al. (2) also observed independently similar relationships, but only for the tobacco, brome, and alfalfa mosaic viruses. Several conserved motifs (subsequence segments about 10 to 20 residues in length) were observed by Kamer and Argos (1); however, one was particularly well preserved and consisted of a central Gly-Asp-Asp triplet (GDD) flanked by pentapeptides composed mostly of hydrophobic residues, hinting at a β-hairpin tertiary structure composed of two hydrogen-bonded antiparallel β-strands separated by a short exposed loop encompassing the GDD amino acids. Such a motif was also found in phage Qβ replicase and in reverse transcriptases (RD pols) from various retroviruses, influenza virus, and the cauliflower mosaic DNA virus. These observations suggested that many viral species, across many hosts, and with great variety in structural and infective features and genomic organization, all could be evolutionarily and divergently related. In fact, scoring criteria based on the degree of similarity for the pol homologies showed that the plant cowpea mosaic virus (CPMV) was more closely related to the animal picornaviruses such as polio or foot-and-mouth disease, than to the other plant viruses such as alfalfa or brome mosaic. The CPMV shares many more structural features and infective mechanisms with the picornaviruses. Thus, viruses are first likely to have developed their strategies along given paths and, once effective, spread across various hosts and species. Apparently, structural and functional development superseded development in and for particular host types, such as plants or animals, and required much more time and evolutionary effort.

Many more RNA pol sequences have been discovered since the early efforts, and subsequent conserved motifs have been added to the ever-expanding list of pols. In 1989, 5 years after the Kamer and Argos (1) observations, three important research reports concentrating on viral evolutionary relationships appeared. Doolittle et al.

(3) examined the origins of retroviruses through sequence alignments of their proteinases, reverse transcriptases, and ribonuclease HH II segments. Habili and Symons (4) concentrated on evolutionary relationships among luteoviruses and other RNA plant viruses, with reliance on conserved sequence motifs in their RNA pols and helicases. They concluded that the positive-stranded RNA plant viruses can be divided into three supergroups based on their structural and functional relationships to certain animal viruses: Sindbis virus-like, luteovirus-like, and picornavirus-like, with the middle group providing the evolutionary link between the former and latter families.

Poch et al. (5) performed the most extensive investigation of RR and RD pol motifs, and their work remains a standard for the delineation and recognition of distant RNA pol sequences. Their study included RR pols from plus-stranded RNA viruses in the picorna-like and Sindbis-like groups, double-stranded RNA viruses, and minus-stranded viruses. In the RD pol category were DNA viruses, retroviruses, and the gypsy-, Ty-, and line-like groups. Figure 1 shows a sampling of aligned sequences from each of these groups over four motifs (A through D), the first three of which were recognized by Kamer and Argos (1) in aligned plus-stranded RNA virus sequences. Each of the subsequences center on a fully conserved residue(s); namely, Asp, Gly, Asp-Asp, and Lys for the respective A to D segments. The invariant residues are often predicted to be in exposed tight-turn regions, which themselves are often flanked by hydrophobic regions predicted as β-strands. The aspartates can easily be imagined as important for catalysis or metal binding, whereas the lysine would be ripe for DNA phosphate interaction. Poch et al. (5) list many experiments focusing on site-directed mutagenesis or side-group chemical modification and pointing to the crucial nature of these sequence spans.

In 1990, Xiong and Eickbush (6) performed an exhaustive study on reverse transcriptase sequences from animals, plants, protozoans, and bacteria. They used seven conserved sequence segments encompassing most of the RD pol primary structures, as well as the motifs of Poch et al. (5), and constructed two separate phylogenetic trees: one for retroviruses and the other for the remaining RD pols. The major groups in the latter tree included plus-stranded RNA viruses, minus-stranded DNA-associated reverse transcriptases, group II introns, retrotransposons lacking long-terminal repeats (LTRs), hepadnaviruses, retrotransposons with LTRs, and caulimoviruses. Xiong and Eickbush (6) suggested that RNA viruses are as old as or even older than retroelements and that non-LTR retrotransposons are the oldest group of retroelements.

In 1991, two important reports, one by Koonin (7) and the other by Bruenn (8), described evolutionary relationships among positive-stranded and double-stranded RNA viruses as viewed through their RR pols. Since new sequences appear at such a fast rate, periodic and frequent updates involving multiple sequence matches and tree constructions are justified. Koonin (7) concluded that these viruses constituted three viral supergroups, yet each had grossly different genome organizations and host specificities. It would thus seem probable that gene module shuffling was important in the evolution of positive-stranded RNA viruses and that transfer of vi-

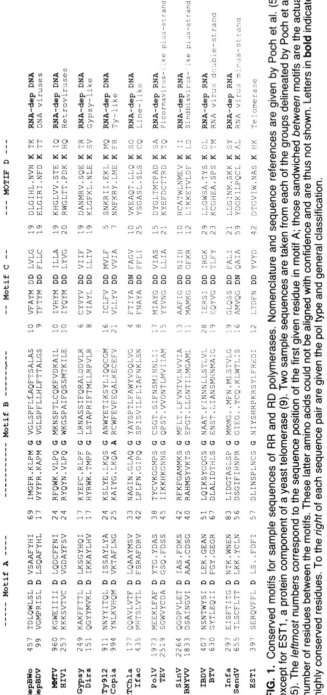

**FIG. 1.** Conserved motifs for sample sequences of RR and RD polymerases. Nomenclature and sequence references are given by Poch et al. (5) except for EST1, a protein component of a yeast telomerase (9). Two sample sequences are taken from each of the groups delineated by Poch et al. (5). The *leftmost* numbers correspond to the sequence position of the first given residue in motif A; those sandwiched *between* motifs are the actual number of residues between the motifs. These latter amino acids could not be aligned with confidence and are thus not shown. Letters in **bold** indicate highly conserved residues. To the *right* of each sequence pair are given the pol type and general classification.

ruses among various and varied species could occur relatively quickly once a viral structural or functional scheme had been developed for facile viral multiplication and interkingdom spreading. Bruenn (8) also found three supergroups, but suggested a different nomenclature [luteo-like, as Koonin (7); potyvirus-like, in lieu of picorna; tobamovirus-like, instead of Sindbis]. Once again the viruses did not cluster according to plant, animal, or fungal cell categories, although grand divisions according to prokaryotic and eukaryotic hosts were maintained. Bruenn (8) goes on to rule out convergent evolution, as helicase sequences resulted in trees similar to those derived from pol sequences, an unlikely event if only convergent evolution were to be responsible for viral development. Furthermore, picornaviral clusters based on VP3 capsid protein primary sequences also yielded phylogenies consistent with the pol and helicase motif results.

Lindblad and Blackburn (9) found the RD pol motifs in a protein component (EST1) of an essential yeast telomerase. Since reverse transcriptases had been universally ascribed to propagation functions for associated retroposons or viruses, their appearance in a cellular gene would also imply their significance in cell growth, once again displaying the generally conservative modes of natural development and the tightly knit scheme of evolutionary divergence.

## DNA POLYMERASES: MOTIFS

Many primary structures have appeared for DNA pols (DD and DR polymerases), and their alignments and grouping into various families have followed a developmental path similar to that for the RNA pols. The most recent and comprehensive alignment collection over entire DD pol sequences has been given by Ito and Braithwaite (10). An extensive set of matched DR pol sequences has not been reported according to the knowledge of the authors. The DD pols that can be classified were placed into four families (A, B, C, and X). Table I, which is largely a repeat of that given by Ito and Braithwaite (10), lists the families and their constituents; for a few new members that we have added, sequence references are explicitly given; for the others, Ito and Braithwaite (10) can be consulted. The family A pols are also referred to as pol I's and have now been found in bacteria, bacteriophage, and yeast mitochondria. The family B pols are often called the pol α-types (and sometimes pol δ) and cover various species, including bacteria, bacteriophages, viruses, yeast, mammals, and plasmids. Family C comprises the α-subunit of pol III's from bacteria, whereas family X is represented by mammalian pol β's and terminal transferases (TDT). The family A sequences include two domains with respective 3'-5' exonuclease and polymerase activity; family B constituents add further nucleotide-editing capability at their $NH_2$-terminus, with the incorporation of a 5'-3' exonuclease domain. It is clear from the complete sequence alignments of Ito and Braithwaite (10) that there are several well-conserved subsequence regions (motifs).

In 1988, Argos (11) noted that a strongly conserved motif in DD pol αs characterized by a central oligopeptide YGDTD bore resemblance to the core of motif C

**TABLE 1.** *DNA-dependent DNA polymerase families*[a]

| Family | Organism and enzyme |
|---|---|
| **FAMILY A** | |
| 1. Bacterial | *E. coli* pol I |
| | *Streptococcus pneumoniae* pol I |
| | *Thermus aquaticus* pol I |
| 2. Bacteriophage | T3 (29) |
| | T5 |
| | T7 |
| | Spo2 |
| 3. Mitochondrial | Yeast (MIP1) |
| **FAMILY B** | |
| 1. Bacterial | *E. coli* pol II |
| 2. Bacteriophage | PRD1 |
| | PZA (30) |
| | Φ29 |
| | M2 |
| | T4 |
| 3. Eukaryotic | Human pol $\alpha$ |
| | Yeast pol I |
| | Yeast pol II |
| | Yeast pol III $\delta$ (cdc2) |
| | Yeast rev 3 |
| | Bovine pol $\delta$ (31) |
| 4. Viral | Herpes simplex-1 |
| | Human cytomegalovirus |
| | Epstein–Barr |
| | Varicella–zoster |
| | Fowlpox pol $\alpha$ |
| | Vaccinia pol II |
| | *Autographa californica* nuclear polyhedrosis (AcMNPV) |
| | Adenovirus-2, -7, -12 |
| 5. Eukaryotic linear DNA plasmid-encoded | S-1 maize |
| | *Kluyveromyces lactis* plasmid pGKL1 |
| | *Kluyveromyces lactis* plasmid pGLK2 |
| | *Claviceps purpurea* plasmid pCLK1 |
| | *Ascobolus immersus* plasmid pAI2 |
| **FAMILY C** | |
| 1. Bacterial | *E. coli* pol III $\alpha$-subunit |
| | *Salmonella typhimurium* pol III $\alpha$-subunit |
| | *Bacillus subtilis* pol III |
| **FAMILY X** | |
| 1. Mammalian | Rat pol $\beta$ |
| | Human pol $\beta$ |
| | Human terminal transferase |
| | Bovine terminal transferase |
| | Mouse terminal transferase |

[a]References for most of the sequences are given by Ito and Braithwaite (10). References for additional primary structures used here are given by a numerical designation in parentheses.

(YGDD) from the plus-strand picorna-like viruses (see Fig. 1). Since both central segments were flanked by several hydrophobic residues, it was suggested that a β-hairpin structure existed for both pol types, with the two aspartates on an exposed loop that accommodated a threonine insertion for the DD pol α-types. The aspartates are presumably used for catalysis or metal binding. Since this one-motif relation was small relative to the length of the entire pol sequences, it was only hypothesized that some DD, RR, and RD pols could adopt the same overall tertiary fold through divergent evolution.

In 1990, Delarue et al. (12) expanded considerably the possible pol connections through typical motif delineation and matching. By aligning distant pol I (family A) sequences, they narrowed the conserved subsequences to five. Earlier alignments by Wang et al. (13) limited the motifs for pol αs to seven regions. Delarue et al. (12) observed that three motifs designated A, B, and C, in sequential order, could be matched over the DD pol α and pol I sequences. They went on to find the same three motifs in DR pols from yeast mitochondria and from several bacteriophages. Motifs A and C were also suggested to be in DD family X primary structures. If Delarue et al. (12) are correct, and divergent evolution rules, DD pols from families A, B, and X, as well as one type of DR pol may possess related tertiary folds with similarly constructed active sites and nucleic acid-binding regions.

Figure 2 shows four 3'-5' exonuclease motifs found in family A and B DD pols. The sequence spans are designated exo-I, -II, -II', and -III. The motif-II' has not been previously observed and is suggested here as a potential area for site-directed mutagenesis tests where a conserved aspartate is a clear target. This subsequence was delineated from a visual examination of the overall alignments given by Ito and Braithwaite (10). Since the three-dimensional structure of one family B pol (Klenow fragment of *E. coli* DD pol I) is known (14), it was found that the exo-II' aspartate mainchain Cα atom was 8.5 Å from that of the conserved active site aspartic acid Cα in the exo-III motif.

Figure 3 shows the aligned and conserved motifs in DD and DR pols. Delarue et al. (12) had not previously found a motif B substructure for family X sequences, whereas in Fig. 3 such a motif is suggested, although a well-conserved tyrosine in other pols is not present. However, the crucial lysine and glycine residues do appear with the proper sequential spacing to preserve an α-helical structure, as suggested by the equivalent region in the *E. coli* Klenow fragment fold.

There are several experiments that support the importance of many of the exonuclease and pol motifs in DD and DR pols. A recent and excellent review by Joyce (15) describes and references many of them; Delarue et al. (12) also report on several of the empirical efforts. As expected, the conserved aspartates are likely to be involved in catalysis or metal binding, whereas conserved lysines and tyrosines are most probably useful for nucleic acid interactions. Site-directed mutagenesis and chemical labeling of specific groups provide the basic techniques for the bulk of the empirical investigations. Most significantly, results from tertiary structural studies also support the DD and DR pol relationships. The exo and pol motif regions, as located on the known fold of *E. coli* pol I (14) invariably fall within clefts

Conserved sequence alignment for the 3'–5' exonuclease domains (columns: **Exo I**, **Exo II**, **Exo II'**, **Exo III**):

```
                     |---- Exo I ----|              |---- Exo II ----|           |- Exo II' -|          |---- Exo III ----|

PRD1            7  KKVEYKIAAF DFE TDPFKHDRIP   30 IERPHVIYAH N GGK FD FLFMLYFRG   11 EVEHGIHKFR D SYAILPVPLA   17 REQHKAEILE YLKGD CVTLHKMVSL
PZA             1  ..MPRKMYSC DFE TTTKVEDCRV   27 LKVQADLYFH N .LK FD GAFINWLER    34 GKRKIHTVIY D SLKKLPFFVK   23 GYKITPDEYA YIRND IQIIAEALLI
ø29             2  KIHMPRKMYSC DFE TTTKIDDCRV  36 LKVQADLYFH N .LK FD GAFINWLER    25 GKRKIHTVIY D SLKKLPFFVK   23 GHEITPEEYE YIRND IEIIARALDI
M2              1  .MSRKMFSC DFE TTTKIDDCRV    35 MEIQADLYFH N .LK FD GAFINWLEQ    25 GKRKIHTVIY D SLKKLPFPVK   23 GHEITPEEYE YIRND IEIIARALDI
Adeno-2       131  PRTERLFVTY DVE TYTWMGAFGK  112 RFLELYIVGH N ING FD EIVLAAQVIN   14 MPRAGKILFN D VTFALPNPRS  101 KYDIKETLD YCALD VQVTAELVNK
Adeno-12      129  PRTERLFITY DVE TYTWMGAFGK  112 RFVELYIVGH N ING FD EIVLAAQVIN   14 IPRAGKILFN D ITFALPNPHY  101 NYNLIQETLN YCALD VLVTASLVEK
Adeno-7       198  PRTERLFITY DVE TYTWMSFGK   112 RFLELYIVGH N ING FD EIVLAAQVIN   14 MPRAGEILFN D VTFALPNPAY  101 RYDIIQETLN YCALD VLVTAELVAK
S-1           203  MKNKTIFFVA DLE TLLLKRRDTD   62 RGSSMVVYFH N LSQ FD GIMLSFLTK    27 LPSIRDSLT D SYLLKVKLA   101 LPSIRDSLT SYLLKVKLA ILITAAVMQR
pA12          536  TVNLKSISTL DLE TRMDTNNRLI   35 TNHGKKFYVH N LAH FD SVFIDTLSK    24 KNTEYSISFL D SLLMLPNSLD  61 NWDFKNELKN YCEID CLALHDILTL
PCLK1         377  VLANPNLGTI DLE TTTGAQPKCY   37 KYRNVLFYAH N LGK FD AVFLLKELLR   33 DWDFKDETLE D YLNLD LISLHQVLVK 47 DWDFKDETLE YLNLD LISLHQVLVK
PGLK2         354  SDVKNITVCF DIE SYFDPEKESN   44 NISSVELIAH N GGG YD FHYILSSMYN   19 AHDGVKFSVR D SYSFLCSLA   66 SKTLIEWSKD YCRND VIVLSKVWLE
PGLK1         353  ENEVKEVFF DIE SFSDETKHQI    34 YNVDILLYAW Y GSG YD YOHVLPYMKS   19 YYENALILTR D PYLFLTSLD   63 YNTILENAIE YCRVD VLAMEKVWIK    DNA-dep DNA
HSV-1         358  DLPAYKLMCF DIE CKAGGEDELA   74 QYGPEFVTGY N IIN FD WPFLLAKLTD   33 KIKVNGMVNI D MYGIITDKIK  32 GPAQRGVIGE YCIQD SLLVGQLFFK    family B
HCMV          288  SWPRYRCLSF DIE CMSGEGFPC    87 RYAPAFVTGY N INS FD LKYILTRLEY   51 KVYIAGSVVI D MPVCMAKTN   32 NAEGRAQVGR YCLQD AVLVRDLFNT
EBV           283  SWPSYQALAF DII CL.GEGFPT    62 DLSVEIVTGY N VAN FD WPYILRRARH   36 KVRITGLIPI D MYAVCRDKLS  32 GPEGRRRLGM YCVQD SALVMDLLNH
VZV           339  SWDYKLLCF DII CKSGGSNELA    75 QYAPEFATGY N IVN FD WAFIMEKLNS   32 KVKINGLISL D MYAIATEKLK  32 GPNTRGIIGE YCIQD SALVCKLFFK
FPV           148  FEVKFTYLLF DII CQFDKKFPSV   70 EHRFDEVITF N GNN FD IRYISGRLEI   37 INNNNGVMFF D LYAFIQKTER 109 NLEIALDMER YCIHD ACLCKYIWDY
Vac V         125  FDIPRSYLFL DII CHPDKKFPSV   79 ELTPDVVVTF N GHN FD LRYITNRLEL   42 NLNIALDMAR D LYSFIQKFEK 107 NLNIALDMAR YCIHD ACLCQIWEY
AcMNPV        186  SGIMPVLSCY DII THSDGINMSK   46 ITNPDVILDF N VVQ FD LYPIIGRLNK   28 TYFNYYIHI D LYKFFSSDSN  33 NTRRLDVIAK YNVQD CMLPIKLFVK
E.C. pol T4   146  YRPPLKKWSI DII TTRHGELYCI   62 NVDDVIIGW N VVY FD LRMLQMRAEM   25 FAQAKGRLLI D GIEALKSAFW  34 FAQAKGRLLI YNLKD CELVTQIFHK
               97  DRKFVRANC DII VTGDKFDDPM    85 QKRPAIFTGW N IEG FD VPYIMNRVKM  28 IYSIDGVSIL D YIDLYKKFAF  30 RETNHQRYIS YNIID VESVQAIDKI
Yeast cdc2    306  HTAPIRIMSF DIE CAGRIGVFPQ   62 ADYICRAGTY N IIQ FD VGIILFINSG 34 NVNIDGRLQL D LLQFIQREYK  32 DSETRRRIAV YCLKD AYIPLRIMEK
Yeast pol II  275  AFADPVVMAF DIE TKRPPLKFPD   81 DVRPTVISTF N CDF FD WPFIHNRSKI  14 HMDCFRWVKR D SYLPQGSQGL  23 AFEKQPHLSE YSVSD AVATYYLYMK
Yeast pol I   511  ASHCAVESV DKP QNTRPVTSKT    94 VEDPDVLIGH N LQN VT LDVLAHRMHD  34 SDICSGRLIC D IANEMGQSLI  91 ILLHEFSRNG FIVPD KEGNRSRAQK
Human pol I   511  CKVEAMALKP DLV NVIKDVPPP    94 KIDPDIIVGH N IYG FE LEVLLQRINV  28 RNATCGRMIC D VEISAKELIR  89 LLLHAFYENN YIVPD KQIFRKPQQK
Yeast Rev 3   469  KYHNTINFSM DCA MTQNMASKRK    91 OFGGETVSYY N KPT FD MFSSWKYALK  32 HSKFLYKFAS D YGMQKRKKKS  74 EIPVMFYESE FEMFE ALIDLVLLLD
Bovine pol d  305  RIAPLRVLSF DIE CAGRKGIFPE   58 IMDPDVITGY N IQN FD LPYIISRAQT  35 VVSMGVRVQM D MLQVLLREYK  30 NDQTRRRLAV YCLKD AFIPLRLLER

E.c. pol I    345  KLEKAPVFAF DTE TDSLDNISAN   42 EDEKALKVGQ N L.K YD RGILANYGIE  14 LNSVAAGRHM D SLAERWLKHK  17 NQIALEEAGR YAAED ADVTIQLHLK    DNA-dep DNA
S.p. pol I    260  GLEDLVYSGP DVE NLGKFYDEMG   35 SIFHFELFGE N Y.H YD NLVGFAWSCG   8 LELLQDPIFK D FLEKTSLRVY  38 ATIASLYGGT YIVDD ETFYCKGVKK
Taq pol I     212  GSEAILKNL DRL KPAIR.....   0 EKILARMDDL N L.S WD LAKVRTDLPL  45 VLSRKEPMWA D LLALAAARGG  47 PGGDPMLLA YLLDP SNTPEGVAR    family A
T5            128  YNMVIGPVAF DSE TSALYCRDGY   33 DSENHTIVFH N L.K YD MHFYKYHLGL   2 DKAHKERRLH D TMHQHYVLDE  44 DLPPDIWAP YAAKD TPATIRLHNF
T3              1  .....MIVS DIE ANNLLEKVTK   32 VKRGGLIVFH N GHK YD VPALTKLAKL  20 LRIHSNLKDT D MGLLRSGKLP  43 WWNFNEEMQD YNVQD VVVTKALLEK
T7              1  .....LKTLSI DIE TFSSVDLLKA   32 VARGGLIVFH N GHK YD VPALTKIAKL   4 EFHLPRENCI D TLVLSRLIHS  59 WWNFNEEMQD YNVQD VVVTKALLEK
Spo2            1  .....MIVS DIE TFSSVDLLKA   45 TSSKVIKTAY N A.N FE RTCLAKHFNL  13 ATTLGLPGNL D GVAKALKISA  33 DPEKWEKFKV YCIQD VEVERAJKNR
Mip1          161  AYPDEELVVF DVE TLYNVSDYPT   41 LNKEQVVIGH N V.A YD RARVLEEYNF  41 SEVHPEISIE D YDDPWLNVSA  30 IIENFQKLAN YCATD VIATSQVFDK
```

**FIG. 2.** Conserved sequence motifs for the 3'–5' exonuclease domains in DD polymerases. Only subsequence exo-II' has been given here. Table 1 lists the full names of the species to which the sequences are attached. The alignment format is as given in Fig. 1.

or pockets that could well bind nucleotides or even the duplex DNA strand itself. Joyce (15) provides a good illustration (see Fig. 7 in his article) of these sites on the Klenow architecture. Recently, the tertiary fold of bacteriophage T7 DR pol has been determined from an x-ray crystallographic electron-density map (16,17). Many Cα atoms from one of the two T7 domains could be largely superimposed onto equivalent main-chain atoms of the Klenow pol domain. The T7 primary sequence (18) is related to that from bacteriophage T3, the motif subsequences of which are presented in Fig. 3. Site-directed mutagenesis and mechanistic studies on T7 DR pol also support the pol motif identifications (19,20). Finally, a large cleft has been observed by electron crystallography at reasonable resolution in *E. coli* DR polymerase containing several subunits; the overall size and shape of the duplex DNA-binding cleft is similar to that found in *E. coli* pol I (21).

## DNA POLYMERASES: EVOLUTIONARY RELATIONSHIPS

Several phylogenetic trees were constructed with the exonuclease and polymerase motifs. Separate trees were built for each of the DD pol exo motifs and then for all the exo subsequences combined as a single input string. The latter was also repeated for the individual and combined pol segments. The distance scores were calculated as a sum of the Dayhoff mutation weights (22) over each aligned residue pair. The Dayhoff PAM250 values reside in a symmetric $20 \times 20$ matrix in which each element corresponds to a particular amino acid substitution and represents the preference of the exchange as observed in 71 aligned protein sequence families. Several clustering techniques were employed to effect the evolutionary trees and to test for their consistency: complete linkage (23), single linkage (23), group averaging (23), neighbor-joining (24), and Ward's averaging (25). The BIOPAT program package (26) provided the means to apply and analyze the various tree-building routines. Only motif sequences could be used, as overall alignments are impossible to effect with any accuracy for such distant families. Trees were not constructed across the DNA and RNA pols because the two motifs suggested to be held in common by the two major pol types were considered too distant to reach any reasonable conclusion.

It must be expected that the DD and DR pol trees would at least show major overall clustering according to the basic families listed in Table 1 and in Figs. 2 and 3. Surprisingly, none of the exonuclease subsequences, taken individually or in combination and applied over all the clustering procedures listed, resulted in any interpretable tree and, often, stood in opposition to the required major classifications. In contrast, the polymerase motifs, again taken individually or in concert and over all the tree-constructing methods, almost always produced consistent results that correlated well with the major classifications of Table 1.

Figure 4 illustrates the evolutionary relationships made salient by the combined pol motifs and group-averaging construction. To the right of each DD pol tree entry is shown the family designation according to Table 1 (e.g., B1 for the first listed member in family B). The DR pol constituents that do not appear in Table 1 are

| | ------- Motif A ------ | | ------ Motif B ------ | | ------ Motif C ------ | | |
|---|---|---|---|---|---|---|---|---|
| PRD1 | 210 | GIIEDDIKVY D VNSMYPHAMR | 99 | AGDLFHNIFY K LILNS.S YG KFAQNPENYK | 57 | LAQERPLYC DTD SIICRDLKNV | | |
| PZA | 236 | EKEIEGEMVF D VNSLYPAQMY | 113 | TSEGAIKQLA K LMLNS.L YG KFASNPDVTG | 50 | QACFDRIIYC DTD SIHLTGTEIP | | |
| ø29 | 239 | EKEIEGEMVF D VNSLYPAQMY | 113 | TSEGAIKQLA K LMLNS.L YG KFASNPDVTG | 42 | QACYDRIIYC DTD SIHLTGTEIP | | |
| M2 | 236 | EKEIEGEMVF D VNSLYPSQMY | 113 | HEEGAKKQLA K LMLNS.L YG KFASNPDVTG | 42 | QACYDRIIYC DTD SIHLTGTEVP | | |
| Adeno-2 | 531 | GILREPLYVY D ICGMYASALT | 131 | DKNQTLRSIA K LLSNA.L YG SFATKLDNKK | 144 | EDRPLKSVYG DTD SLFVTERGHR | | |
| Adeno-12 | 531 | GVLKEPIVYY D ICGMYASALT | 131 | DKNQTLRSIA K LLSNA.L YG SFATKLDNKK | 144 | EDRPIKSVYG DTD SLFVTEKGRR | | |
| Adeno-7 | 600 | GILEEPLYVY D ICGMYASALT | 130 | EKNQTMRSIA K LLSNA.L YG SFATKLDNKK | 144 | EDRPIKSVYG DTD SLFVTQRGHE | | |
| S-1 | 475 | KPYGENLYYY D VNSLYPSSML | 111 | KGEKALDFIA K ITMNS.L YG RFGISPESTT | 69 | FISRDDCYYT DTD SVVVERELPE | | |
| pAI2 | 812 | KPEGKNIHSY D INSLYPSAMA | 107 | TPDDDMYFIA K LLMNS.L YG RFGMDPITIK | 63 | IKYADNLYAV DTD GIKVDTEIDK | | |
| pCLK1 | 661 | ADSTKSXYYY D VNSLYPFASI | 104 | PNNVTEKNIA K LILNS.L IG RFGMNINKIK | 90 | LENNGTLYYT DTD SIVTDLKLPE | | |
| pGLK2 | 639 | NGIYKDVLCL D VKSLYPSAMA | 114 | QNNKVKRNVI K IIMNS.L WG KFAQKWVNFE | 58 | AKEGAECIYS DTD SIFVHKEHFK | | |
| HSV-1 | 626 | GIYEENIVYA D VVSLYPSAMK | 112 | EQPCPIRMVA K IALNGGG YG KFVQKPIDKE | 78 | QFENIDIIYS DTD SIFVKQKSVD | | |
| HCMV | 707 | GFHVNPVVVF D FASLYPSIIQ | 73 | VLLDKQQAAI K VVCNS.V YG FTGVQHGLLP | 44 | GEYSMRIIYG DTD SIFVLCRGLT | | |
| VZV | 707 | GYYNDPVAVF D FASLYPSIIM | 73 | MLLDKEQMAL K VTCNA.F YG FTFVVNGMMP | 68 | RRVEARVIYG DTD SVFVRFRGLT | DNA-dep DNA | |
| BBV | 574 | GFYNSPVLVV D FASLYPSIIQ | 76 | TILDKQQLAI K CTCNA.V YG FTGVANGLFP | 40 | PEGQLRVIYG DTD SLFIECRGFS | | family B |
| FPV | 672 | GFYIDPVVVL D FASLYPSIIQ | 72 | VLLDKQQAAI K VVCNS.V YG FTGVAQGFLP | 45 | KAYEVKVIYG DTD SVFIRFKGVS | | |
| Vac V | 525 | KTFERNVMIF D YNSLYPVNCI | 91 | TLYDSLQYIY K IIANS.V YG LMGFSNSTLY | 57 | ESFKFRSVYG DTD SIFSEISTKD | | |
| AcMNPV | 514 | KMFSNNVLIF D YNSLYPNVCI | 91 | AIYDSMQYTY K IVANS.V YG LMGFRNSALY | 59 | YRFRFRSVYG DTD SVFTEIDSQD | | |
| [unclear] | 520 | AGIYKNAFSL D FNSLYLTIMI | 55 | DLYDQKQNSV K RTANS.I YG YYGIFYKVLA | 31 | GSITFKVVYG DTD STFVLPTFNY | | |
| B.C. pol II | 409 | PGLYDSVLVL D YKSLYPSIIR | 53 | GNKPLSQ.AL K IMNA.F YG VLGTTACRFF | 20 | EAQGYDVIYG DTD STFVWLKGAH | | |
| T4 | 398 | PIARRYIMSF D LTSLYPSIIR | 128 | TLANTNQLNR K ILINS.L YG ALGNIHFRYY | 20 | TNDEDFIAAG DTD SVYVCVDKVI | | |
| Yeast Cdc2 | 596 | GYYDVPIATL D FNSLYPSIMM | 71 | DVLNGRQLAL K ISANS.V YG FTGATVGKLP | 30 | YKHDAVVVYG DTD SVMVKFGTTD | | |
| Yeast pol II | 630 | RNELPLIYHV D VASMYPNIMT | 163 | VLYDSLQLAH K VILNS.F YG YVMRKGSRWY | 20 | VERVGRPLEL DTD GIWCILPKSF | | |
| Yeast pol I | 854 | GLHKNYVLVM D FNSLYPSIIQ | 59 | VQCDIRQQAL K LTANS.M YG CLGYNSRFY | 21 | ESMNLLVVYG DTD SVMIDTGCDN | | |
| Human pol α | 850 | FGYDKFILLL D FNSLYPSIIQ | 69 | LQYDIRQKAL K LTANS.M YG CLGFSYSRFY | 30 | QKMNLEVIYG DTD SIMINTNSTN | | |
| Yeast Rev 3 | 965 | AFYKSPLIVL D FQSLYPSIMI | 90 | RLLNNKQLAI K LLANV.T YG YTSASFSGRM | 25 | ETWNAKVVYG DTD SLFVIPGKT | | |
| Bovine pol δ | 591 | GYYDVPIATL D FSSLYPSIMM | 71 | QVLDGRQLAL K VSANS.V YG FTGAQVGRLP | 30 | YSTSAKVVYG DTD SVMCRFGVSS | | |

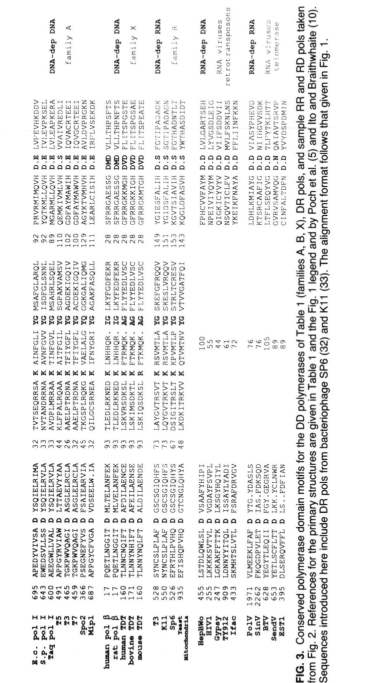

**FIG. 3.** Conserved polymerase domain motifs for the DD polymerases of Table 1 (families A, B, X), DR pols, and sample RR and RD pols taken from Fig. 2. References for the primary structures are given in Table 1 and the Fig. 1 legend and by Poch et al. (5) and Ito and Braithwhaite (10). Sequences introduced here include DR pols from bacteriophage SP6 (32) and K11 (33). The alignment format follows that given in Fig. 1.

named family R. All the A, B, X, and R family members are clustered appropriately from an overall perspective. Families A, B, and R are separated from family X, and A and R are more closely related than A is to B or to X. It is thus noteworthy that all the DNA-dependent DNA pol families (A, B, X) do not all cluster, and that a DNA-dependent RNA polymerase (family R) can evolve from a particular DD pol (family A) for which, presumably, more specific catalytic mechanisms are shared. At a higher-resolution level, not all the subgroups within given families seem properly aligned; for example, the B4 of adenoviruses are separated from the B4 of other viruses. The motif subsequences are apparently too short to allow proper distinction. Ten-residue lengths were taken as flanks for the conserved motif cores, as they represent the average length (eight residues) of a secondary structural element ($\alpha$-helices and $\beta$-strands) and half of the length of a connecting tight turn (two residues). It would be normally expected that the two-flanking secondary structures of a functionally essential loop would likely be conserved without deletions and insertions.

The almost general loss of evolutionary memory in the exonuclease motifs over these distant structures appears arcane. The exo motifs are as likely to be crucial for function as are the pol motifs and are also likely to involve metal and nucleotide binding. Perhaps the exo motifs are older and the cutting of nucleic acid chains preceded polymerization in importance for biological development and survival. In any event, caution must be exercised in using motifs to reconstruct evolutionary development paths.

## A COMMON FOLD FOR ALL POLYMERASES?

It is possible that there are at least two basic pol folds: one for DNA pols, be they of the DD or DR types, and the other for RNA pols, again of the RR or RD types. If this hypothesis were correct, then the dominant factor in the formation of the two ancestors would be the type of template nucleic acid, one pol fold developed to bind DNA in the likely B-conformation and the other to associate with RNA in the probable A-form. The catalytic mechanism and active site for polymerizing the DNA or RNA products in each of the DNA or RNA pols would, in contrast, be expected to be similar. In the tree of Fig. 4, the DR pols, as a group, are closely associated with family A DD pols and are not distinctly separated from all three of the DD families A, B, and X.

It must be emphasized that there are many polymerase sequences, presumably with polymerase activity, that do not contain motif characteristics of the two folds just discussed and, therefore, may well display different three-dimensional structures; for example, the authors (unpublished work) recently examined all DR pol

**FIG. 4.** A phylogenetic tree constructed from the group-averaging technique (23) using the BIO-PAT computer program package (26) for the DD and DR pol motifs given in Fig. 3. To the *right* are indicated, according to Table 1, the families (A, B, X, R) and subfamily classes (numbers) to which the tree entries belong.

| | |
|---|---|
| Phage PRD1 | B2 |
| Phage Phi29 | B2 |
| Phage PZA | B2 |
| Phage M2 | B2 |
| Maize mitochondrial S-1 | B5 |
| *Ascobolus immersus* pA12 | B5 |
| *Claviceps purpurea* pCLK1 | B5 |
| Adenovirus-2 | B4 |
| Adenovirus-12 | B4 |
| Adenovirus-7 | B4 |
| *Kluyveromyces lactis* pGLK2 | B5 |
| *Kluyveromyces lactis* pGLK1 | B5 |
| Herpes simplex virus (HSV-1) | B4 |
| Varicella–zoster virus (VZV) | B4 |
| Epstein–Barr virus (EBV) | B4 |
| Human cytomegalovirus (HCMV) | B4 |
| Yeast pol III δ (Cdc2) | B3 |
| Bovine pol δ | B3 |
| Yeast rev3 | B3 |
| Fowlpox virus (FPV) | B4 |
| Vaccinia virus pol II (VacV) | B4 |
| *E. coli* pol II | B1 |
| Yeast pol I | B3 |
| Human pol α | B3 |
| *Autographa californica* nuclear polyhedrosis virus (AcMNPV) | B4 |
| Phage T4 | B2 |
| Yeast pol II | B3 |
| *E. coli* pol I | A1 |
| *Streptococcus pneumoniae* pol I | A1 |
| *Thermus aquaticus* pol I | A1 |
| Phage Spo2 | A2 |
| Phage T5 | A2 |
| Yeast mitochondrial Mip1 | A3 |
| Phage T7 | A2 |
| Phage T3 | A2 |
| Phage T3 (DNA-dep RNA pol) | R |
| Phage PK11 (DNA-dep RNA pol) | R |
| Phage SP6 (DNA-dep RNA pol) | R |
| Yeast mitochondrial (DNA-dep RNA pol) | R |
| Human pol β | X |
| Rat pol β | X |
| Human TDT | X |
| Bovine TDT | X |
| Mouse TDT | X |

sequences in the present-day data banks, and at least two large families could be found that did not ostensibly possess the three motifs of the DNA pols listed in Fig. 3. Nonetheless, it is still possible that the tertiary architectures could be similar, but the motifs not recognizable owing to large evolutionary distances.

Delarue et al. (12) suggested that two (motifs A and C) of the three DNA pol regions (see Fig. 3) and the four RNA pol segments (see Fig. 1) could be shared (see Fig. 3 for examples). Delarue et al. (12) also show an alignment of a hepatitis B reverse transcriptase (RD pol) sequence with that from the DD pol of herpes simplex virus, for which a near 30% residue identity can be achieved. These observations could imply that many DNA and RNA pols share a common fold and ancestry for at least a catalytic subdomain; however, caution must be exercised. The two common motifs rely primarily on conserved central aspartates, which frequently occur in active sites from various enzymes (27), although it is notable that motif C contains double aspartates.

A recent report on the three-dimensional structure of the viral HIV-1 reverse transcriptase (28) considerably clarifies the polymerase tertiary relations. The HIV RD pol (sometimes also a DD pol) is a heterodimer with one subunit (p66) containing ribonuclease H and polymerase domains and the other (p51) consisting of only a pol segment that is identical in sequence with that of the former domain. The HIV tertiary structure shows that both the p66 and p51 pol domains have four conformationally similar subdomains (palm, thumb, connection, and fingers); however, their spatial relationships are very different in the two subunits. Kohlstaedt et al. (28) suggest that p51 forms part of the binding site for the tRNA and template primers, whereas p66 directly catalyzes the pol reaction. The "palm" subdomain, with about 110 residues, contains sequence motifs A and C (see Fig. 3), has a very similar fold to the corresponding region in the Klenow pol I segment (Fig. 5), and is probably responsible for polymerization by a two-metal, ion-catalyzed phosphoryl transfer mechanism. A second subdomain ("thumb") of the HIV pol segment bears some structural resemblance to a corresponding helical region in the Klenow pol I, whereas the "finger" and "connection" subdomains are structurally very different in the two pol types. Given the overall similarity of fold for the T7 DR and *E. coli* DD pol domains and the shared palm catalytic subdomain in T7 DR, Klenow DD, and HIV RD pols (see Fig. 5), it becomes realistic to think that many DNA and RNA pols, if not all, will share a common tertiary architecture in their catalytic polymerization substructures, which are likely to have divergently evolved from an ancestor. It is noteworthy that the finger subdomains display a similar subsequence (motif B, see Fig. 3) in the T7 DD and *E. coli* DR pols, whereas the HIV RD pol primary structure does not contain the motif and, consistently, its finger subdomains do not display a tertiary structure similar to those found in the DNA pols. This allows the suggestion (28) that the finger subdomains are responsible for binding the primer template, which would be in different conformations for the respective DNA (B-form) and RNA (A-form) polymerases.

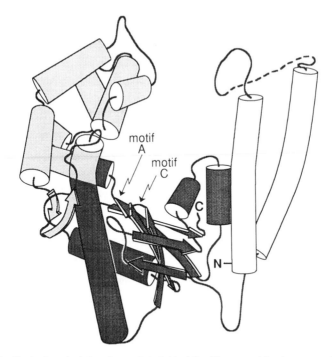

**FIG. 5.** An illustration depicting the protein fold of the Klenow pol I polymerase domain. *Cylinders* represent α-helices and *arrows* refer to β-strands; loops are shown as *connecting ribbons*. The *lightest-shaded* regions correspond to the thumb subdomain, the *intermediate* to the finger region, and the *darkest* to the palm or catalytic polymerization substructure. The structural sites of the conserved aspartates contained in sequence motifs A and C (Fig. 3) are noted; their positioning in the exposed and putative DNA-binding cleft is appropriate for activity.

## SUMMARY

Sequence fragment motifs conserved in many viral and host polymerase primary structures are listed and discussed. These sequence spans have been shown critical for polymerase structural stability and functional attributes. Such residue conservations have allowed suggestions that many polymerases, both DNA- and RNA-directed, share the same overall tertiary fold. Phylogenetic trees based on the motif relationships can be used to trace possible divergent evolutionary development among these very critical biomacromolecules.

## ACKNOWLEDGMENTS

The authors are deeply grateful to Nelly van der Jagt-González whose assistance was invaluable in preparing this manuscript with its many complex sequence alignments. The authors also wish to acknowledge the use of the BIOPAT program package as generously supplied by Professor Pauline Hogeweg.

## REFERENCES

1. Kamer G, Argos P. Primary structural comparison of RNA-dependent polymerases from plant, animal, and bacterial viruses. *Nucleic Acids Res* 1984;12:7269–7282.
2. Haseloff J, Goelet P, Zimmern D, Ahlquist P, Dasgupta R, Kaesberg P. Striking similarities in amino acid sequence among nonstructural proteins encoded by RNA viruses that have dissimilar genomic organizations. *Proc Natl Acad Sci USA* 1984;81:4358–4362.
3. Doolittle RF, Jeng DF, Johnson MS, McClure MA. Origins and evolutionary relationships of retroviruses. *Q Rev Biol* 1989;64:1–30.
4. Habili N, Symons RH. Evolutionary relationship between luteoviruses and other RNA plant viruses based on sequence motifs in their putative RNA polymerases and nucleic acid helicases. *Nucleic Acids Res* 1989;17:9543–9555.
5. Poch O, Sauvaget I, Delarue M, Tordo N. Identification of four conserved motifs among the RNA-dependent polymerase encoding elements. *EMBO J* 1989;8:3867–3874.
6. Xiong Y, Eickbush TH. Origin and evolution of retroelements based upon their reverse transcriptase sequences. *EMBO J* 1990;9:3353–3362.
7. Koonin EV. The phylogeny of RNA-dependent RNA polymerases of positive-strand RNA viruses. *J Gen Virol* 1991;72:2197–2206.
8. Bruenn JA. Relationships among the positive strand and double-strand RNA viruses as viewed through their RNA-dependent RNA polymerases. *Nucleic Acids Res* 1991;19:217–226.
9. Lundbald V, Blackburn EH. RNA-dependent polymerase motifs in EST1: tentative identification of a protein component of an essential yeast telomerase. *Cell* 1990;60:529–530.
10. Ito J, Braithwaite DK. Compilation and alignment of DNA polymerase sequences. *Nucleic Acids Res* 1991;19:4045–4057.
11. Argos P. A sequence motif in many polymerases. *Nucleic Acids Res* 1988;16:9909–9916.
12. Delarue M, Poch O, Tordo N, Moras D, Argos P. An attempt to unify the structure of polymerases. *Protein Eng* 1990;3;461–467.
13. Wang TSF, Wong SW, Korn D. Human DNA polymerase α: predicted functional domains and relationship with viral DNA polymerases. *FASEB J* 1989;3:14–21.
14. Ollis DL, Brick P, Hamlin R, Xuong NG, Steitz TA. Structure of large fragment of *Escherichia coli* DNA polymerase I complexed with dTMP. *Nature* 1985;313:762–766.
15. Joyce CM, Can DNA polymerase I (Klenow fragment) serve as a model for other polymerases? *Curr Opin Struct Biol* 1991;1:123–129.
16. Sousa R, Chung YJ, Rose JP, Wang B-C. Crystal structure of bacteriophage T7 RNA polymerase at 3.3 Å resolution. *Nature* 1993;364:593–599.
17. Chung YJ, Sousa R, Rose JP, Lafer E, Wang BC. Crystallographic structure of phage T7 RNA polymerase at resolution of 4.0 Å. In: Wu FYH, Wu CW, eds. *Structure and function of nucleic acids and proteins.* New York: Raven Press; 1990:55–59.
18. Dunn JJ, Studier FW. Complete nucleotide sequence of bacteriophage T7 DNA and the location of T7 genetic elements. *J Mol Biol* 1983;166:477–535.
19. Sousa R, Patra D, Lafer EM. Model for the mechanism of bacteriophage T7 RNAP transcription initiation and termination. *J Mol Biol* 1992;224:319–334.
20. Osumi-Davis PA, de Aguilera MC, Woody RK, Woody AYM. Asp537, Asp812 are essential and Lys631, His811 are catalytically significant in T7 RNA polymerase activity. *J Mol Biol* 1992;226: 37–45.
21. Darst SA, Kubalek EW, Kornberg RD. Three-dimensional structure of *Escherichia coli* RNA polymerase holoenzyme determined by electron crystallography. *Nature* 1989;340:730–732.
22. Dayhoff MD, Barker WC, Hunt LT. Establishing homologies in protein sequences. *Methods Enzymol* 1983;91:524–545.
23. Lance GN, Williams WT. General theory of classificatory sorting strategies I. Hierarchical systems. *Comput J* 1967;9:373–380.
24. Saitou N, Nei M. The neighbor-joining method: a new method for reconstructing phylogenetic trees. *Mol Biol Evol* 1987;4:406–425.

25. Ward JH. Hierarchical grouping to optimize an objective function. *Am Stat Assoc* 1963;58:236–244.
26. Hogeweg P, Hesper B. The alignment of sets of sequences and the construction of phyletic trees: an integrated method. *J Mol Evol* 1984;20:175–186.
27. Zvelebil MJJM, Sternberg MJE. Analysis and prediction of the location of catalytic residues in enzymes. *Protein Eng* 1988;2:127–138.
28. Kohlstaedt LA, Wang J, Friedman JM, Rice PA, Steitz TA. Crystal structure at 3.5Å resolution of HIV-1 reverse transcriptase complexed with an inhibitor. Science 1992;256:1783–1790.
29. Beck PJ, Gonzalez S, Ward CL, Molineux IJ. Sequence of bacteriophage T3 DNA from gene 2.5 through gene 9. *J Mol Biol* 1989;210:687–701.
30. Paces V, Vlcek C, Urbanek P, Hastomsky Z. Nucleotide sequence of the major early region of *Bacillus subtilis* phage PZA, a close relative of Φ 29. *Gene* 1985;38:45–56.
31. Zhang J, Chung DW, Tan CK, Downey KM, Davie EW, So AG. Primary structure of the catalytic subunit of calf thymus RNA polymerase δ: sequence similarities with other DNA polymerases. *Biochemistry* 1991;30:11742–11750.
32. Katani H, Ishizaki Y, Hiraoka N, Obayashi A. Nucleotide sequence and expression of the cloned gene of bacteriophage SP6 RNA polymerase. *Nucleic Acids Res* 1987;15:2653–2664.
33. Dietz A, Weisser HJ, Koessel H, Hausmann R. The gene for *Klebsiella* bacteriophage K11 RNA polymerase: sequence and comparison with the homologous genes of phage T7, T3, and SP6. *Mol Gen Genet* 1990;221:283–286.

The Evolutionary Biology of Viruses,
edited by Stephen S. Morse.
Raven Press, Ltd., New York © 1994.

# 6

# RNA Viral Supergroups and the Evolution of RNA Viruses

Rob Goldbach and Peter de Haan

*Department of Virology, Wageningen Agricultural University, Binnenhaven 11, 6709 PD, Wageningen, The Netherlands*

During the past 10 years, advances in molecular technologies and in virology have led to an explosive accumulation of information about the genetic organization and nucleotide sequence of viral RNA genomes. This has created the possibility of determining evolutionary relationships between many taxonomic groups, which in other aspects, such as host range, pathology, and virion morphology, may show a wide variation. Comparisons of genome structures and nonstructural viral proteins have revealed that many plant RNA viruses are genetically related to animal-infecting counterparts. The majority of eukaryotic RNA viruses can thus be placed into a limited number of *supergroups* or *superfamilies*; each of these groups comprises viruses that share several relevant molecular and genetic properties. With more viral sequences having been published over the past years, the concept of supergrouping has gained increasing support (1–5).

This chapter will discuss some new insights in the genetic functions and evolution of viral RNA genomes. Since it is not possible to discuss all the relevant data and advances for all RNA viruses in a single contribution, we will focus mainly on three different supergroups, the picorna-like viruses, the alpha-like viruses, and the negative-stranded RNA viruses.

## RNA VIRUS SUPERGROUPS

Viruses with an RNA genome and parasitic on eukaryotic cells form a very diverse group, with a wide variation in morphologic appearance and biological properties. On the basis of their genomic form, they have been traditionally divided into the positive-stranded, the negative-stranded, and the double-stranded RNA viruses. They have been further subdivided into a large number of families and genera (6). So far, only one order has been approved by the International Committee on Taxonomy of Viruses (ICTV) in virus taxonomy, the order Mononegavirales, comprising all *mono*partite, *nega*tive-stranded RNA viruses. Despite all differences, the genetic

**TABLE 1.** *Supergroups of RNA viruses*

| Supergroup | Taxonomic group | Common characteristics |
|---|---|---|
| *Picorna-like* | Picorna<br>Calici<br>Como<br>Nepo<br>PYFV<br>Poty | Plus-stranded RNA<br>5'-VPg, 3'-poly(A)<br>No subgenomic mRNA<br>Polyprotein processing<br>Conserved gene set |
| *Alpha-like* | Toga<br>Bromo<br>Carla<br>Clostero<br>Cucumo<br>Furo<br>Hordei<br>Ilar<br>Potex<br>Tobamo<br>Tobra<br>Tymo | Plus-stranded RNA<br>5'-cap<br>Subgenomic mRNA<br>Often read-through<br>Conserved gene set |
| *Flavi-like* | Flavi<br>HCV<br>Pesti | Plus-stranded RNA<br>5'-cap, no poly(A)<br>No subgenomic mRNA<br>Conserved gene set |
| *Carmo-like* | Carmo<br>Diantho<br>Machlomo<br>Necro<br>Tombus | Plus-stranded RNA (small)<br>5'-cap, no poly(A)<br>Subgenomic mRNA<br>Read-through or frame-shift<br>Conserved polymerase gene |
| *Sobemo-like* | Sobemo<br>Luteo | Plus-stranded RNA (small)<br>5'-VPg, no poly(A)<br>Conserved gene set |
| *Corona-like* | Corona<br>Toro<br>Arteri | Plus-stranded RNA (small)<br>5'-cap, 3'-poly(A)<br>Nested set of mRNAs<br>Enveloped particle |
| *Minus-stranded* | Paramyxo<br>Rhabdo<br>Orthomyxo<br>Bunya<br>Arena | Minus-stranded RNA<br>Complementary RNA termini<br>5'-pyrophosphate, no poly(A)<br>Similar gene set<br>Enveloped particle |
| *Double-stranded* | Birna<br>Crypto<br>Cysto<br>Partiti<br>Reo<br>Toti | Double-stranded RNA<br>Segmented genome<br>5'-cap, no poly(A) |

Data from Strauss et al., ref. 2 and Goldbach et al., ref. 9.

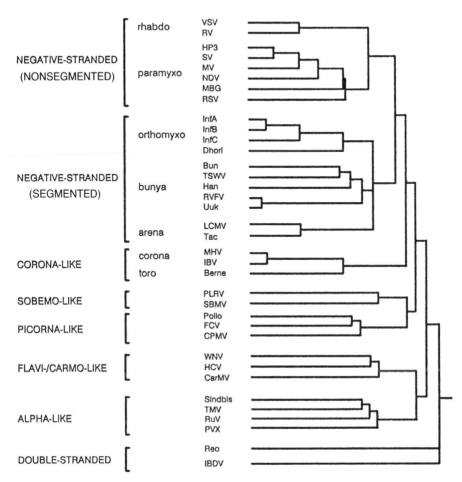

**FIG. 1.** Tentative phylogenetic tree for the RNA-dependent RNA polymerases of eukaryotic RNA viruses. The tree is partly based on the cluster analysis published by Koonin (5), extended with the polymerases of negative-stranded and double-stranded RNA viruses. The degree of homology between amino acid sequences is inversely proportional to the length of the branches. The various supergroups are indicated.

and molecular data now available reveal that most of the eukaryotic RNA viruses, irrespective of whether they infect plants or animals, can be clustered into a limited number of supergroups (Table 1) (1,2). Although each supergroup may contain viruses that are very diverse in virion architecture, host range, and pathology, they share a number of molecular and genetic properties (i.e., RNA strandedness, genomic structure, conserved genes), which make us believe that they are evolutionarily interrelated.

One of the molecular characteristics on which the supergrouping, as presented in Table 1, has been based represents the polymerase gene. All (nondefective) RNA viruses, in spite of their variation in, for example, gene number, invariably possess a gene for an RNA-dependent RNA polymerase and, therefore, this gene is a good candidate to try to reconstruct phylogenetic relations between RNA viruses. All viral RNA-dependent RNA polymerases share several (six to eight) short, conserved sequence motifs of which the so-called Gly-Asp-Asp (GDD) motif (SDD or GDN in negative-stranded viruses) is the most obvious (5–7). Various authors have published parsimony trees, in which RNA viruses have been clustered and arranged using these conserved polymerase sequence motifs (5,8–10). The most accurate and detailed polymerase sequence dendrograms yet published (5), however, contained only positive-stranded RNA viral genes. Starting from this dendrogram, the polymerase genes of negative-stranded and double-stranded RNA viruses have now been included into a more complete, overall phylogenetic scheme also comprising the negative-stranded and double-stranded RNA viruses (Fig. 1).

For simplicity, for each taxonomic group, only a limited number of representative viruses (including a plant- and an animal-infecting virus when a group bridges the plant and animal kingdom) have been included in the scheme of Fig. 1. Inspection of this putative phylogenetic tree of polymerase genes reveals that the clustering of RNA viruses solely based on these genes agrees rather well with the supergrouping in Table 1, which is merely based on the strandness, on the structure and organization of the respective genomes, and on additional conserved genes.

Of the various supergroups listed in Table 1 the supergroups of picorna-like viruses, alpha-like viruses, and minus-stranded viruses are among the most important for our understanding of virus evolution and, therefore, only these groups will be further discussed in more detail. These three supergroups each embrace RNA viruses of plants and animals and allow us to compare related viral genomes that, during evolution, have successfully adapted to very different ecological niches.

## The Picorna-like Viruses

The supergroup of picorna-like viruses comprise the Picornaviridae, the Caliciviridae (both families exclusively infecting animals), and several groups of plant viruses [e.g., the family Potyviridae, the floating genera *Comovirus* and *Nepovirus*, and the parsnip yellow fleck virus (PYFV) "group"; see Table 1)]. A comparison of the genetic maps of some members of the supergroup is shown in Fig. 2, to illustrate their genetic similarities and differences.

The genetic interrelation among the picorna-like viruses are demonstrated by the following shared properties.

1. They all have positive-stranded RNA genomes, provided with a small protein (VPg) covalently linked to the 5'-end, and poly(A) tail.
2. Their translation strategy is similar, involving the synthesis of polyproteins (one or two, depending on whether or not the genome is split), from which the functional proteins are derived by proteolytic cleavages.
3. They encode several nonstructural proteins that exhibit significant sequence homology [i.e., the viral polymerase, a proteinase, and a (putative) helicase].

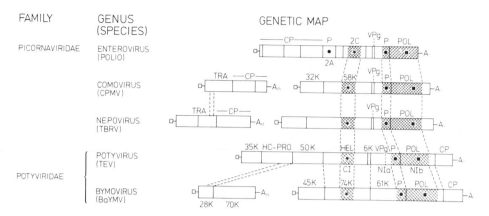

**FIG. 2.** Comparison of the genetic maps of members of picorna-like supergroup [shown here: poliovirus, cowpea mosaic virus (CPMV), tomato black ring virus (TBRV), tobacco etch virus (TEV), and barley yellow mosaic virus (BaYMV)]. Coding regions in the genomes are indicated as *open bars*; regions of amino and sequence homology are indicated by *similar shading*. Other symbols: *CP*, coat protein; *TRA*, transport (cell-to-cell movement) protein; *P*, proteinase; *POL*, polymerase; *, nucleotide-binding domain (in putative helicases); ●, conserved trypsin-like cysteine proteinase motif; ■, conserved polymerase domain.

4. Moreover, these conserved proteins are, together with VPg, encoded by a similarly arranged gene set (5'—helicase, VPg, proteinase, polymerase -3') located in the 3' halves of their respective genomes (see Fig. 2).

5. Since these conserved nonstructural proteins have all been demonstrated or suggested to be involved in the RNA replication process (11,12), all viruses belonging to the picorna-like supergroup will share, in principle, a similar RNA replication strategy. This statement is further strengthened by the fact that the genomes of these viruses have similar terminal structures (VPg, poly(A) tail).

Despite the fact that all viruses belonging to the supergroup of picorna-like viruses are very similar in genomic structure, gene arrangement (3' half), and in the two basic processes of protein and RNA synthesis, there are also several major differences among the various members of the supergroup. For instance, although the plant-infecting comoviruses (e.g., CPMV, see Fig. 2) superficially look like a "split" picornavirus, they possess an extra gene in the middle (M-)RNA, encoding a 58–48-kDa protein doublet involved in cell-to-cell movement during plant tissue infection (13). The acquisition of this gene is obviously an adaptation to plants as hosts. Conversely, there is an additional gene in the (true) picornaviral genome, that is obviously absent in the plant comoviral genome (i.e., the 2A proteinase cistron). This function is indeed not required during expression of the split comoviral genome, since the precursor for the structural proteins is encoded by a distinct polyprotein (encoded by M-RNA) and need not to be released from the P2/P3-like polyprotein encoded by the bottom (B-)RNA. It is tempting to assume that the genomic differences between comoviruses and picornaviruses reflect adaptations to their respective (plant and animal) hosts. The genome comparisons provide (indirect) evidence that these adaptations are realized by capturing extra genes (e.g., the cell-to-cell transport protein gene of comoviruses) by recombinational events.

Another good example of creating diversity by recombination and gene exchange is represented by the Potyviridae, the genomes of which seem to be (successful) products of a series of recombinational events. Potyviruses are very distinct from the other picorna-like viruses in having a rod-shaped particle structure, instead of having a pseudo T = 3 isometric particle. On the genetic level this difference is illustrated by the linkage—at the 3′-side—of a unique coat protein gene to the set of conserved genes (see Fig. 2). Furthermore, upstream from the conserved replication gene module (helicase, VPg/proteinase, and polymerase genes) the genome of the potyviruses belonging to the genus *Potyvirus* contains three genes for which there seem to be no counterparts present in the genome of the other members of the picorna-like supergroup (see Fig. 2) This linkage of unique and conserved genes, together with other lines of evidence, demonstrates that, in addition to common ancestry, gene shuffling by interviral recombination has been a major mechanism on which RNA virus evolution is based (14–16). As a result, each member of the supergroup not only will be related evolutionarily, by the conserved replicatory protein genes, to the other picorna-like viruses, but also, by their additional genes, to other viruses, not belonging to this supergroup. Thus, the 5′ proximate-encoded protein in the potyviral genome (see Fig. 2), suggested to represent the cell-to-cell movement protein required for systemic invasion of plant tissues, has sequence homology to the tobacco mosaic virus (TMV) 30-kDa movement protein (17). Therefore, the conclusion may be drawn that, through the 5′ proximate gene, the genus *Potyvirus* might be distantly related to tobamoviruses. It should be noted, however, that this gene seems to be lacking in the bipartite genome of the genus *Bymovirus* (see Fig. 2), also belonging to the Potyviridae.

The second unique gene of potyviruses encodes the helper component (HC-PRO), a two-domain protein required for aphid transmission (HC) and possessing proteolytic activity (PRO). The protein exhibits some sequence homology to the aphid transmission factor (gene II product) of caulimoviruses, which suggests a genetic interrelationship between poty- and caulimoviruses. Barley yellow mosaic virus (BaYMV) is transmitted by a fungus (*Polymyxa graminis*) and not by aphids. In concert with this, in the BaYMV genome only the PRO-domain is conserved, whereas the HC domain is lacking (see Fig. 2).

The protein encoded by the third unique gene of potyviruses (i.e., the 50-kDa protein of TEV) does not show homology to any other known viral or cellular protein.

In addition to the three unique genes in the 5′-terminal part of the potyviral genome, a fourth gene is exclusively found in the *Potyviridae* (i.e., the coat protein gene located at the 3′-end of the genome). Whereas picorna- and comoviruses (and probably also nepoviruses) encode coat proteins that are folded into three globular domains, (or "β-barrels") (18), building up isometric virions with the same basic geometry, members of the Potyviridae encode a single capsid protein species (30 to 35 kDa) that folds into a core, similar to the TMV coat protein, with extended $NH_2$- and COOH-termini, and builds a tobamo-like rod-shaped particle (19). The structure of the coat protein, the resulting rod-shaped architecture of the potyviral particle, and the deviant position of the coat protein gene within the genome set the Potyviridae distinctly apart from the other picorna-like viruses, and links this virus family to rod-shaped plant viruses such as the tobamoviruses (belonging to the alpha-like supergroup).

Last but not least, there is a genetic link between Potyviridae and the animal virus family Flaviviridae. Although the helicase (CI protein) of members of the Potyviridae shows significant sequence homology to the (putative) helicases of the como-, nepo-, and picornaviruses (see Fig. 2), this protein is even more closely related to the nucleotide-binding motif (NTP motif)-containing protein (NS3) of Flaviviridae (20–22). On the basis of sequence homology, the NTP–motif-containing proteins encoded by a considerable number of positive-stranded RNA viruses can be placed into three main groups (alphavirus-like, picornavirus-like, and poty/flavivirus-like), each of them revealing closer relationships to distinct groups of cellular NTP-binding motif-containing proteins (both prokaryotic- and eukaryotic) than to the NTP-binding proteins of the other viral groups (20,23).

## The Alpha-like Supergroup

Most of the remarks made for the picorna-like supergroup also hold for the supergroup of alpha-like viruses. On one hand, they share several genetic properties to place them sensefully into a supergroup; on the other hand, the individual members all have unique properties. The genetic interrelationship among these viruses is demonstrated by the following shared properties (Fig. 3):

1. They all have plus-stranded RNA genomes provided with a cap.
2. They all produce at least one subgenomic mRNA, encoding the structural protein(s).
3. Often read-through translation (at a suppressible stop codon), is involved for the expression of a downstream cistron.
4. They encode several nonstructural genes that exhibit significant sequence homology [i.e., the viral polymerase, a (putative) helicase, and a (putative) methyltransferase (domain)].
5. Moreover, these conserved proteins are encoded by a similarly arranged gene set (5'—methyltransferase–helicase–polymerase—3').
6. Also for this supergroup, the conserved genes are all likely to be involved in the RNA replication process; therefore, again, it is tempting to assume that all members share a similar RNA replication strategy.

As noted for the picorna-like supergroup, the various members of this group are dissimilar in having unique genes, which are not shared by the group, in their

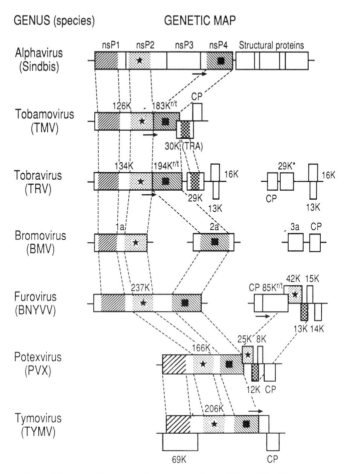

**FIG. 3.** Comparison of the genetic maps of some members of the alpha-like supergroup [including Sindbis virus, tobacco mosaic virus (TMV), tobacco rattle virus (TRV), brome mosaic virus (BMV), beet necrotic yellow vein virus (BNYVV), potato virus X (PVX), and turnip yellow mosaic virus (TYMV)]. Coding regions in the genomes are indicated as *open bars*; regions of amino acid sequence homology in the gene products by *similar shading*. Other notations: *CP*, coat protein; *TRA*, transport (cell-to-cell) protein; *, nucleotide-binding motif (putative helicase); ■, conserved polymerase domain; →, leaky termination codon, *r/t*, read-through.

respective genomes. For instance, alphaviruses have a number of extra genes (encoding nonstructural protein nsP3 and some additional structural proteins), whereas the related plant virus TMV has acquired an extra gene encoding a 30-kDa protein involved in cell-to-cell movement through infected plant tissue (see Fig. 3). Also, a considerable number of the plant-infecting supergroup members have split genomes, a property hardly known for animal-infecting positive-stranded viruses.

Therefore, the conclusion seems justified that at least a part of the genomic variation among the various members of the supergroup, as also observed for the picorna-like supergroup, reflects adaptations of ancestral viruses to different (plant versus animal) hosts, adaptations that seem to have been partly fulfilled by acquiring extra genes through recombinational events.

## The Supergroup of Negative-Stranded Viruses

The negative-stranded supergroup forms a rather homogeneous group, also in terms of structure. In principle, all have nucleoprotein particles surrounded by a lipid envelope containing one or more glycoproteins. Although only limited homology among the various viral-encoded proteins is present (mainly the polymerase proteins), they share several common features of genomic organization and replication strategy (Fig. 4). The monopartite Rhabdoviridae and Paramyxoviridae are so similar that they have been placed into an order, the Mononegavirales (6). Indeed, their genetic maps are almost colinear. The multipartite members Bunyaviridae, Orthomyxoviridae, and Arenaviridae are less similar, both in number of genome segments, number of genes, and genetic map. Still, they all share similar proteins (and genes), that is, membrane glycoproteins, a nucleocapsid protein, and polymerase proteins, the latter being a single large (L) protein or a set of different smaller proteins (PA, PB1, and PB2 for Orthomyxoviridae).

The clustering of all negative-stranded RNA viruses in a single supergroup is further supported by the phylogenetic polymerase tree of Fig. 1, although it shows a clear subclustering of the nonsegmented and segmented viruses.

Of the five families of negative-stranded viruses two families bridge the gap between the animal and plant kingdom: the Rhabdoviridae and the Bunyaviridae. Comparing the genomes of a plant-infecting rhabdovirus—sonchus yellow net virus (SYNV)—and an animal-infecting rhabdovirus—vesicular stomatitis virus—reveals that the plant-infecting counterpart has acquired an additional gene, encoding a protein called SC4. Although the function of this additional gene is still unknown, it is tempting to assume that the capturing of this extra gene represents an adaptation to plant hosts. All other rhabdoviral genes, however, seem unaltered in both genomes.

Comparison of the genome of tomato spotted wilt virus (TSWV), a plant-infecting bunyavirus (24), with those of the animal-infecting bunyaviruses (e.g., Bunyamwera, see Fig. 4) reveals that they have very similar genetic maps, with, again, the tomato spotted with virus genome containing one additional gene, NSm.

It, therefore, seems that the negative-stranded viruses have been under much less differentially selective pressures in their plant and animal hosts, than have the positive-stranded viruses. One explanation could be that the negative-stranded viruses have invaded the plant kingdom much more recently than the positive-stranded ones.

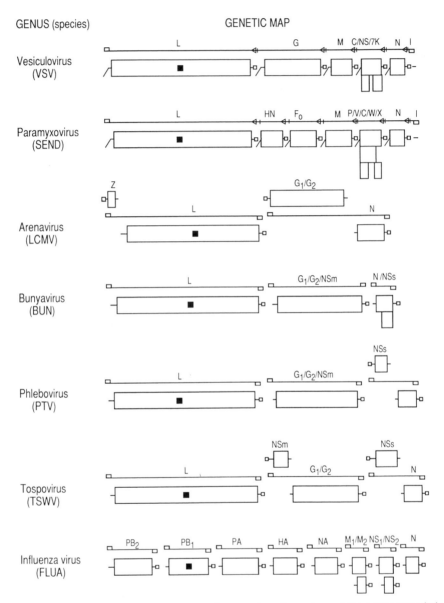

**FIG. 4.** Comparison of the genetic maps and expression of members of the minus-stranded virus supergroup [shown here: vesicular stomatitis virus (VSV); Sendai virus (SEND); lymphocytic choriomeningitis virus (LCMV); influenza A virus (FLUA); Bunyamwera virus (BUN); punta toro virus (PTV); tomato spotted wilt virus (TSWV)]. Genomic RNAs are indicated by *single lines*, with the complementary termini as *open squares*; open-reading frames in the transcripts are indicated as *open bars*. Other symbols: ■, conserved polymerase domain.

## EVOLUTIONARY MECHANISMS

From comparisons made in the foregoing and summarized in Table 1, the conclusion seems justified that the eukaryotic RNA viruses for which the genome sequence have now been determined can be clustered in several supergroups, each group comprising viruses that share a number of relevant molecular and genetic properties. The question now arises of how present day RNA viruses, which infect eukaryotic kingdoms that diverged $1.2 \times 10^9$ years ago, can be genetically so closely related that they share similar gene sets encoding conserved proteins (e.g., picorna-like and alpha-like viruses of plants and animals). In theory four different evolutionary pathways may be considered to account for the interviral relations that have been detected so far:

1. Convergent evolution
2. Transduction of (conserved) host genes
3. Common ancestry
4. Interviral recombination

The first two mechanisms seem difficult to imagine as the major explanations, since they do not explain the colinearity of the genetic maps of the various viruses. Hence, the third and fourth mechanisms (divergency from common ancestors and interviral recombination) are the only plausible possibilities that remain, and evidence is presented in the following that evolution of RNA viruses is probably based on both.

### Divergency as an Evolutionary Mechanism

Why consider *divergent evolution*, that is, the evolution of related viruses (e.g., plant and animal viruses) from common ancestors? This option can be best illustrated by comparing all similarities between the comovirus CPMV and the picornavirus poliovirus, one of the best studied combinations of a plant and animal virus (see Fig. 2). The CPMV and poliovirus, apart from the properties characteristic of all picorna-like viruses mentioned in Table 1, also share their capsid architecture. Although CPMV has two capsid proteins, these proteins fold into three β-barrel domains that mimic, in many details, the three independent β-barrel domains of the major proteins in the picornavirus capsid (18). Moreover, for both viruses the arrangement of these domains within the protein reflects the alignment of the gene order. In other words, by placing M-RNA of CPMV left of B-RNA, the general gene order in the (divided) genome of this virus is colinear with that of poliovirus. Hence, CPMV may be regarded as a split picornavirus, with the P1 region (M-RNA) physically separated from the P2 and P3 regions (B-RNA). Obviously, common ancestry seems to be the only plausible evolutionary mechanism to underlie the genetic relation between como- and picornaviruses. In this context, a number of plant viruses (e.g., dandelion yellow mosaic, anthriscus yellows, and parsnip yellow fleck; 25) have been identified as putative picornaviruses. Hence, this virus group may bridge the gap between the plant and animal kingdom, thereby making common ancestry an even more credible explanation.

The high mutation rates generally observed for RNA genomes make it very improbable that common ancestral viruses predated the evolutionary separation of plant and animal cells. It seems more likely that when divergency has been a major mechanism during RNA virus evolution, common ancestors of plant and animal viruses existed in much more recent times. If this is true, then there are several valid arguments for supposing that insects may have played a major role in RNA virus evolution, either as sources of ancestral viruses into plants or vice versa: (a) The host ranges of plant and animal viruses overlap in insects. Numerous plant viruses are transmitted by insects, and a number of them even multiply in their respective insect vectors (26). On the other hand, some insect viruses use plants as reservoirs for horizontal transmission (e.g., *Rhopalosiphum padi* virus in barley; 27). (b) There are many RNA viruses in insects, with a great variety in morphology and genomic properties. Hence, insects are indeed potential sources of ancestral viruses. (c) Some positive-stranded RNA viruses of insects have divided genomes (the Nodaviridae). Since plus-stranded RNA viruses of plants often have divided genomes, whereas the Nodaviridae are, in this aspect unique among the animal viruses, this insect virus group may represent an important link in the evolution of plant and animal RNA viruses.

If we consider the differences between positive-stranded and negative-stranded RNA viruses, it is notable that the corona-like viruses obviously take a somewhat intermediate evolutionary position. On one hand, they share properties with the (other) positive-stranded viruses (e.g., genome polarity, conserved proteinase and helicase motifs); on the other hand, they have affinities to the negative-stranded viruses (particle morphology, gene order, genomic termini, and transcriptional expression). This intermediate position is also reflected in the phylogenetic tree of polymerases (see Fig. 1).

### Recombination as Evolutionary Mechanism

Divergency from common ancestors as the sole evolutionary mechanism, however, cannot explain all similarities and, maybe even more importantly, the differences between RNA viruses belonging to the same supergroup. For instance, the unique gene order in the potyviral genome, with a coat protein gene downstream from the conserved gene set typical for all picorna-like viruses and, moreover, several genes upstream from this set for which no counterparts are found for the como- and nepoviruses (see Fig. 2), make this virus group very distinct from the others. Also, in addition to their conserved genes, other viruses appear to have one or more extra genes not found in the genomes of related viruses, for which no counterpart is found in the genomes of related animal viruses: Sindbis virus nsP3 and envelope protein genes, CPMV 58-kDa gene, and others. The coupling of common genes to unique genes in various viral genomes suggests that, in addition to common ancestry, there is at least a second major evolutionary mechanism, namely, interviral recombination, by which sets of genes can be exchanged among viruses belonging to different taxonomic groups. This has led to the concept of *modular evolution* (i.e., the idea that viral RNA genomes are constructed by the mixing and joining of gene modules; 14–16). Indeed, recombination has been shown to occur for several virus families or groups [e.g., picornaviruses (28,29), coronaviruses (30), and bromoviruses (31)]. For poliovirus, evidence has been presented that this process occurs by a copy-choice mechanism during RNA replication (32), and this may also hold for the other viruses. For a considerable number of plant RNA viruses, there is an accumulating amount of evidence that present-day RNA genomes have actually arisen by recombination events. One of the best examples represents the Potyviridae, which can be considered (in view of their widespread occurrence and economic impact in agriculture) successful products of several interviral recombination events. As a result of such recombination events, potyviruses appear to possess genes that originate from at least four different genetic sources [i.e., the proteinase (*NIa*) and polymerase (*NIb*) genes, which are both related to genes of picornaviruses; the *CI* (helicase) gene, most closely related to the *NS3* gene of flaviviruses; the *HC-PRO* gene, distantly related to gene II of caulimoviruses; and a coat protein gene, related to other rod-shaped viruses such as tobamoviruses]. But there are more examples. For instance the RNA-2 segment of the tobacco rattle virus (TRV; see Fig. 3) may be extended at its 3′ end with sequences of RNA-1 of variable length (depending on the strain) (33,34). Moreover, some TRV isolates appear to represent true, natural recombinants of TRV and pea early browning virus (35). And, last but not least, some luteoviruses [e.g., potato leafroll virus (PLRV) and beet western yellow virus (BWYV)] specify a polymerase homologous to that of sobemoviruses, whereas another luteovirus (e.g., BYDV) specifies a very distinct polymerase homologous to that of carmoviruses (36,37).

Two important notes can be made with reference to these examples. First, recombination not only can occur within a single virus, but also between different viruses belonging to the same taxonomic group, and even between viruses belonging to different taxonomic groups. Second, recombination occurs not only under the selective pressure of artificial, laboratory conditions, but it also occurs in nature, some recombinant genomes obviously being more advantageous under certain conditions and ecological niches than those of the parental viruses.

## CONCLUSIONS

With the current flood of genetic and molecular data, it has become possible to compare groups of eukaryotic RNA viruses in terms of nucleotide and protein sequences, genetic maps, and replication strategies. This has led to supergrouping of distinct viruses that were previously not suspected to be evolutionarily related. The data now available indicate that RNA virus evolution is mainly based on two mechanisms: divergency from common ancestors and interviral recombination. As a result of these two mechanisms the evolutionary dynamics and flexibility of viral RNA genomes may be expected to be so great that RNA viruses will always be able to retain their important position among the most successful parasites of eukaryotic organisms.

## REFERENCES

1. Goldbach RW. Molecular evolution of plant RNA viruses. *Annu Rev Phytopathol* 1986;24:289–310.
2. Strauss EG, Strauss JH, Levine AJ. Virus evolution. In: Fields BN, Knipe DM, et al, ed. *Virology*. New York: Raven Press; 1990.
3. Habili N, Symons RH. Evolutionary relationship between luteoviruses and other RNA plant viruses based on sequence motifs in their putative RNA polymerases and nucleic acid helicases. *Nucleic Acids Res* 1989;17:9543–9555.
4. Bruenn JA. Relationships among the positive strand and double-strand RNA viruses as viewed through their RNA-dependent RNA polymerases. *Nucleic Acids Res* 1991;19:217–226.
5. Koonin EV. The phylogeny of RNA-dependent RNA polymerases of positive-strand RNA viruses. *J Gen Virol* 1991;72:2197–2206.
6. Francki RIB, Fauquet CM, Knudson DD, Brown F. Fifth report of the International Committee on Taxonomy of Viruses. *Arch Virol* 1991;2:1–450.
7. Kamer G, Argos P. Primary structural comparison of RNA-dependent polymerases from plant, animal and bacterial viruses. *Nucleic Acids Res* 1984;12:7269–7282.
8. Candresse T, Morch MD, Dunez J. Multiple alignment and hierarchical clustering of conserved amino acid sequences in the replication-associated proteins of plant RNA viruses. *Res Virol* 1990; 141:315–329.
9. Goldbach R, Le Gall O, Wellink J. Alpha-like viruses in plants. *Semin Virol* 1991;2:19–25.
10. Poch O, Sauvageut L, Delarue M, Tordo N. Identification of four conserved motifs among the RNA-dependent polymerase encoding elements. *EMBO J* 1989;8:3867–3874.
11. Goldbach R, Van Kammen A. Structure, replication and expression of the bipartite genome of cowpea mosaic virus. In: Davies JW, ed. *Molecular plant virology*. Boca Raton: CRC Press; 1985; 2:83–120.
12. Takegami T, Kuhn RJ, Anderson CW, Wimmer E. Membrane-dependent uridylation of the genome-linked protein VPg of poliovirus. *Proc Natl Acad Sci USA* 1983;80:7447–7451.
13. Van Lent J, Storms M, Van der Meer F, Wellink J, Goldbach R. Tubular structures involved in cell-to-cell movement of cowpea mosaic virus are also formed in infected cowpea protoplasts. *J Gen Virol* 1991;72:2615–2623.
14. Zimmern D. Evolution of RNA viruses. In: Holland J, Domingo E, Ahlquist P, eds. *RNA genetics*. Boca Raton: CRC Press; 1987:211–240.
15. Gibbs A. Molecular evolution of viruses: "trees," "clocks," and "modules." *J Cell Sci* 1987;7:319–337.
16. Goldbach R, Wellink J. Evolution of plus-strand RNA viruses. *Intervirology* 1988;29:260–267.
17. Domier LL, Shaw JG, Rhoads RE. Potyviral proteins share amino acid sequence homology with picorna-, como- and caulimoviral proteins. *Virology* 1987;158:20–27.
18. Rossmann MG, Johnson JE. Icosahedral RNA virus structure. *Annu Rev Biochem* 1989;58:533–573.
19. Shukla DD, Strike PM, Tracy SL, Gough KH, Ward CW. The N and C termini of the coat proteins of potyviruses are surface located and the N terminus contains major virus-specific epitopes. *J Gen Virol* 1988;69:1497–1508.
20. Gorbalenya AE, Koonin EV, Donchenko AP, Blinov VM. A novel superfamily of nucleoside triphosphate-binding motif containing proteins which are probably involved in duplex unwinding in DNA and RNA replication and recombination. *FEBS Lett* 1988;235:16–24.
21. Hodgman TC. A new superfamily of replicative proteins. *Nature* 1988;333:578.
22. Lain S, Riechmann JL, Martin MT, Garcia JA. RNA helicase: a novel activity associated with a protein encoded by a positive strand RNA virus. *Nucleic Acids Res* 1991;18:7003–7006.

23. Lain S, Riechmann JL, Martin MT, Garcia JA. Homologous potyvirus and flavivirus proteins belonging to a superfamily of helicase-like proteins. *Gene* 1989;82:357–362.
24. De Haan P, Kormelink R, De Oliveira Resende R, Van Poelwijk F, Peters D, Goldbach R. Tomato spotted wilt virus L RNA encodes a putative RNA polymerase. *J Gen Virol* 1991;71:2207–2216.
25. Murant AF, Hemida SK, Mayo MA. Plant viruses that resemble picornaviruses. *Abstr 7th Int Congr Virol*; 1987,R24.2.
26. Matthews REF. *Plant virology*, 3rd ed. New York: Academic Press; 1991.
27. Gildow FE, D'Arcy CJ. Barley and oats as reservoirs for an aphid virus and the influence on barley yellow dwarf virus transmission. *Phytopathology* 1988;78:811–816.
28. Cooper PD. Genetics of picornaviruses. *Compr Virol* 1977;9:133–207.
29. King AMQ, McCahon D, Slade WR, Newman JWI. Recombination in RNA. *Cell* 1982;29:921–928.
30. Makino S, Keck JG, Stohlman SA, Lai MMC. High-frequency RNA recombination of murine coronaviruses. *J Virol* 1986;57:729–733.
31. Bujarski JJ, Kaesberg P. Genetic recombination between RNA components of a multipartite plant virus. *Nature* 1986;321:528–531.
32. Kirkegaard K, Baltimore D. The mechanism of RNA recombination in poliovirus. *Cell* 1986;47:433–443.
33. Angenent GC, Linthorst HJM, Van Belkum AF, Cornelissen BJC, Bol JF. RNA2 of tobacco rattle virus strain TCM encodes an unexpected gene. *Nucleic Acids Res* 1986;14:4673–4682.
34. Cornelissen BJC, Linthorst HJM, Bredero FT, Bol JF. Analysis of the genome structure of tobacco rattle virus strain PSG. *Nucleic Acids Res* 1986;14:2157–2169.
35. Robinson D, Hamilton WDO, Harrison DB, Baulcombe DC. Two anomalous tobravirus isolates: evidence for RNA recombination in nature. *J Gen Virol* 1987;68:2552–2562.
36. Veidt I, Lot H, Leiser M, Scheidecker D, Guilley H, Richards K, Jonard G. Nucleotide sequence of beet western yellows virus RNA. *Nucleic Acids Res* 1988;16:9917–9932.
37. Van der Wilk F, Huisman MJ, Cornelissen BJC, Huttinga H, Goldbach R. Nucleotide sequence and organization of the potato leafroll virus genomic RNA. *FEBS Lett* 1989;245:51–56.

The Evolutionary Biology of Viruses,
edited by Stephen S. Morse.
Raven Press, Ltd., New York © 1994.

# 7

# Origin and Evolutionary Relationships of Retroelements

## Thomas H. Eickbush

*Department of Biology, University of Rochester, Rochester, New York 14627*

"Natural selection operating within genomes will inevitably result in the appearance of DNAs with no phenotypic expression whose only function is survival within genomes" (1). This nonphenotypic model for the origin of selfish, or parasitic DNAs (1,2) remains the most viable explanation for the remarkable abundance, in virtually every genome that has been studied, of interspersed repeated sequences with no apparent function (3,4). Many of these repeated sequences in prokaryotic and eukaryotic genomes are autonomous mobile elements, termed *transposable elements* (5,6). Although highly variable in their structure, distribution, and abundance within a genome, transposable elements can be divided into two major classes, based on the mechanism by which they propagate.

One class of transposable elements uses entirely DNA-mediated mechanisms. These elements do not encode their own polymerase and are simply replicated along with the chromosomal DNA in which they are inserted. Their mobility is made possible by enzymatic activities that are able to precisely excise a copy from one chromosomal location for insertion elsewhere within the host genome. The copy number of the DNA-mediated transposable elements can increase in a genome by excision of copies from one of the two chromatids behind a DNA replication fork and insertion into a chromosomal site in front of a replication fork, or by gene conversion mechanisms using the element on one homologue to replace the element that was excised from the other homologue (7,8). The second major class of transposable elements are not able to excise themselves from the chromosome. These elements encode their own replication machinery, an RNA-directed DNA polymerase (reverse transcriptase), which uses the sequence of an RNA transcript from the element to make a new DNA copy for insertion into the genome (9,10). These elements have been termed *retrotransposable elements*, or *retrotransposons*.

In addition to transposable elements a variety of other intra- or intercellular DNA parasites are known (5,6). For example, certain plasmids, introns, and a variety of viruses propagate by DNA-mediated or RNA-mediated mechanisms that are directly related to those of transposable elements. Thus, an understanding of the evolutionary origins of viruses, which is the focus of this section of the book, requires an understanding of their relation to the intracellular transposable elements. Indeed, one of the most frequently raised questions in the study of these elements is did viruses evolve from transposable elements by acquiring specialized mechanisms to exit and enter cells, or did transposable elements evolve from viruses by the loss of these functions?

In this chapter I will discuss the possible origins and evolutionary relation of the retrotransposable elements to viruses and other genetic elements that encode reverse transcriptase. As suggested by Temin (11) I will use the term *retroelement* for any DNA insertion element that encodes reverse transcriptase. As shown in Table 1, besides the retrotransposons, a current list of retroelements includes RNA viruses (12), DNA viruses (13,14), and a number of more recently discovered genetic elements found in mitochondria, plastids, and bacteria (15–18). Because the reverse transcriptases encoded by each of these diverse elements contain the same set of conserved amino acid sequences (16,19), it appears likely that these elements, or at least their reverse transcriptase genes, have had a common origin. Here, the structure of each class of retroelement and what is known of their ability to retrotranspose will be briefly reviewed, followed by a description of attempts to use the sequence of the reverse transcriptases to trace their evolutionary relations.

The relationship of retroviruses, hepadnaviruses, caulimoviruses, and retrotransposons has been previously reviewed (13,14,20,21). The intent here is to offer an updated view of newly discovered subclasses of retrotransposons as well as the new classes of retroelements from organellar and prokaryotic genomes. It is hoped that by deriving the evolutionary relations among the different groups of retroelements, new clues to their mechanism of retrotransposition can be found that will help guide experiments to determine their life cycle and their effect on the formation and utilization of genomes. Major emphasis throughout this chapter will be on the retrotransposons, because their abundance, broad distribution in all eukaryotes, and variety of replication machinery indicate that they are not only the most successful of the retroelements, but quite possibly the oldest.

**TABLE 1.** *List of retroelements and their distribution*

| Class | Distribution |
|---|---|
| Retroviruses | Mammals, birds |
| Hepadnaviruses | Mammals, birds |
| Caulimoviruses | Plants |
| Retrotransposons | |
|   LTR retrotransposons | Animals, plants, fungi, protozoa |
|   Non-LTR retrotransposons | Animals, plants, fungi, protozoa |
| Group II introns | Fungi and plant mitochondria, algae plastids |
| Mauriceville plasmid | *Neurospora* mitochondria |
| RTL gene | *Chlamydomonas* mitochondria |
| msDNA-associated reverse transcriptase | Purple bacteria, other bacteria |

## RETROVIRUSES

Because of their role in cancers and immunodeficiency diseases, the most extensively studied retroelements are, without question, the mammalian and avian retroviruses. Any discussion of retrotransposition naturally begins with an analysis of these well-characterized viruses. Extensive descriptions of the life cycles and biological properties of retroviruses can be found in numerous reviews (12,22,23). Only the common aspects of retroviral genomes and their replication mechanisms are described here for comparison with those of the retrotransposons.

All retroviruses have a relatively homogeneous structure, with three genes, termed *gag*, *pol*, and *env*, encoding the structural proteins and enzymes needed for the replication cycle. As shown in Fig. 1, these three genes are flanked by terminal repeats several hundreds of base pairs (bp) in length called long terminal repeats (LTRs). All three genes are transcribed from one promoter located in the left (upstream) LTR. The *gag* gene encodes a polyprotein that is processed into a series of low molecular weight structural proteins that are the major proteins of the nucleocapsid and matrix of the virion. The most highly conserved amino acid sequences of the *gag* protein are a series of characteristic cysteine and histidine residues with the sequence $CX_2CX_4HX_4C$ (where C, cysteine; H, histidine; and X, any amino acid). This sequence, located near the COOH-terminus of the polyprotein, is responsible for the tight association of this protein with the RNA of the nucleocapsid (24). Retroviral *gag* proteins usually contain one or two of these nucleic acid-binding (NB) domains.

The *pol* gene of retroviruses also encodes a polyprotein, the processed products of which supply the catalytic components found in the mature virions. In many retroviruses, the beginning of this gene encodes an aspartate proteinase (PR) that is responsible for processing the primary translation product of the *gag* and *pol* genes into the various structural and enzymatic components of the virion. In certain other retroviruses, this proteinase domain is encoded at the end of the *gag* gene, or represents yet a fourth separate gene between *gag* and *pol*. The remaining catalytic components encoded by the *pol* gene include DNA polymerase (RT), which can function as both a RNA-directed polymerase (reverse transcriptase) and a DNA-directed polymerase; RNase H activity (RH), which destroys the RNA strand of the RNA:DNA heteroduplex made by reverse transcription, thereby enabling second-strand synthesis; and finally, integrase (IN), which is responsible for the integration of the double-stranded DNA intermediate made from the RNA transcript into the chromosomes of the host cell. The *pol* gene of the various retroviruses is either in a different reading-frame from the *gag* gene (as it is shown in Fig. 1) or in the same frame as *gag*, but separated by a termination codon. The *pol* gene does not contain its own initiation codon, thus its expression in the different retroviruses is dependent on either frame-shifting between the *gag* and *pol* genes during translation, or readthrough of the *gag* termination codon (25,26). These mechanisms result in expression of the enzymes encoded by the *pol* gene at a level 10- to 20-fold lower than the structural proteins of the *gag* gene.

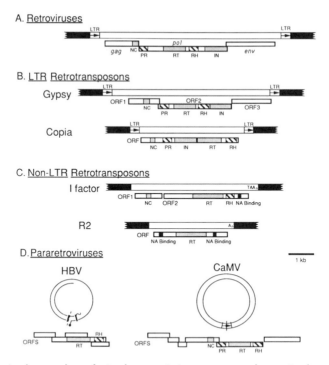

**FIG. 1.** Structural comparison of retroviruses, retrotransposons, and pararetroviruses: The integrated DNA form is shown for the retrovirus and retrotransposons (*black bars* representing flanking DNA sequences) and as the circular genome for the pararetroviruses. The 3′ end of the integrated non-LTR retrotransposons is a variable number of TAA repeats (I factor) or a poly(A) tail (R2). The *black dot* at the end of one strand of the hepatitis B virus (HBV) genome corresponds to the protein primer used to prime first-strand (cDNA) synthesis. The *wavy line* at the end of the other, incomplete strand of HBV corresponds to the RNA primer used to prime second-strand synthesis. *Thick bars* in HBV represent short direct repeats, involved in the strand-transfer reactions. *Dotted lines* indicate the initiation site for the full-length RNA template in both HBV and cauliflower mosaic virus (CaMV). The open-reading frames (ORFs) for each element is shown *below* the DNA diagram as a series of *horizontal boxes*. *Boxes at the same level* represent ORFs separated by termination codons, *boxes at different levels* represent ORFs in different reading frames. A consensus retrovirus is shown depicting only the *gag, pol,* and *env* genes. Additional genes following the *pol* or *env* genes are found in several retroviruses. Identified domains within the ORFs are indicated by *shading*. NC, nucleocapsid protein; PR, asparate protease; RT, reverse transcriptase; RH, RNase H; IN, integrase; NA binding, an identified cysteine–histidine motif with an unknown function in RNA binding or DNA binding.

The third gene found in all retroviruses is the *env* gene. This gene encodes one protein that interacts with the host cell receptors, mediating virus entry, and a second, transmembrane protein, that is believed to mediate fusion of the viral and host membranes. The *env* protein is translated from an mRNA that is generated by a splicing reaction in which the *gag* and *pol* gene sequences are removed from the full-length retroviral RNA transcript. More recently, it has been found that retroviruses sometimes encode proteins other than the *gag*, *pol*, and *env* genes (22). The genes for these proteins are located downstream from the *pol* or *env* genes and are expressed by differential splicing of the same RNA transcripts that give rise to the expression of the *gag*, *pol*, and *env* proteins. These additional proteins, which are highly variable among retroviruses, appear to regulate proviral expression by controlling the expression of both viral and cellular genes.

## RETROTRANSPOSABLE ELEMENTS (RETROTRANSPOSONS)

Transposable elements similar in structure to the retroviruses were first discovered in *Saccharomyces cerevisiae* and *Drosophila melanogaster* (27–29). Today, over 40 retrotransposable elements have been characterized. They are remarkably widespread and abundant in eukaryotes, with putative examples found in virtually every species of protists, fungi, plants, and animals for which the genomes have been extensively analyzed. However, with the exception of Ty1 of *S. cerevisiae* (30), characterization of the precise mechanisms of their retrotransposition lags considerably behind that of the retroviruses. The location of various structural and enzymatic domains within each retrotransposon has usually been determined only by sequence similarity to the retroviruses.

The structures of retrotransposons are more variable than those of retroviruses. Although the gene organization of some elements is remarkably similar to retroviruses, other retrotransposons contain structures that have little resemblance to retroviruses, except for the sequence similarity of one or two encoded proteins. The single best indicator that divides the retrotransposable elements into those that resemble retroviruses and those that do not is the presence of long terminal repeats. This simple taxonomic framework for placing retrotransposable elements in either the LTR or non-LTR groups has been confirmed by molecular phylogenetic studies of their encoded reverse transcriptases (21,31,32).

Many of the seminal studies of retrotransposons were of elements originally discovered in the fruit fly, *D. melanogaster* (27,28,33–35). Today, this species still has the most extensive and thoroughly described set of retrotransposons of any eukaryote (36). Shown in Fig. 1 are four *D. melanogaster* elements, two with LTRs and two without LTRs, which are representative of the many structural differences that exist between various retrotransposons. Most retrotransposons that contain LTRs have gene organizations similar to that of either the gypsy (37) or copia elements (28). Gypsy elements are remarkable for the extent to which their gene organization resembles that of retroviruses. Three open-reading frames (ORFs) are present that are of sizes and arrangement similar to the *gag*, *pol*, and *env* genes of retroviruses. Cysteine-histidine motifs identical with the nucleic acid-binding motifs of the *gag* gene are present in the first ORF, and sequence similarity to the conserved motifs of the proteinase, reverse transcriptase, RNase H, and integrase domains of the *pol* gene are found in ORF2. The third ORF of gypsy, although reminiscent of the location of the *env* of retroviruses, contains no sequence similarity to the *env* genes. However, ORF3 does contain a hydrophobic domain near its COOH-terminus analogous to the transmembrane portion of retroviral *env* proteins (38). There is no current evidence to suggest that the protein encoded by ORF3 confers on this element the ability to form a virion and leave the cell. Retrotransposons similar to gypsy can be found in a wide array of eukaryotes including the fungal element Ty3 of *S. cerevisiae* (39) and the plant element del of *Lilium henryi* (40). Most of these gypsy-like elements have only two ORFs that are similar to the first and second ORF of gypsy. Only three gypsy-like elements, all present in insects, contain the third ORF (27,38,41).

The second major subgroup of LTR-containing retrotransposons is represented in *D. melanogaster* by the element copia. Copia has a single ORF that contains a nucleic acid-binding motif similar to that of *gag* genes near its $NH_2$-terminal end and the four enzymatic domains found in the *pol* gene. Thus, the single ORF of copia encodes the same products as the first and second (*gag* and *pol*) ORFs of retroviruses and gypsy-like elements. The order of the enzymatic domains is somewhat different in copia, however, in that the integrase domain precedes the reverse transcriptase domain, rather than following the RNase H domain, as it does in retroviruses and gypsy. Retrotransposons similar to copia can be found in a wide array of eukaryotes including, Ty1 of *S. cerevisiae* (29) and Ta1 of *Arabidopsis thaliana* (42). The inverted order of the integrase and reverse transcriptase domains is found in all copia-like elements, and most encode a single ORF. Only Ty1 differs from the remaining copia-like elements in containing separate ORFs for the *gag* and *pol*-like functions. Both copia and Ty1 have been shown to make virus-like particles within cells; however, these particles do not appear to be infectious (30,33).

The second major group of retrotransposons contains those elements that lack LTRs. These elements have only recently been recognized as autonomous retrotransposons, and the number of characterized elements is expanding rapidly. Although I will use the term non-LTR retrotransposons for all elements of this group, because it emphasizes the key feature that they lack terminal repeats, these elements have been referred to by several different names. They have been called retroposons (11), and nonviral retrotransposable elements (9), because of their lack of structural similarity with the retroviruses. Many elements end with short poly(A) tails at their 3′ end; thus, they have also been called the poly(A)-type retrotransposable elements (10). Finally, because the best-known elements within this group are the long interspersed nucleotide elements (LINE) 1 (L1) elements found in all mammalian genomes (43), they have been called the LINE-like elements.

Two examples of the non-LTR retrotransposons found in *D. melanogaster*, the I factor (44) and R2 (45), are shown in Fig. 1. The non-LTR elements are probably the most varied group of retroelements, and I factor and R2 are not representative of any major subgroups. Most non-LTR retrotransposons share at least a few of the features associated with the *gag* and *pol* genes of retroviruses. Similar to the I factor, most contain two ORFs that are in different frames or separated by a termination codon. Other non-LTR elements are like R2 in containing a single ORF. The I factor contains a series of three putative RNA-binding motifs within the first ORF, with a spacing of cysteine and histidine residues highly similar to the motifs found in retroviral *gag* genes. R2, on the other hand, contains a single cysteine–histidine motif near the 5′ end of its single ORF, with a spacing of cysteine and histidine residues, $CX_2CX_{12}HX_4C$, identical with the zinc finger motif found in TFIIIa (reviewed in 46). The structure of this motif suggests that the product of the $NH_2$-terminal domain of the R2 ORF could bind DNA. Still other non-LTR elements [e.g., L1 elements of mammals (47) and ingi of *Trypanosoma brucei* (48)], contain two overlapping ORFs, but the first ORF contains no apparent nucleic acid-binding motifs. Consequently, it is unclear whether all non-LTR retrotransposons encode proteins with *gag*-like functions.

The reverse transcriptase domain of the non-LTR retrotransposons are usually centrally located in the element, but occasionally are closer to either the $NH_2$-terminal or COOH-terminal ends (48,49). Direct demonstration of the reverse transcriptase activity encoded by non-LTR elements has been obtained for Jockey (50), CRE1 (51), L1 (52), and R2 (53). Of the three additional enzymatic domains associated with the *pol* gene of retroviruses, the aspartate proteinase domain is absent from all non-LTR retrotransposons, and the RNase H and integrase domains are found in only a few elements. Weak sequence similarity to the RNase H domain has been found downstream from the reverse transcriptase domain in only the I factor and ingi of *T. brucei* (44,54), and only three of the non-LTR elements, CRE1, SLACS, and CZAR (55) contain an integrase domain similar to retroviruses. The integrase domain in these three related non-LTR elements is located upstream from the reverse transcriptase domain, as it is in the copia-like elements.

Many non-LTR retrotransposons, including both I and R2, encode a conserved cysteine–histidine motif, $CX_{1-3}CX_{7-8}HX_4C$, COOH-terminal of the reverse transcriptase domain (45,56). This domain is likely to correspond to a nucleic acid-binding domain; however, the spacing of cysteine and histidine residues in this motif is clearly different from that of the integrases identified in retroviruses and LTR retrotransposons, $HX_3HX_{22-32}CX_2C$ (57,58). Instead the COOH-terminal nucleic acid-binding motif found in many non-LTR retrotransposons is more like motifs that bind RNA (24,46). The only enzymatic activity, other than reverse transcriptases, that is known to be encoded by a non-LTR retrotransposon, is an endonuclease encoded by R2 (59). It is not known whether this endonuclease is part of the nucleic acid-binding motifs located either upstream or downstream from the reverse transcriptase domain (see Fig. 1).

In summary, the structure of the LTR retrotransposons is quite uniform, with the only major differences being the presence or absence of a third *env*-like ORF, and the location of the integrase either upstream or downstream from the reverse transcriptase and RNase H domains. The ORF structure of the non-LTR retrotransposons, on the other hand, is quite varied. Other than the reverse transcriptase domain and a lack of terminal repeats, there are no uniform characteristics in their gene organization. The non-LTR elements are missing many of the conserved domains present in all retroviruses and LTR retrotransposons. All non-LTR elements lack the proteinase domain, most lack the RNase H and integrase domains, and some may be missing the *gag*-like functions. Therefore, other than their ability to encode reverse transcriptase, there is strikingly little to suggest that these non-LTR retrotransposons have any relation to the LTR retrotransposons. There is no question that they are true retrotransposons, however, because both L1 and I elements precisely excise introns inserted in their genomes when they undergo transposition (34,35,60). These experiments directly demonstrate the I and L1 transpose by means of an RNA intermediate. In addition R2 specifically recognizes and reverse transcribes its RNA template at the chromosomal integration site (53).

## Mechanism of Retrotransposition

Retroviruses and LTR retrotransposable elements use their terminal repeats to initiate and terminate transcription of the RNA template, for strand-transfer steps in the synthesis of the DNA intermediate made from that template, and eventually for the integration of the DNA intermediate into new chromosomal sites. This complete dependence of the replication cycle on the presence of LTRs indicates that those retrotransposons that lack these terminal repeats must employ a fundamentally different mechanism (9,47,61). In this section, these two mechanisms of retrotransposition will be compared. The LTR mechanism has been extensively reviewed (12,22,23) and will only be summarized here. Insights into non-LTR methods of retrotransposition have appeared only recently and will be described in greater detail.

A summary of the various retrotransposition steps employed by retroviruses and LTR retrotransposons is shown in Fig. 2. The RNA genomic sequence used in the replication process is the same full-length RNA transcript beginning in the left LTR and ending in the right LTR, which is translated to yield the *gag* and *pol* protein products. Reverse transcription of this RNA is primed by a host tRNA molecule. Priming typically involves 15 to 20 nucleotides at the 3' end of the tRNA, base-pairing to the viral RNA sequences immediately downstream from the 5' LTR. On reaching the 5' end of the RNA, the presence of identical sequences at the 5' and 3' ends of the RNA template enables the reverse transcriptase to undergo strand-transfer to the 3' end of another mRNA molecule. This transfer enables completion of the first DNA strand, the negative strand. In retroviruses, this strand transfer is believed to be an intermolecular event; however, for simplicity, in Fig. 2 it is drawn as an intramolecular event. During reverse transcription, the RNA component of the synthesized RNA:DNA heteroduplex is removed by the RNase H domain of the polymerase. Second-strand (or plus-strand) synthesis is primed by oligoribonucle-

# LTR-based Integration

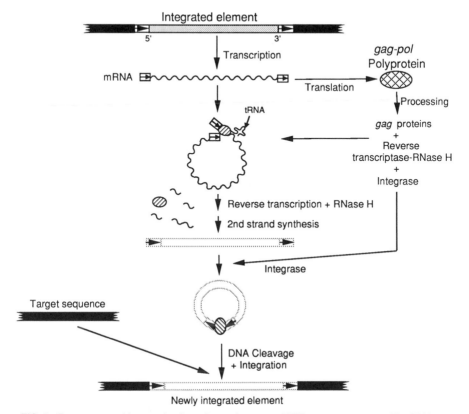

**FIG. 2.** Retrotransposition mechanism of retroviruses and LTR retrotransposons: The RNA template is depicted as a *wavy line*, with the terminal redundancy indicated by *boxed arrows*. First- and second-strand synthesis involves strand transfers between the ends of the RNA template. First-strand synthesis is shown as an intramolecular reaction to simplify the drawing, but at least for retroviruses, is an intermolecular reaction. The result of first- and second-strand synthesis is a linear double-stranded DNA molecule. The integrase binds to the LTR sequence at the ends of this linear molecule, cleaves a target site on chromosomal DNA and inserts the linear intermediate. More detailed descriptions of the retrotransposition of retroviruses and Ty1 can be found in several recent reviews (22,23,30).

otides generated by this RNase H activity. A polypurine stretch of nucleotides primes second-strand synthesis immediately upstream from the 3' LTR sequences. Again the presence of identical sequences at the 5' and 3' ends of the molecule allows strand transfer, this time from the 3' end to the 5' end of the same molecule and completion of the second-strand synthesis. Both first- and second-strand syntheses displace the DNA strand already synthesized in the region of the terminal repeat. The result of this complex series of events is a double-stranded linear DNA intermediate that has a complete LTR sequence at either end. This DNA intermediate migrates to the nucleus where the integrase, bound to the LTR sequences at its ends, is capable of making a staggered cut in chromosomal DNA and covalently joining the ends of the linear intermediate to the chromosomal DNA.

Only a few clues are now available about the mechanism of non-LTR retrotransposition. Indeed, considering the extensive differences in their ORFs, it is unclear whether one mechanism will apply to all elements. Even transcription of non-LTR elements is somewhat of a mystery. If transcription of the RNA template is controlled, as in retroviruses, by a conventional RNA polymerase II promoter with downstream initiation, the promoter sequence itself would not be present in the RNA template. (This is overcome in the LTR elements by the terminal redundancy and the regeneration of the upstream LTR.) A solution to this problem has recently been found for several non-LTR elements. Jockey (62), F (63), and I (64) of *D. melanogaster* and L1 of mouse (65) have been shown to contain promoter sequences near their 5' ends that can initiate transcription upstream at the first nucleotide of the element, thereby preserving the promoter in the course of replication by reverse transcription.

An alternative mechanism to ensure transcription of all nucleotides of a non-LTR retrotransposon has been suggested by the remarkable number of non-LTR elements that are site-specific in their insertion within the host genome (66). Of the 19 known non-LTR retrotransposons, 9 exhibit high levels of target-site specificity. R1 (67), R2 (68), and RT1 (69) insert at specific sites in the 28S gene of insects; G inserts into the spacer region of the ribosomal DNA unit (70); CRE1 (71), SLACS (72), and CZAR (55) insert in the spliced leader exons of species of *Trypanosomatidea*; Tx1 inserts into another mobile element in *Xenopus laevis* (73); and, finally, DRE specifically inserts 5' of tRNA genes in *Dictyostelium discoideum* (74). In contrast, of the 23 known LTR retrotransposons only 1, Ty3 of *S. cerevisiae*, exhibits similar target-site specificity (75). The high percentage of non-LTR elements that have evolved specificity in their insertion may be a result of their need to insert adjacent to a reliable exogenous promoter sequence.

The first comprehensive model for a non–LTR-based method of retrotransposition was proposed for the cin4 element of *Zea mays* (56). This model was subsequently expanded to include the retrotransposition of I factor (61), and L1 elements (47), and eventually all non-LTR retrotransposons (66). Key to the model was the observation that, for most non-LTR elements, all copies of the element have identical 3' ends, whereas their 5' ends are subject to truncation. For example, over 90% of the L1 elements in humans are truncated at their 5' ends (47). A summary of this model for the retrotransposition of non-LTR elements is shown in Fig. 3. The reverse transcriptase from the non-LTR element associates with the full-length, polyadenylated mRNA transcript. Because an aspartate proteinase is not encoded by non-LTR elements, the entire polyprotein encoded by the ORF(s) of the element is shown associated with the RNA template. This polyprotein may associate with the

# Non-LTR Integration

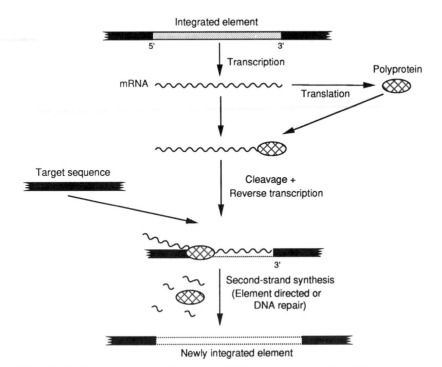

**FIG. 3.** Model for the retrotransposition of non-LTR retrotransposons: Non-LTR elements do not appear to encode a proteinase; thus, a single protein is shown to encode both reverse transcriptase and endonuclease activities. The protein binds near the 3′ end of a full-length RNA transcript. The protein cleaves the target DNA and uses the 3′ end of one of the DNA strands to prime first-strand synthesis. Although shown as a double-stranded cleavage, this cleavage could be a nick in only one of the chromosomal DNA strands, as in R2 (53). Most non-LTR elements do not encode RNase H activity; hence, it is not clear if second-strand synthesis is conducted by an enzymatic activity encoded by the non-LTR element or by DNA repair enzymes healing the break in the chromosome. This non-LTR model for retrotransposition is still speculative and may show significant variation between elements. Direct demonstration of the critical step in this model, reverse transcription of the template at the target site, has been obtained for only the R2 element of *B. mori* (53,59).

RNA at several locations along the molecule, but at a minimum, involves the region at or near the 3′ end. The RNA–protein complex migrates to the chromosomal DNA and generates a chromosomal break. This break is shown in Fig. 3 as a double-stranded cleavage, but could be a single-stranded nick. The 3′ end of one of the DNA strands generated by this break is then used to prime reverse transcription of the first DNA strand (negative-strand). For I factor and ingi, which appear to encode an RNase H activity, the RNA strand of the heteroduplex formed by the reverse transcriptase is removed and second-strand synthesis can take place, again primed by a 3' hydroxyl group of the chromosomal DNA. However, the reverse transcriptases of most non-LTR elements do not appear to contain an associated RNase H activity; consequently, removal of the RNA may be brought about by cellular RNase H activity or by strand displacement. It is also possible that formation of the second strand and the remaining steps in the integration, are conducted by host DNA repair mechanisms.

This model of non-LTR retrotransposition easily explains why the 3' ends of all integrated copies are identical, whereas the 5' ends can contain major truncations. The reverse transcriptase responsible for integration must bind in a specific manner to the 3' end of the RNA to initiate first-strand synthesis. However, once the reverse transcription process is initiated, if polymerization is aborted before reaching the 5' end of the RNA, or if the RNA template itself is truncated, the remaining steps can still occur, since they are nonspecific and may be controlled by DNA repair processes.

The only direct support for the insertion model shown in Fig. 3 comes from studies of the R2 element of *Bombyx mori* (53). This element inserts in a sequence-specific manner into 28S ribosomal (rRNA) genes of most insects (76). The single ORF of R2 (see Fig. 1) encodes a 120-kDa protein that binds nucleotide sequences near the 3' end of the R2 RNA transcript, and after generating a single-stranded nick at the insertion site of the 28S rRNA gene, is able to use the 3' end of the DNA to prime reverse transcription. In vitro, only after reverse transcription is initiated does the enzyme complete the double-stranded break at the target site (53).

Given the variability in the coding capacity of different non-LTR retrotransposons, it is difficult to predict how universal this mechanism of integration will be for all members of the non-LTR class. It is significant that all non-LTR elements contain precise 3' ends, and most are subject to the 5' truncations that are readily explained by this model. For those non-LTR elements that are known to be site-specific in their integration, it is likely that highly specific endonucleases initiate the retrotransposition process, whereas those non-LTR elements that integrate at multiple locations in the genome employ endonucleases that are not sequence-specific. One finding, however, argues against all non-LTR elements encoding their own endonuclease. Many of the non-LTR elements that insert at multiple locations do not generate a precise target-site duplication on insertion (66). It has been suggested (61) that these elements may use random DNA breaks in chromosomes to prime the reverse transcription step, rather than encode their own endonucleases.

The non-LTR mechanism of retrotransposition has several features in common with earlier suggestions for the mechanism of short interspersed nucleotide elements (SINE), and Alu element insertions in mammals (77,78). Since SINES do not encode their own reverse transcriptases, it is likely that the reverse transcriptases encoded by the non-LTR elements are responsible for the insertion of SINEs in eukaryotes. Clearly, more investigations of the mechanisms by which various non-LTR elements are able to integrate into their host genome are needed, if the universality of this mechanism and its role in the generation of SINEs is to be understood. Finally, comparison of the LTR-based and non-LTR methods of retrotransposition suggests that the non-LTR method is both simpler and less efficient, particularly, if certain elements depend on random breaks in the chromosome and DNA repair for second-strand synthesis. This would support the view that the non-LTR elements represent a more primitive form of retrotransposition, as suggested by the sequence phylogeny of reverse transcriptase sequences described latter in this chapter.

## OTHER RETROELEMENTS

Space does not permit a detailed description of the remaining retroelements listed in Table 1. Only the general features of their structure and what is known of their potential retrotransposition cycle will be summarized here.

### Hepadnaviruses

The hepadnaviruses primarily infect liver cells of mammalian and avian species (reviewed in 79). The best-studied member of this group is human hepatitis B virus (HBV), shown in Fig. 1. Although certain of the proteins encoded by hepadnaviruses and various aspects of their life cycle are similar to that of retroviruses (80), hepadnaviruses differ from retroviruses in two important respects. First, they do not require integration into the host genome for completion of their life cycle. Second, synthesis of viral DNA from an RNA template occurs immediately after the RNA is encapsidated into the virus particle; accordingly, the preponderant extracellular form is DNA, rather than RNA. Most hepadnaviruses encode four genes, the largest of which shows similarity to the *pol* genes of retroviruses, in that it contains reverse transcriptase and RNase H domains. The replication cycle of hepadnaviruses is known in considerable detail. Reverse transcription begins at the 3' end of a terminally redundant RNA template, primed by a hydroxyl group on the reverse transcriptase itself (81). The primer for second-strand synthesis is an oligomer of the template RNA that initiates DNA synthesis near the 3' repeat sequence, as in retroviruses. Both first- and second-strand syntheses undergo intramolecular strand transfers using the direct repeats (DR sequences) at the ends of the same RNA template. Second-strand synthesis is not complete, resulting in a gapped open-circular DNA molecule that is part double-stranded and part single-stranded.

### Caulimoviruses

Caulimoviruses are a group of plant viruses the best-studied member of which is cauliflower mosaic virus (CaMV; reviewed in 82). Caulimoviruses are similar to hepadnaviruses in that the preponderant extracellular form is DNA, and they do not require integration into the genome for completion of the life cycle. Of the variety of genes encoded by CaMV, two appear similar to the *gag* and *pol* genes of retroviruses (see Fig. 1). These genes contain an RNA-binding domain, proteinase, reverse transcriptase, and RNase H domains in the same order as retroviruses and gypsy-like elements. Formation of the DNA genome from a terminally redundant RNA template occurs by a mechanism that is nearly identical with that of retroviruses (83). A specific host tRNA primes first-strand synthesis near the 5' end of the RNA template, and on reaching the end the reverse transcriptase, strand-transfers to the 3' end of the same RNA template. Second-strand synthesis is primed by means of a polypurine tract upstream from the 3' end of the RNA. Unlike retroviruses, neither first- nor second-strand synthesis of the CaMV genome displaces the DNA strand already synthesized in the region of the terminal repeat; thus, the result is a double-stranded circular DNA molecule, with one repeat sequence, rather than the linear DNA molecule with complete terminal repeats on each end, as in retroviruses.

## Group II Introns

Group II introns are found inserted in the mitochondrial genomes of fungi and plants, and in the plastid genomes of algae (reviewed in 84). They are spliced from the primary transcription product by a mechanism involving a transesterification reaction that results in the formation of an intron lariat, similar to that produced during the splicing of nuclear mRNA introns. This similarity in splicing mechanisms has been used to suggest that the group II introns are the ancestors of present-day nuclear mRNA introns. Several properties of group II introns suggest that they are not simply introns, but are, in fact, mobile elements. First, group II introns have conserved, complex secondary structures that enable them to self-splice. Second, a fraction of the group II introns encode ORFs approximately 800 amino acids in length, containing sequence similarity to reverse transcriptase. Third, related introns have been found in different genes in different organisms.

Recently, evidence has been obtained demonstrating that two group II introns of the yeast mitochondrial gene *cox1*, al1 and al2, encode functional reverse transcriptase (85). The reverse transcriptase from al2 was shown to use either excised intron RNA or *cox1* pre-mRNA as template and initiate cDNA synthesis near the 3' end of the intron or within the downstream exon. This data strongly argues that the mechanism for the mobility of certain group II introns will be retrotransposition. Because most group II introns do not encode ORFs, yet are fully self-splicing, it is not clear whether the ancestral introns were self-splicing and the ORF was acquired subsequently, or whether the ancestral mobile element contained an ORF that was subsequently lost from most elements. Alternatively, a completely different mechanism of intron mobility has been suggested by experiments that demonstrate that the reverse of the intron-splicing step may be possible (86). Mobility of the group II intron in this reverse-splicing model would involve the integration of the intron RNA template into a target RNA molecule, reverse transcription of a cDNA copy of the intron and flanking sequences, followed by recombination of the intron sequences into the genome by means of these flanking sequences.

Group I introns are also mobile elements, but their mechanism of self-splicing is completely different from that of the group II introns, and none of the ORFs detected in the group I introns have a reverse transcriptase domain. A number of group I introns have been shown to encode a site-specific endonuclease that can initiate a transposition event by gene conversion (reviewed in 84,87). Thus, like the mobile elements of the nucleus, the organellar introns may be divided into two groups, the mobility of which is directed by RNA- and DNA-mediated mechanisms.

## Mitochondrial Plasmids

The Mauriceville and Varkud plasmids are closely related double-stranded closed circular DNA elements found in mitochondria of *Neurospora crassa* (15). The plasmids encode a 710-amino acid ORF, with a reverse transcriptase domain. The major transcript from these plasmids is a full-length linear RNA, with 5′ and 3′ ends that are immediately adjacent on the circular plasmid DNA such that every nucleotide of the plasmid is represented (88). Replication of the plasmid has been shown to occur through this full-length RNA template, with minus-stranded DNA synthesis beginning precisely at the 3′ end of the transcript. The 3′ end of the Mauriceville plasmid RNA can assume a tRNA-like structure that is recognized by the reverse transcriptase encoded by the plasmid. This tRNA structure does not directly prime reverse transcription by formation of RNA:RNA duplex as in retroviruses; rather, short cDNA fragments associated with the reverse transcriptase prime synthesis (89). Although the plasmids are not functional as introns, they do contain conserved sequence elements characteristic of group I introns, and they appear to insert into mitochondrial DNA in a site-specific manner (90).

## The *RTL* Gene

A third retroelement identified in mitochondria is the *RTL* (reverse transcriptase-like) gene of *Chlamydomonas reinhardtii* (17). The gene is expressed as a 1.2-kb mRNA that encodes a 368-amino acid ORF, with sequence similarity to reverse transcriptases. The *RTL* gene is flanked by two of the "scrambled" rRNA gene pieces that are a unique feature of the *C. reinhardtii* mitochondrial DNA (91). Extreme levels of codon bias are found in genes of these mitochondria, and the *RTL* gene employs several codons that are not used by any other mitochondrial gene. This codon bias suggests that *RTL* may have been recently acquired by the genome; however, there is as yet no information to suggest the mechanism of its mobility.

## Multicopy Single-Stranded DNA-Associated Elements

For some time, it was believed that prokaryotes did not contain retroelements. However, in 1989, observations in the myxobacterium, *Myxococcus xanthus* (92) and in *Escherichia coli* (93), led to the discovery of an abundant (500 to 700 copies per cell) single-stranded DNA molecule generated by reverse transcription. These DNAs, which are named multicopy single-stranded DNAs (msDNAs), are 60 to 160 nucleotides in length and are linked at their 5′ end to a short RNA molecule by an unconventional 2′,5′-phosphodiester linkage. From six to eight nucleotides at the 3′ end of the DNA are base-paired to the same RNA molecule. The msDNAs from different species show no sequence homology, but do share a common secondary structure. Synthesis of the single-stranded DNA has been shown to occur from an RNA template in which the 2′ hydroxyl group of an internal guanine nucleotide self-primes reverse transcription of a portion of the RNA template (reviewed in 18). The gene encoding the reverse transcriptase for this reaction is present on the RNA template used to generate msDNA, but does not become part of msDNA itself, indicating that msDNA is not an intermediate of a retrotransposition process. The msDNA with its associated reverse transcriptase gene is present in all strains of different species of myxobacteria, but is found in only a small fraction of clinical or wild-type strains of *E. coli*. In *E. coli*, these sequences appear to be associated with phage genomes.

## USE OF REVERSE TRANSCRIPTASE TO TRACE RETROELEMENT PHYLOGENY

The origin of retroviruses from intracellular mobile DNA elements was first proposed by Temin in his protovirus hypothesis, and he discussed structural features common to retroviruses Ty1 and copia, even before these transposable elements were completely sequenced (94). Today, given the similarity in their structures and mechanisms of retrotransposition a common origin for retroviruses, LTR retrotransposons, hepadnaviruses, and caulimoviruses is abundantly clear (13,14,20,21). Both hepadnaviruses and caulimoviruses have been described as pararetroviruses (20), and the LTR retrotransposons as members of the retroviral superfamily (9). However, which of these retroelements was the progenitor of the others is much less obvious. The presence in some retrotransposons of what appear to be remnants of an *env*-like ORF, whereas others lack this third ORF, would seem to suggest that the retrotransposons are retroviruses that have lost the ability to leave a cell. Indeed, endogenous retrovirus-like elements (e.g., VL30 and IAP), are abundant in mammals and appear to be ancient exogenous retroviruses now existing only as intracellular mobile elements (20). Yet retrotransposons are present in a much wider diversity of organisms than retroviruses, suggesting that they may be older. Why do mammalian and avian retroviruses have fewer features in common with the hepadnaviruses, present in the same mammals and birds, than they do with the caulimoviruses which are present in plants? Because of these uncertainties, no generally accepted phylogeny for the evolution of these retroelements has appeared.

The evolutionary relation of the non-LTR retrotransposons, the various organellar retroelements, and the msDNA-associated elements of bacteria is even more poorly understood. These classes of retroelements have virtually no structural features in common, nor is there similarity in what is known of their mechanism of replication. Indeed the *only* property they have in common is the ability to encode reverse transcriptase. The reverse transcriptase domains of these various retroelements contain the same highly conserved series of amino acid residues, indicating that they are more related to each other than to any other known DNA or RNA polymerase. Thus, at a minimum, this enzyme component of all retroelements appears to have a common ancestor, rather than being independently evolved from different DNA or RNA polymerases.

How can we derive the evolutionary relation of the retroelements? The only rigorous approach is to use the information stored in the nucleotide or amino acid sequences of the elements as a record of their evolution. Furthermore, because the only protein encoded by all retroelements is reverse transcriptase, this sequence is our only choice for such a molecular analysis. Can the sequence of the reverse transcriptase domain be expected to reconstruct this phylogeny? To be useful in tracing the evolutionary history of these elements the reverse transcriptases should first evolve slowly, such that their sequences can be unambiguously aligned over an extended region and, second, these polymerases must serve the same function in all elements. Both of these requirements appear to be met by the reverse transcriptases of retroelements. With only the exception of the RTL gene, reverse transcriptases from each group of retroelements synthesize DNA from an RNA template in the same assays [poly(rC) primed with a short oligonucleotide of dG]. The major difference between the reverse transcriptases of the various retroelements is in the nature of the priming reaction. As will be described in the following section, the region of the enzyme responsible for this priming is outside the region of the sequence comparisons. The region used for the phylogenetic analysis contains the active domain responsible for the polymerization of DNA from the RNA template. This domain is conserved in three-dimensional structure with the cellular DNA polymerases (DNA-directed DNA polymerases), as well as certain RNA polymerases (RNA-directed RNA polymerase) (95).

The use of molecular sequence data of the reverse transcriptase to reconstruct the evolutionary history of retroelements can be viewed as analogous to attempts to use rRNA sequences to trace the phylogeny and origins of various groups of organisms (96,97). However, our attempts to reconstruct the phylogeny of retroelements is on much shakier grounds than attempts to trace the phylogeny of organisms. In organism phylogeny, a variety of genes are available to be used to trace the evolutionary relation in addition to the rRNA genes (98,99). If these independent sequences give the same phylogeny, confidence in that phylogeny can be high. For retroelements, we have no other sequence data to serve as an independent check of the relationships. Attempts have been made to use the RNase H, proteinase, or integrase domains to independently determine the relations between certain classes of retroelements (21). In general, although these analyses support the relationships determined from the reverse transcriptase sequences, they involve smaller, less-conserved protein domains and, consequently, are less reliable. More importantly, these domains are absent in over half of all retroelements.

A second limitation of our ability to use molecular data to trace the evolutionary history of the retroelements is that it is difficult to place even approximate dates on when various elements diverged. In organism phylogeny, approximate dates can be established for the origin of certain taxa from the fossil record. No fossil record exists for the retroelements, and it is not certain whether one can use the distribution of these retroelements in various groups of organisms as an indicator of their approximate age. Because these retroelements are mobile, it is possible that they have been transferred across species barriers. Putative examples of cross-species transfers of mobile elements and viruses have been reported (21,32,100–103). However, as yet, convincing examples deal only with transfers between closely related species (100,103). No convincing evidence has been presented for such transfers between major taxa of organisms. Thus, if the same group of retroelement (e.g., the copia group) is present in both plants and animals, it remains unclear whether the elements of that group have been vertically transmitted in the germline of plants and animals since their divergence, or whether there has been more recent horizontal transfer between plants and animals.

Even with these limitations, the analysis of reverse transcriptase sequences is worthwhile as it constitutes the only method available to study the evolutionary relationships of the retroelements. The only major assumption of this approach is that elements with a common ancestor will have greater sequence similarity to each other than they both have to any other class of elements. Given this assumption, the question then becomes: Is there enough information in the reverse transcriptase sequences to resolve this phylogeny?

## Reverse Transcriptase Sequences and Their Alignment

The level of amino acid sequence identity of the reverse transcriptase domain from different elements can be quite low, whether one is comparing sequences from the same or from different classes of retroelements. For example, the level of amino acid sequence identity between the murine leukemia retrovirus (MuLV) and the human immunodeficiency retrovirus (HIV-1) is only 25%. Sequence identity between HIV-1 and the I factor is only 15%. Because of these low levels of sequence identity, current alignment of reverse transcriptase sequences is based on an identification of a series of highly conserved amino acid residues. Conserved reverse transcriptase residues were originally identified by Toh et al. (19,104) by comparison of sequences from retroviruses, hepadnaviruses, caulimoviruses, and several LTR-containing retrotransposable elements. A related group of conserved residues was subsequently identified in the reverse transcriptase domains of group II introns

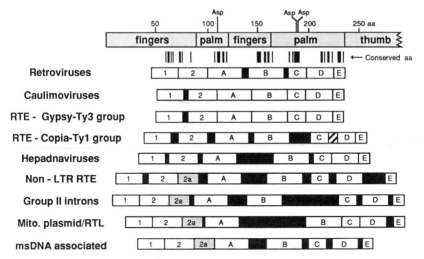

FIG. 4. Conserved regions of all reverse transcriptases: Shown at the *top* are the location of the various subdomains of the active subunit determined for the reverse transcriptase of human immunodeficiency virus (HIV)-1 (95). *Vertical lines* below these subdomains correspond to the 42 conserved positions found in all reverse transcriptase sequences (32). The reverse transcriptase domain of each class of retroelement is indicated by the *horizontal bar*. *Open regions*, conserved peptide regions in all reverse transcriptase used in the subsequent phylogenetic analysis; *solid regions*, expansions regions found in certain classes; *stippled region*, conserved expansion segment found in all retroelements that lack terminal repeats; and *diagonal shading*, segment of region D missing in retroelements of the copia group. Regions A–E correspond to the nomenclature of Poch et al. (106) and Kohlstaedt et al. (95), and were indicated as regions 3–7 in earlier reports (31,32).

and mitochondrial plasmids (16), the non-LTR retrotransposons (67) and in the ms-associated DNAs (92). At present 42 amino acid positions have been identified as containing largely unvaried or chemically similar residues in most reverse transcriptase sequences (32). These 42 positions are conserved, on average, in 88% of reverse transcriptase sequences, and define seven, short (6 to 16 residues) conserved peptide segments that can be identified in all reverse transcriptase sequences (Fig. 4). Alignment of the reverse transcriptase domain from the *RTL* gene, the only retroelement in which reverse transcriptase activity has not been directly demonstrated, is as good as all other groups of retroelements, thus it is likely that reverse transcriptase activity will eventually be demonstrated for the *RTL* gene.

Additional support for the alignment of the reverse transcriptase sequences based on these conserved residues has come from independent attempts to combine the amino acid sequence similarities of the reverse transcriptases with predictions of secondary structures. Webster et al. (105) identified four conserved peptide motifs corresponding to the second through fifth regions. Poch et al. (106) identified the third through seventh regions in all reverse transcriptases, which they termed regions A through E. In Fig. 4, this terminology is used for the five regions identified by Poch et al. (106), whereas the two regions nearer the $NH_2$-terminal end of the protein are numbered 1 and 2. Regions A through E are also found in the RNA-directed RNA polymerases of RNA viruses (106), and regions A and C are found in both DNA-directed DNA polymerases and DNA-directed RNA polymerases (107).

The number of amino acids separating the seven groups of conserved residues varies considerably between the different groups of retroelements (see Fig. 4). For example, the non-LTR retrotransposons and mitochondrial sequences have approximately 100 amino acid residues that have no homology in the retroviral sequences. Most of these additional sequences occur between the conserved peptides of regions 2, A, B, and C. The unitary matrix algorithm (108) has been used to align the sequences between the seven groups of conserved residues (32). This algorithm assigns a penalty for the insertion of gaps in aligning a sequence, thus the low levels of sequence similarity in these regions resulted in the localization of the additional sequences to the same positions between the sets of conserved peptide sequences. In a few instances, these expansion regions do contain sequence similarity between different groups of retroelements. This was particularly true for the expansion segment between 2 and A that is shared by the non-LTR retrotransposons, the mitochondrial sequences, and ms-associated DNA. This expansion segment has therefore been termed region 2a. Most expansion segments, such as the sequences between regions A and B in the mitochondrial plasmids or the sequences between regions B and C in the group II introns are unique to particular groups of retroelements.

Recently, x-ray crystal structure determinations have revealed the three-dimensional structure of the reverse transcriptase from the HIV-1 retrovirus (95). The active enzyme from HIV-1 is a heterodimer with one subunit, p66 retaining the RNase H domain, and the second subunit, p51, having its RNase H domain removed by proteolysis. The p66 subunit contains the active site for the polymerization reactions, and the p51 subunit is in a significantly different tertiary structure that appears to function predominantly as part of the tRNA primer–template-binding site. The catalytic subunit encoded by the HIV-1 retrovirus resembles a right hand, therefore the various subdomains of this subunit have been named "fingers," "palm," and "thumb." Three aspartic acid residues that have been shown by mutagenesis experiments to be part of the active site of the enzyme (109) are located within the palm subdomain. The regions of the enzyme that contain sequence similarity to all reverse transcriptases corresponds to the palm subdomain, and all but the first 40 amino acids at the $NH_2$-terminal end of the finger subdomain (see Fig. 4). No sequence similarities between all classes of retroelements have been found in the thumb subdomain of the reverse transcriptases. From this three-dimensional structure, the expansion regions that are detected in certain of the retroelements appear to be located in regions extending away from the active site; thus, they would not be expected to have a major affect on the activity of the enzyme.

It is likely that the structure surrounding the active site of all reverse transcriptases is highly similar to that of the HIV-1 enzyme, because the three-dimensional structure of this region is remarkably similar to even that of DNA-directed DNA polymerases and RNA-directed RNA polymerases (95). Additional evidence to suggest that this hand represents the basic core of all reverse transcriptases is provided by the observation that the ORF of the *RTL* gene (17) and most of the reverse transcriptase genes associated with msDNA (18) are large enough to encode only the fingers, palm, and thumb subdomains.

## Phylogenetic Relations of Retroelements Based on the Sequence of Their Reverse Transcriptases

The amino acid sequences of the seven regions identified as common to all reverse transcriptase sequences (see Fig. 4) have been used to estimate the evolutionary relationships within and between the different classes of retroelements (31,32). The expansion segments found in some retroelements were not included in the analysis. The number of identical amino acid residues in the 178 positions of the common regions were determined for all pairwise comparisons of 91 reverse transcriptase sequences. This matrix of pairwise distances was then used to construct a phylogenetic tree using the neighbor-joining (NJ) method of Saitou and Nei (110). We are currently (Lathe WC and Eickbush TH, in preparation) using a different method for the construction of this phylogeny, the phylogenetic analysis using parsimony (PAUP) method of Swofford (111). In general, the topology of the trees obtained by the distance and parsimony methods are similar, testifying to the reliability of the trees. However, differences do exist between these distance and parsimony trees, and various statistical tests (112) clearly indicate that certain branch points are not well-defined. The precise nature of these uncertainties and their effect on the conclusions that can be drawn concerning the evolutionary relations of the retroelements will be indicated as each segment of the analysis is discussed. As sequences from a greater number of retroelements and more sophisticated algorithms become available, greater reliability in the branch positions should be obtained.

The evolutionary relationship of retroelements based on the NJ analysis of their reverse transcriptase sequence divergence is shown in Fig. 5. The analysis is shown as an unrooted tree with the length of each branch directly proportional to the amount of sequence divergence that has occurred on that branch. To root this tree requires independent information or assumptions about the ages of the elements. Two alternative rootings of the tree are shown in Fig. 6, and their implications for the origins of retroelements will be discussed in the next section.

The analysis contains all distinct retroelements, the reverse transcriptase domain of which has been completely sequenced. For the hepadnaviruses and retroviruses, for which many highly similar elements have been sequenced, it was necessary to select a smaller number of elements, representative of the sequence diversity present within each of these groups of viruses. For the non-LTR retrotransposons,

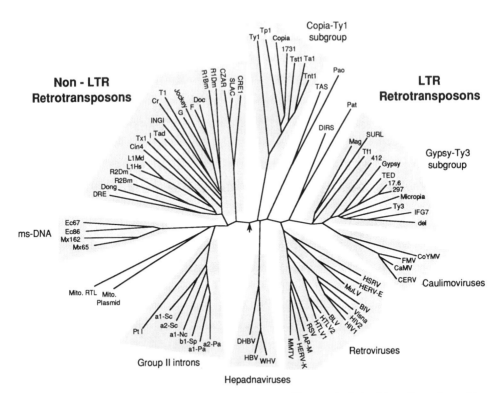

**FIG. 5.** Unrooted phylogenetic tree of retroelements: The tree is derived from the amino acid sequences of the reverse transcriptase domain encoded by each retroelement. The amino acid sequence divergence of the 178 positions common to all enzymes (*open bars* in Fig. 4) was used (31,32) to derive the tree using the NJ method (110). All elements that define a particular class or subclass of retroelements are indicated by *shading*. Identification of the host species and publication reference for each element can be found in Xiong and Eickbush (31,32), except for the following: CZAR (55), TST1 (137), Tp1 (138), TAS (119), PAT(120), SURL(139), Tf1 (140), TED (38), DRE (74), DOC (141), TAD (Kinsey, J, personal communication), Cr1 (Burch, J, personal communication). *Arrow* indicates the location of the branch leading to the RNA-directed RNA polymerases used to root the tree in Fig. 6B.

several examples of the same element sequenced from distant species have been included. The analysis includes five recently sequenced non-LTR retrotransposons and seven LTR retrotransposons, but is otherwise the same as in a previous report (32). The topology of this tree has several differences from that originally reported by Doolittle and coworkers in their review on the evolution of retroviruses (21). In that report the hepadnaviruses were located closer to the retroviruses than the cau-limoviruses, and the copia subgroup of retrotransposable elements was found to be more distant from the retroviruses than the non-LTR retrotransposable elements. Part of this discrepancy was a result of the incorrect alignment of the reverse tran-scriptase sequences in that study by the progressive alignment method (113). More

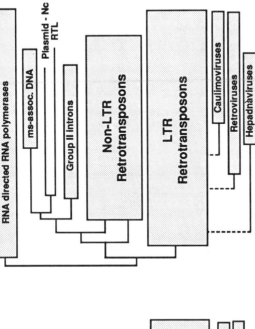

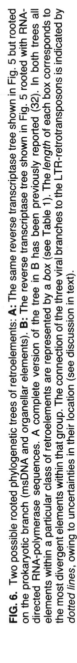

**FIG. 6.** Two possible rooted phylogenetic trees of retroelements: **A:** The same reverse transcriptase tree shown in Fig. 5 but rooted on the prokaryotic branch (msDNA and organellar elements). **B:** The reverse transcriptase tree shown in Fig. 5 rooted with RNA-directed RNA-polymerase sequences. A complete version of the tree in B has been previously reported (32). In both trees all elements within a particular class of retroelements are represented by a *box* (see Table 1). The *length* of each box corresponds to the most divergent elements within that group. The connection of the three viral branches to the LTR-retrotransposons is indicated by *dotted lines*, owing to uncertainties in their location (see discussion in text).

recently, McClure (54) has reported a tree that is nearly identical with the one shown here.

Several significant conclusions concerning the evolution of retroelements are suggested from the analysis in Fig. 5. First, all examples of elements from the same class are clustered on the tree. Sequences from the three classes of viruses are each located on separate branches, as are the ms-associated elements and the group II introns, indicating a monophyletic origin for each of these classes of retroelements. The retrotransposons are distributed over several branches, but there is no interspersion of the branches containing the LTR and non-LTR retrotransposons, again suggesting a single, separate origin for each of these two groups. Second, the reverse transcriptase tree can be divided into two major branches. One branch (corresponding to the right half of the tree) contains the three types of viruses and the LTR retrotransposons. I will refer to this branch as the LTR branch because all elements on this branch use terminal repeats and a reverse transcriptase that is dependent on strand transfer to complete replication. The second branch (corresponding to the left half of the tree) contains the non-LTR retrotransposable elements, the organellar elements, and the ms-associated DNAs. All elements on this branch lack any type of terminal repeat in their inserted genome, and there is no evidence to suggest that their RNA templates are associated with terminal repeats. I will refer to this branch as the non-LTR branch. The recently reported DRE element of *Dictyostelium discoideum* has been described as containing a complex series of direct repeats (74). However, the protein-encoding region of DRE ends with a poly(A) tail, suggesting that like Tx1 (73), DRE is a non-LTR element that has specialized for insertion into a second mobile element.

Without question, the two classes of retroelements with the highest degree of sequence diversity are the retrotransposons. Non-LTR retrotransposons are located on two branches. One branch contains CZAR, SLACS, and CRE1. These three elements were independently found in different species, but appear to represent the same element that is specialized for insertion into leader exon sequences of Trypanosomatidea (55). These elements differ from all other non-LTR elements in that they encode a putative integrase domain, suggesting they may employ a mechanism of replication that is intermediate between that of the LTR retrotransposons and most non-LTR elements.

Located on a second branch are the 16 remaining non-LTR elements found in insects, vertebrates, maize, fungi, and protozoans. Most of these elements are connected to the tree by extremely long branches, reflecting their high levels of sequence divergence. This level of divergence is nearly equal between these elements, resulting in little confidence in their specific branching order. Indeed, only Jockey, Doc, F, and G elements of *D. melanogaster* exhibit significant clustering. In three separate cases, the same non-LTR element has been sequenced from distantly related species. L1 elements have been sequenced from a number of mammals, including humans and mouse, species separated by an estimated 80 million years (114, 115), and both R1 and R2 elements have been sequenced from *B. mori* and *D.*

*melanogaster*, species separated by an estimated 240 million years (45). If one uses these divergence times to date the corresponding branch points, it would suggest that most of the different non-LTR elements represent ancient divergences. (This conclusion is tentative because it requires that these elements have not undergone cross-species transfer.)

The LTR retrotransposons also have a broad range of sequence divergence that divides them into a number of subgroups. One subgroup has been termed the Copia-Ty1 subgroup. The seven elements of this subgroup have been characterized in *D. melanogaster*, several plants, *S. cerevisiae* and *Physarum polycephalum*. These elements are remarkably distinct from other LTR elements in the sequence of their reverse transcriptase domain and share the unique feature that the integrase domain is located upstream from the reverse transcriptase domain (see Fig. 1). Several recent studies using PCR methodology have shown that members of this subgroup are also present in vertebrates (116), and that most species of plants have multiple families of these elements (117,118).

A second subgroup of LTR retrotransposons is the Gypsy-Ty3 subgroup. The 12 members of this subgroup are present in insects, sea urchins, plants, and fungi. The four elements within this group (gypsy, TED, 17.6, and 297) that have a third, potential *env*-like ORF are located on a distinct subbranch, as are the two elements found in plants, del and IFG7. The branching order for the remaining elements within this subgroup, 412, micropia, and mag of insects, Tf1 and Ty3 of fungi, and SURL of sea urchins, is not well supported.

Four LTR retrotransposons, TAS (119) and Pat (120) from nematodes, Pao from *B. mori* (121), and DIRS from *Dictyostelium discoideum* (122), do not appear to fit cleanly into either the Copia-Ty1 or Gypsy-Ty3 subgroups. The integrase domain of these elements is downstream from the reverse transcriptase domain, suggesting that they are more related to the Gypsy-Ty3 subgroup. Three of these elements contain unusual terminal repeats. Pao elements contain LTRs that have a series of internal tandem repeats, DIRS contains inverted terminal repeats, and Pat contains unusual split direct repeats. These elements may represent new subgroups of the LTR retrotransposons that have evolved modified methods of replication.

Interspersed with the LTR retrotransposons are the retroviruses, hepadnaviruses, and caulimoviruses. These groups of viruses are located on separate branches, suggesting that they have had independent origins. Their branching within the LTR retrotransposons clearly supports the suggestions that they arose from cellular transposable elements (94). The exact locations of these viral branches within the LTR retrotransposons are unfortunately major uncertainties of the tree. Doolittle et al. (21) have placed the hepadnaviruses much closer to the retroviruses. We have found that maximum parsimony methods place retroviruses along with the caulimoviruses within the Gypsy-Ty3 subgroup of elements (Lathe WC and Eickbush TH, in preparation). Statistical tests (112,123) suggest that neither the distance nor parsimony methods can now reliably place the retroviral and caulimoviral branches either outside or within the Gypsy-Ty3 subgroup.

The inability to place these viral sequences on the tree is probably a result of their more rapid evolution. It has been suggested that retroviral genomes can evolve at rates that are over a million times faster than those of nuclear genomes (124). This rapid rate of evolution is a result of two processes: the number of replication events per year, which for an actively spreading virus is many orders of magnitude greater than an element that is replicating in synchrony with the host genome, and the higher error rates of their reverse transcriptases as an adaptation for avoiding the immune system of the host. However, as discussed by Doolittle and coworkers (21), the incredibly rapid rates of evolution possible in retroviruses, as seen for example in the current evolution of HIV-1 in humans, are probably short-lived. They suggest that the only model consistent with what is known about the long-term evolution within the retroviral group is to assume that every infectious retrovirus has recently evolved from an endogenous virus (e.g., HERV-E, HERV-K, IAP-M). Thus, although the long-term rate of evolution of retroviruses is considerably slower than that of HIV-1 in its current pandemic progression, these short-lived bursts of evolution would accelerate the rate of evolution of these viruses over that of their host's genome. The LTR retrotransposons, on the other hand, may be evolving at a rate more similar to the host's genome, because they are believed to remain quiescent within an organism for generations, retrotransposing only rarely when their copy number is reduced by recombination to low levels.

Currently, the faster rate of retroviral evolution makes it impossible to determine if these viruses, like the caulimoviruses, are located inside or outside the Gypsy-Ty3 subgroup of retrotransposons. One intriguing location for the retroviral branch within the Gypsy-Ty3 subgroup is on the branch containing gypsy, TED, 17.6, and 297, the only retrotransposons with a third ORF. The positioning of both retroviruses and caulimoviruses within the Gypsy-Ty3 group would help explain a troubling point in the evolutionary tree shown in Fig. 5. The exclusive location of retroviruses in mammals and birds has been used to suggest that they evolved from intracellular transposable elements at about the time of the mammalian–avian divergence (21). If the location of the retroviral branch in Fig. 5 is correct, then the entire Gypsy-Ty3 subgroup also arose at about this time. Because the Gypsy-Ty3 elements are present in fungi, animals, and plants, their distribution can be explained only by assuming that these retrotransposons have been horizontally transferred across major taxonomic groups (32). However, if the retroviral branch in Fig. 5 is misplaced and, along with the caulimoviruses, actually falls within the Gypsy-Ty3 group of elements, then the Gypsy-Ty3 group is much older, and no such cross-species transfer is necessary to explain their distribution. The continued analysis of the available sequence data and the recovery of the LTR retrotransposons from greater numbers of organisms should one day resolve these uncertainties. If the phylogeny of the Gypsy-Ty3 elements is identical with the phylogeny of their hosts, it would suggest that these elements are quite old and cross-species transfers have not occurred. On the other hand, if the phylogeny of the Gypsy-Ty3 elements violate the phylogeny of their hosts, their widespread distribution is a result of cross-species transfers, and they need not be older than retroviruses.

## Rooting the Reverse Transcriptase Tree

The reverse transcriptase tree in Fig. 5 reveals the sequence relation between the different retroelements; but because it is unrooted it does not indicate which retroelements are the oldest. Which group of known retroelements most resembles the original retroelement? Although any branch can potentially be considered as the root from which all other elements evolved by the gain or loss of various functions, only two of the possible roots are consistent with accepted methods of rooting phylogenetic trees and will be considered here.

One method of rooting the tree uses the phylogeny of the host organisms. This method is reliable only if cross-species transfer of retroelements has not occurred. As discussed earlier, this is not a particularly safe assumption for mobile elements and viruses. However, convincing evidence is now lacking to suggest that such transfers have occurred across major taxa, and so for now, at least, using the distribution of retroelements remains a viable method of rooting the tree. Non-LTR retrotransposons and LTR retrotransposons of both the Copia-Ty1 and Gypsy-Ty3 subgroups are present in fungi, protozoans, plants, and animals, suggesting that each of these branches dates back to the origin of eukaryotes. The remaining retroelements, the msDNA-associated elements and the several classes of retroelements found in mitochondria, are found in prokaryotes or in organelles that are derived from prokaryotes. Because prokaryotes are known to predate eukaryotes, these prokaryotic elements could be considered the root of the retroelement tree. Although it is possible that the organellar retroelements have been derived from nuclear retroelements present in the same cells, this appears unlikely. The msDNA elements and the organellar elements are more related to each other than to any other retroelement, suggesting that the origin of the organellar retroelements dates back to a common prokaryotic origin. Therefore, the branch containing msDNA, the Mauriceville plasmid, the *RTL* gene and the group II introns should be considered the prokaryotic branch and can be used to root the retroelement tree. A simplified version of this tree drawn with its root on this prokaryotic branch is shown in Fig. 6A. The individual classes of retroelements in this tree are indicated with a box, with the length of each box corresponding to the most divergent elements within that class. The three viral branches are shown as being derived from the LTR retrotransposons, with their exact location on this branch unknown. One can also follow the evolution of the different retroelement classes from the prokaryotic elements by moving across the tree in Fig. 5 from left to right.

The nature of the first prokaryotic retroelements is subject to only speculation at this time. The Mauriceville mitochondrial plasmid can replicate independently of the host genome, and has been suggested to be an example of one of the earliest DNA replication mechanisms (88,125). Soon after their discovery, it was suggested that msDNA-associated elements, as the only true prokaryotic retroelements, were the progenitors of all retroelements (11,126).

The tree rooted as in Fig. 6A suggests that the first true eukaryotic (nuclear) retroelements were the non-LTR retrotransposons. Two possible origins for the non-LTR retrotransposons can be suggested. They were originally present in the genome that was to become the first eukaryotic nucleus, or they entered the first eukaryotic nucleus along with other genes from the endosymbiont genomes that became the mitochondria and chloroplasts. The most convincing argument for a specific prokaryotic retroelement being the likely progenitor of the non-LTR retrotransposons would be that element's use of the 3' end of DNA at a nick or break to prime reverse transcription of its RNA template. The progenitor non-LTR retrotransposons evolved additional protein functions not found in the prokaryotic elements, including RNA-binding proteins (progenitors of the *gag* gene) as well as an endonuclease that aided the integration of the elements. The structure of present day non-LTR retrotransposons suggests that this group has had considerable flexibility in gene organization and in the nature of its encoded proteins.

The next step in the evolution of retroelements was the addition of terminal repeats to a non-LTR retrotransposon and the origin of the LTR retrotransposons. These elements had an entirely new mechanism of replication, independent of the chromosomal insertion site, using terminally redundant RNA templates and a reverse transcriptase capable of strand transfer. Interestingly, CZAR, CRE1, and SLACs are located on the retroelement tree between the major non-LTR branch and the LTR retrotransposons. These three elements are the only non-LTR retrotransposons with an integrase domain (54), suggesting that they are a potential intermediate step in the evolution of the LTR retrotransposons from the non-LTR retrotransposons. These elements specifically insert into the leader exons of Trypanosomatidae. It is possible that these leader exon sequences, which are transspliced onto the mRNAs of these species, served as primordial LTRs. Recent expression of CRE1 proteins in a Ty1-based expression system should permit characterization of its mode of retrotransposition (51).

Since the origin of the LTR retrotransposons, their enzymatic and structural proteins have remained relatively stable. All LTR retrotransposons contain similar *gag* and *pol*-like genes, with only the integrase domain shifting its location between the copia-like and gypsy-like elements. With few exceptions (e.g., DIRS), the LTRs have also remained highly similar, including the mechanisms to prime first- and second-strand synthesis. Indeed, of all the elements on the LTR branch of the retroelement tree, only the hepadnaviruses use a different method of priming reverse transcription.

Rooting of the retroelement tree on the prokaryotic branch, particularly the msDNA branch, is satisfying in that it predicts a logical progression from the simpler to the more complex retroelements. Many of these prokaryotic retroelements have been discovered recently, thus it is likely that additional classes of retroelements will be identified in bacteria or organellar genomes that will provide new insights into the origin of retroelements. Interestingly, if prokaryotic elements more closely approximating non-LTR retrotransposons are identified, then the presence of non-LTR retrotransposons in both prokaryotes and eukaryotes would suggest these elements are the most ancient, a conclusion consistent with the following, alternative method of rooting the retroelement tree.

A second independent method of rooting the retroelement tree is to determine the relation of the reverse transcriptase sequences to other nucleic acid polymerases. It has been suggested that reverse transcriptase dates back to an RNA world in which it was the enzyme responsible for turning information stored in the RNA of the progenote to the more stable form of DNA, as it is in cells today (127, 128). In this model, the RNA-directed DNA polymerase activity of reverse transcriptase evolved from the first nucleotide polymerase, an RNA-directed RNA polymerase. Several studies have attempted to find the relation between the RNA and DNA polymerases (107). These studies have revealed that the enzymes most related in sequence to reverse transcriptases are the RNA-directed RNA polymerases identified in various prokaryotic and eukaryotic RNA viruses (106). Each of the seven peptide regions described in Fig. 4 for the reverse transcriptases can also be found in the RNA-directed RNA polymerases. Regions A through E were originally identified in these RNA polymerases by Poch et al. (106), and domains 1 and 2 were subsequently identified by Xiong and Eickbush (32). Of the 42 conserved amino acid positions found in reverse transcriptase, on average, 60% are also found in these RNA polymerases (compared with an average of 88% in reverse transcriptases). The polio group of animal RNA viruses have the highest conservation of these conserved residues, with 72% of the conserved sites. We have repeated the analysis of the reverse transcriptase tree including 15 RNA-directed RNA polymerase sequences (32). The location of the RNA polymerase branch on the retroelement tree in Fig. 5 is shown by the arrow. A simplified version of the tree using these RNA polymerase sequences as the root is shown in Fig. 6B. Because total sequence identify between the reverse transcriptases and these RNA polymerases is low, the analysis has also been conducted using only the sequence data from regions A through E. These regions contain the highest levels of sequence identity, corresponding mostly to segments surrounding the active site of both enzymes (95). The topology of the tree containing this reduced data set is the same as that shown in Fig. 6B.

Rooting the retroelement tree using RNA polymerase sequences does not support the model that retroelements on the prokaryotic branch are the oldest. It suggests, instead, that the root is located between what was previously described as the two major branches of the tree, the LTR and non-LTR branches. If this rooting of the tree is correct, then the question becomes: did the original retroelements have LTRs and employ a mechanism of retrotransposition similar to that in Fig. 2, or did they lack terminal repeats and use a retrotransposition mechanism similar to that in Fig. 3? Several arguments suggest that the reverse transcriptases of non-LTR retroelements are more like the RNA-directed RNA polymerases than are those of the LTR elements. First, the expansion segments of the RNA-directed RNA polymerases have more similarities to the reverse transcriptases of the non-LTR branch. In particular, most RNA-directed RNA polymerases have an expansion segment 2A (see Fig. 4). Second, the most highly conserved amino acid motif, the YXDD sequence in region C, is more similar between the non-LTR reverse transcriptases and the RNA polymerases. For most RNA-directed RNA polymerases, X corresponds to a glycine residue. The reverse transcriptases of all non-LTR elements have a chemically similar alanine residue at this position, whereas the reverse transcriptases of all elements on the LTR branch have a hydrophobic residue at this position. Finally, the third reason to suggest that the reverse transcriptases of non-LTR retrotransposons are more similar to the RNA-directed RNA polymerases is that neither of these polymerases is known to have the strand-transfer ability associated with the reverse transcriptase of the LTR elements (129).

Rooting the reverse transcriptase phylogenetic tree using RNA-directed RNA polymerase assumes that both enzymes are extremely old. In this model, the retroelements must be viewed as relics, or molecular fossils of the first primitive replication systems in the progenote (125). Retroelements in this scenario were present in the first cells, and were most similar to present-day non-LTR retrotransposons. The first major division of these elements took place with the capture of terminal repeats (LTRs), which enabled the retrotransposition process to be independent of the chromosomal insertion site. Subsequent to this first division, members of the prokaryotic branch (e.g., msDNA, group II introns) evolved from the primitive non-LTR retrotransposons (32). Whether such primitive LTR and non-LTR retrotransposons are still present in bacteria will be revealed by the continued search for reverse transcriptase-like sequences in more prokaryotes.

In summary, the correct rooting of the retroelement tree remains unknown because each method of rooting the tree is based on potentially erroneous assumptions. Using the distribution of retroelements to root the tree on the prokaryotic branch assumes that we have an adequate understanding of the types of retroelements present in prokaryotes, and that cross-species transfers between major kingdoms have not occurred. Using RNA-directed RNA polymerase to root the tree, assumes that the origin of retroelements goes back to the progenote appearing at a time when the RNA world was changing to a DNA world. It is possible that the original enzymes that catalyzed these events are no longer in existence with both the RNA-directed RNA polymerase of RNA viruses and the RNA-directed DNA polymerases of retroelements evolving at a later date from typical cellular DNA or RNA polymerases. Thus, both methods of rooting the tree are highly speculative, and it is unclear if the true origin of retroelements can ever be determined. One possible resolution of this problem would be obtained by the discovery of non-LTR retrotransposons in bacteria. With such a discovery, both methods used to root the tree would suggest that non-LTR retrotransposons are at the root of the tree, and we will have taken a significant step in uncovering the origin of retroelements.

## CONCLUDING REMARKS

This chapter has attempted to summarize the variety of known retroelements and describe their evolutionary relation based on their reverse transcriptase sequences. With the discovery of retroelements in bacteria and organelles and of retrotransposons in virtually all eukaryotes, few today would argue with Temin's original suggestion that retroviruses evolved from cellular DNA elements. However, the great variety of retroelements now known leads to questions of how and when these many types of elements evolved. Advances in the following three areas are needed if we are to eventually determine their origins and evolutionary relationships. First, there must be further developments in the use of molecular data to derive sequence phylogenies. The major problem is that the various algorithms currently available are not designed for the low levels of sequence similarity found between these reverse transcriptase sequences. Perhaps the three-dimensional structure that is now available for the HIV-1 enzyme can be extrapolated to all reverse transcriptases not only to confirm the sequence alignment, but also to enable the incorporation of secondary and tertiary structure features in the phylogenetic reconstructions (130).

A second advance would be a better understanding of the stability of retroelements in a species lineage. Are retroelements stable within lineages over hundreds of millions of years, or do they exist as active elements in a species for only relatively short times and require continual reintroduction? Investigations of retroelements need to progress from the occasional discovery of insertion elements within genes that are under study, which is how most of the retroelements have been found to date, to systematic attempts to characterize the distribution and evolution of particular types of retroelements. With the advent of the polymerase chain reaction (PCR) method (131) it is now possible to use degenerate primers directed to conserved regions of the reverse transcriptase domain and amplify from genomic DNA reverse transcriptase sequences from virtually any organism. This has been dramatically shown with the Copia-Ty1 group of elements. The use of degenerate probes to regions B and C of the reverse transcriptase domain of the Copia-Ty1 group has resulted in the recovery of copia-like elements from virtually all plant and animal species studied (117,118). Similar studies should be conducted with the different classes of retroelements to determine if the phylogenies of the individual retroelements agree with those of their host species.

Finally, the use of PCR technology should enable the search for new types of retroelements and reverse transcriptase-like enzymes. Better representation is needed of the retroelements that are present in prokaryotes, particularly the *Archaeobacteria*, and in primitive eukaryotes. The discovery of entirely new types of reverse transcriptase-like enzymes in either eukaryotes or prokaryotes can provide new insights into the origin of retroelements. Indeed, one such reverse transcriptase-like sequence, telomerase, has already been identified. Telomerase synthesizes the GA-rich repeats which form the chromosomal telomeres in most organisms by copying a short RNA template that remains attached to the enzyme (132). Although sequence similarity of telomerase from *S. cerevisae* to reverse transcriptase sequences has been reported (133,134), this similarity is too low to obtain an unambiguous alignment of the seven domains used in our analysis. When telomerase sequences are identified in additional organisms, their conserved residues can be determined, the sequences aligned with reverse transcriptase, and their relation to the other retroelements can be studied.

With advances in these areas, together with progress in our understanding of the enzymatic activities encoded by each retroelement, we will be better able to trace the origin and evolution of these interesting elements. Retroelements are intimately involved with fundamental questions concerning the the evolution of DNA and RNA viruses and of eukaryotic genomes. Retrotransposons and SINEs represent a surprisingly high percentage of many genomes. It is even possible that most introns are derived from retroelement insertions, similar to group II introns (135,136). Clearly, the evolution of retroelements will remain an exciting area of investigation in the field of molecular evolution for many years to come.

## ACKNOWLEDGMENTS

This work was supported by National Institutes of Health Grant GM42790 and American Cancer Society Grant NP-691. The author would like to thank the members of his laboratory for useful discussions and comments on the manuscript.

## REFERENCES

1. Doolittle WF, Sapienza C. Selfish genes, the phenotype paradigm and genome evolution. *Nature* 1980;284:601–604.
2. Orgel LE, Crick FHC, Selfish DNA: the ultimate parasite. *Nature* 1980;284:604–608.
3. Werren JH, Nur U, Wu C-I. Selfish genetics elements. *Trends Ecol Evol* 1988;3:297–302.
4. Charlesworth B, Langley CH. The population genetics of *Drosophila* transposable elements. *Annu Rev Genet* 1989;23:251–287.
5. Shapiro J, ed. *Mobile genetic elements*. New York: Academic Press; 1983.
6. Berg DE, Howe MM, eds. *Mobile DNA*. Washington, DC: American Society of Microbiology; 1989.
7. Kleckner N. Regulation of transposition in bacteria. *Annu Rev Cell Biol* 1990;6:297–327.
8. Engels WR, Johnson-Schlitz DM, Eggleston WB, Sved J. High frequency P element loss in *Drosophila* is homolog dependent. *Cell* 1990;62:515–525.
9. Weiner AM, Deininger PL, Efstratiadis A. Nonviral retroposons: genes, pseudogenes, and transposable elements generated by the reverse flow of genetic information. *Annu Rev Biochem* 1986;55:631–661.
10. Boeke JD, Corces VG. Transcription and reverse transcription of retrotransposons. *Annu Rev Microbiol* 1989;43:403–434.
11. Temin HM. Retrons in bacteria. *Nature* 1989;339:254–255.
12. Varmus HE. Retroviruses. In: Shapiro J, ed. *Mobile genetic elements*. New York: Academic Press; 1983. 411–503.
13. Hull R, Covey SN. Genome organization and expression of reverse transcribing elements: variations and a theme. *J Gen Virol* 1986;67:1751–1758.
14. Mason WS, Taylor JM, Hull R. Retroid virus genome replication. *Adv Virus Res* 1987;32:35–96.
15. Nargang FE, Bell JB, Stohl LL, Lambowitz AM. The DNA sequence and genetic organization of a *Neurospora* mitochondrial plasmid suggest a relationship to introns and mobile elements. *Cell* 1984;38:441–453.
16. Michel F, Lang BF. Mitochondrial class II introns encode proteins related to the reverse transcriptases of retroviruses. *Nature* 1985;316:641–643.
17. Boer PH, Gray MW. Genes encoding a subunit of respiratory NADH dehydrogenase (ND1) and a reverse transcriptase-like protein (RTL) are linked to ribosomal RNA gene pieces in *Chlamydomonas reinhardtii* mitochondrial DNA. *EMBO J* 1988;7:3501–3508.
18. Inouye M, Inouye S. Retroelements in bacteria. *Trends Biochem Sci* 1991;16:18–21.
19. Toh H, Hayashida H, Miyata T. Sequence homology between retroviral transcriptase and putative polymerases of hepatitis B virus and cauliflower mosaic virus. *Nature* 1983;305:827–829.

20. Temin HM. Reverse transcription in the eukaryotic genome: retroviruses, pararetroviruses, retrotransposons, and retrotranscripts. *Mol Biol Evol* 1985;2:455–468.
21. Doolittle RF, Feng DF, Johnson MS, McClure MA. Origin and evolutionary relationships of retroviruses. *Q Rev Biol* 1989;64:1–30.
22. Varmus H, Brown P. Retroviruses. In: Berg DE, Howe MM, eds. *Mobile DNA*. Washington, DC: American Society of Microbiology; 1989:53–108.
23. Whitcomb JM, Hughes SH. Retroviral reverse transcription and integration: progress and problems. *Annu Rev Cell Biol* 1992;8:275–306.
24. Berg JM. Potential metal-binding domains in nucleic acid binding proteins. *Science* 1986;232:485–487.
25. Yoshinaka Y, Katoh I, Copeland TD, Oroszlan S. Murine leukemia virus protease is encoded by the *gag–pol* gene is synthesized through suppression of an amber termination codon. *Proc Natl Acad Sci USA* 1985;82:1618–1622.
26. Jacks T, Madhani HD, Masiarz FR, Varmus HE. Signals for ribosomal frameshifting in the Rous sarcoma virus *gag–pol* region. *Cell* 1988;55:447–458.
27. Saigo K, Kugimiya W, Matsuo Y, Inouye S, Yoshioka K, Yuki S. Identification of the coding sequence for reverse transcriptase-like enzyme in a transposable genetic element in *Drosophila melanogaster*. *Nature* 1984;312:659–661.
28. Mount SM, Rubin GM. Complete nucleotide sequence of the *Drosophila* transposable element copia: homology between copia and retroviral proteins. *Mol Cell Biol* 1985;5:1630–1638.
29. Clare J, Farabaugh P. Nucleotide sequence of a yeast Ty element: evidence for a novel mechanism of gene expression. *Proc Natl Acad Sci USA* 1985;82:2829–2833.
30. Boeke JD. Transposable elements in *Saccharomyces cerevisiae*. In: Berg DE, Howe MM, eds. *Mobile DNA*. Washington, DC: American Society of Microbiology; 1989:335–374.
31. Xiong Y, Eickbush TH. Similarity of reverse transcriptase-like sequences of viruses, transposable elements, and mitochondrial introns. *Mol Biol Evol* 1988;5:675–690.
32. Xiong Y, Eickbush TH. Origin and evolution of retroelements based upon their reverse transcriptase sequences. *EMBO J* 1990;9:3353–3362.
33. Shiba T, Saigo K. Retrovirus-like particles containing RNA homologous to the transposable element copia in *Drosophila melanogaster*. *Nature* 1983;302:119–124.
34. Jenson S, Heidmann T. An indicator gene for detection of germline retrotransposition in transgenic *Drosophila* demonstrates RNA-mediated transposition of the LINE I element. *EMBO J* 1991; 10:1927–1937.
35. Pelisson A, Finnegan DJ, Bucheton A. Evidence for retrotransposition of the I factor, a LINE element of *Drosophila melanogaster*. *Proc Natl Acad Sci USA* 1991;88:4907–4910.
36. Finnegan DJ, Fawcett DH. Transposable elements in *Drosophila melanogaster*. *Oxford Surv Eukaryotic Genes* 1986;3:1–62.
37. Marlor R, Parkhurst S, Corces V. The *Drosophila melanogaster* gypsy transposable element encodes putative gene products homologous to retroviral proteins. *Mol Cell Biol* 1986;6:1129–1134.
38. Friesen PD, Nissen MS. Gene organization and transcription of TED, a lepidopteran retrotransposon integrated within the baculovirus genome. *Mol Cell Biol* 1990;10:3067–3077.
39. Hansen LJ, Chalker DL, Sandmeyer SB. Ty3, a yeast retrotransposon associated with tRNA genes, has homology to animal retroviruses. *Mol Cell Biol* 1988;8:5245–5256.
40. Smyth DR, Kalitsis P, Joseph JL, Sentry JW. Plant retrotransposon from *Lilium henryi* is related to Ty3 of yeast and gypsy group of *Drosophila*. *Proc Natl Acad Sci USA* 1989;86:5015–5019.
41. Inouye S, Yuki S, Saigo K. Complete nucleotide sequence and genome organization of a *Drosophila* transposable element, 297. *Eur J Biochem* 1986;154:417–425.
42. Voytas DF, Ausubel FM. A copia-like transposable element family in *Arabidopsis thaliana*. *Nature* 1988;336:242–244.
43. Singer MF, Skowronski J. Making sense out of LINEs: long interspersed repeat sequences in mammalian genomes. *Trends Biochem Sci* 1985;10:119–122.
44. Fawcett DH, Lister CK, Kellett E, Finnegan DJ. Transposable elements controlling 1-R hybrid dysgenesis in *D. melanogaster* are similar to mammalian LINEs. *Cell* 1986;47:1007–1015.
45. Jakubczak JL, Xiong Y, Eickbush TH. Type I (R1) and type II (R2) ribosomal DNA insertions of *Drosophila melanogaster* are retrotransposable elements closely related to those of *Bombyx mori*. *J Mol Biol* 1990;212:37–52.
46. Berg JM. Zinc fingers and other metal-binding domains. *J Biol Chem* 1990;265:6513–6516.
47. Hutchison CA, Hardies SC, Loeb DD, Shehee WR, Edgell MH. LINEs and related retroposons:

long interspersed repeated sequences in the eukaryotic genome. In: Berg DE, Howe MM, eds. *Mobile DNA*. Washington, DC: American Society of Microbiology; 1989:593–617.

48. Kimmel BE, ole-Moiyoi OK, Young JR. Ingi, a 5.2 kilobase dispersed sequence element from *Trypanosoma brucei* that carries half a smaller mobile element at their end has homology with mammalian LINEs. *Mol Cell Biol* 1987;7:1465–1475.

49. Priimagi AF, Mizrokhi LJ, Ilyin YV. The *Drosophila* mobile element jockey belongs to LINEs and contains coding sequences homologous to some retroviral proteins. *Gene* 1988;70:253–262.

50. Ivanov VA, Melnikov AA, Siunov AV, Fodor II, Illyin YV. Authentic reverse transcriptase is coded by *jockey*, a mobile *Drosophila* element related to mammalian LINEs. *EMBO J* 1991; 10:2489–2495.

51. Gabriel A, Boeke JD. Reverse transcriptase encoded by a retrotransposon from the trypanosomatid *Crithidia fasciculata*. *Proc Natl Acad Sci USA* 1991;88:9794–9798.

52. Mathias SL, Scott AF. Kazazian HH, Boeke JD, Gabriel A. Reverse transcriptase encoded by a human transposable element. *Science* 1991;254:1808–1810.

53. Luan D, Korman M, Jakubczak JL, Eickbush TH. Reverse transcription of R2Bm RNA is primed by a nick at the chromosomal target site: a mechanism for non-LTR retrotransposition. *Cell* 1993;72:595–605.

54. McClure MA. Evolution of retroposons by acquisition or deletion of retrovirus-like genes. *Mol Biol Evol* 1991;8:835–856.

55. Villanueva MS, Williams SP, Beard CB, Richards FR, Aksoy S. A new member of a family of site-specific retrotransposons is present in the spliced leader RNA genes of *Trypanosoma cruzi*. *Mol Cell Biol* 1991;11:6139–6148.

56. Schwarz-Sommer Z, Leclercq L, Gobel E, Saedler H. Cin4, an insert altering the structure of the *A1* gene in *Zea mays*, exhibits properties of non-viral retrotransposons. *EMBO J* 1987;6:3873–3880.

57. Johnson MS, McClure MA, Feng D-F, Gray J, Doolittle RF. Computer analysis of retroviral *pol* genes: assignment of enzymatic functions to specific sequences and homologies with nonviral enzymes. *Proc Natl Acad Sci USA* 1986;83:7648–7652.

58. Khan E, Mack JPG, Katz RA, Kulkosky J, Skalka AM. Retroviral integrase domains: DNA binding and the recognition of LTR sequences. *Nucleic Acids Res* 1991;19:851–860.

59. Xiong Y, Eickbush TH. Functional expression of a sequence-specific endonuclease encoded by the retrotransposon R2Bm. *Cell* 1988;55:235–246.

60. Evans JP, Palmiter RD. Retrotransposition of a mouse L1 element. *Proc Natl Acad Sci USA* 1991;88:8792–8795.

61. Bucheton A. I transposable elements and I-R hybrid dysgenesis in *Drosophila*. *Trends Genet* 1990;6:16–21.

62. Mizrokhi LJ, Georgieva SG, Ilyin YV. Jockey, a mobile element similar to mammalian LINEs, is transcribed from the internal promoter by RNA polymerase II. *Cell* 1988;54:685–691.

63. Minchiotti G, Di Nocera PP. Convergent transcription initiates from oppositely oriented promoters within the 5' end region of *Drosophila melanogaster* F elements. *Mol Cell Biol* 1991;11:5171–5180.

64. Mclean C, Finnegan D. Analysis of the promoter of the I factor—the retrotransposon controlling I-R hybrid dysgenesis in *Drosophila melanogaster*. *J Cell Biochem* 1991;15D:74.

65. Swergold GD. Identification, characterization, and cell specificity of a human LINE-1 promoter. *Mol Cell Biol* 1990;10:6718–6729.

66. Eickbush TH. The non-LTR class of retrotransposable elements. *New Biol* 1992;4:430–440.

67. Xiong Y, Eickbush TH. The site-specific ribosomal DNA insertion element R1Bm belongs to a class of non-long–terminal-repeat retrotransposons. *Mol Cell Biol* 1988;8:114–123.

68. Burke WD, Calalang CC, Eickbush TH. The site-specific ribosomal insertion element type II of *Bombyx mori* (R2Bm) contains the coding sequence for a reverse transcriptase-like enzyme. *Mol Cell Biol* 1987;7:2221–2230.

69. Besansky NJ, Paskewitz SM, Hamm DM, Collins FH. Distinct families of site-specific retrotransposons occupy identical positions in the rDNA genes of *Anopheles gambiae*. *Mol Cell Biol* 1992;12:5102–5110.

70. Di Nocera PP. Close relationship between non-viral retrotransposons in *Drosophila melanogaster*. *Nucleic Acids Res* 1988;16:4041–4052.

71. Gabriel A, Yen TJ, Schwartz DC, Smith CL, Boeke JD, Sollner-Webb B, Cleveland DW. A rapidly rearranging retrotransposon within the miniexon gene locus of *Crithidia fasciculata*. *Mol Cell Biol* 1990;10:615–624.

72. Aksoy S, Williams S, Chang S, Richards FF. SLACS retrotransposon from *Trypanosoma brucei gambiense* is similar to mammalian LINEs. *Nucleic Acids Res* 1990;18:785–792.
73. Garrett JE, Knutzon DS, Carroll D. Composite transposable elements in the *Xenopus laevis* genome. *Mol Cell Biol* 1989;9:3018–3027.
74. Marschalek R, Hofmann J, Schumann G, Gosseringer R, Dingermann T. Structure of DRE, a retrotransposable element which integrates with a position specificity upstream of *Dictyostelium discoideum* tRNA genes. *Mol Cell Biol* 1992;12:229–239.
75. Sandmeyer SB, Hansen LJ, Chalker DL. Integration specificity of retrotransposons and retroviruses. *Annu Rev Genet* 1990;24:491–518.
76. Jakubczak JL, Burke WD, Eickbush TH. Retrotransposable elements R1 and R2 interrupt the rRNA genes of most insects. *Proc Natl Acad Sci USA* 1991;88:3295–3299.
77. Rogers J. The origin and evolution of retroposons. *Int Rev Cytol* 1985;93:187–279.
78. Deininger P. SINEs: short interspersed repeated DNA elements in higher eukaryotes. In: Berg DE, Howe ME, eds. *Mobile DNA*. Washington, DC: American Society of Microbiology; 1989:619–636.
79. Ganem D, Varmus HE. The molecular biology of the hepatitis B viruses. *Annu Rev Biochem* 56;1987:651–693.
80. Summer J, Mason WS. Replication of the genome of a hepatitis B-like virus by reverse transcription of an RNA intermediate. *Cell* 1982;29:403–415.
81. Wang G-H, Seeger C. The reverse transcriptase of hepatitis B virus acts as a primer for viral DNA synthesis. *Cell* 1992;71:663–670.
82. Covey SN, Hull R. Advances in cauliflower mosaic virus research. *Oxford Surv Plant Mol Cell Biol* 1985;2:339–346.
83. Pfeiffer P, Hohn T. Involvement of reverse transcription in the replication of cauliflower mosaic virus: a detailed model and test of some aspects. *Cell* 1983;33:781–789.
84. Lambowitz AM, Belfort M. Introns as mobile genetic elements. *Annu Rev Biochem* 1993;62:587–622.
85. Kennell JC, Moran JV, Perlman PS, Butow RA, Lambowitz AM. Reverse transcriptase activity associated with maturase-encoding group II introns in yeast mitochondria. *Cell* 1993;73:133–146.
86. Jarrell KA, Peebles CL, Dietrich RC, Romiti SL, Perlman PS. Group II intron self-splicing: alternative reaction conditions yield novel products. *J Biol Chem* 1988;263:3432–3439.
87. Lambowitz AM. Infectious introns. *Cell* 1989;56:323–326.
88. Kuiper MTR, Lambowitz AM. A novel reverse transcriptase activity associated with mitochondrial plasmids of *Neurospora*. *Cell* 1988;55:693–704.
89. Wang H, Kennell JC, Kuiper MTR, Sabourin JR, Saldanha R, Lambowitz AM. The Mauriceville plasmid of *Neurospora crassa*: characterization of a novel reverse transcriptase that begins cDNA synthesis at the 3' end of template RNA. *Mol Cell Biol* 1992;12:5131–5144.
90. Akins RA, Kelley RL, Lambowitz AM. Mitochondrial plasmids of *Neurospora*: integration into mitochondrial DNA and evidence for reverse transcription in mitochondria. *Cell* 1986;47:505–516.
91. Boer PH, Gray MW. Scrambled ribosomal RNA gene pieces in *Chlamydomonas reinhardtii* mitochondrial DNA. *Cell* 1988;55:399–411.
92. Lampson RC, Sun J, Hsu M-Y, Vallejo-Ramirez J, Inouye S, Inouye M. Reverse transcriptase in a clinical strain of *Escherichia coli*: production of branched RNA-linked msDNA. *Science* 1989; 243:1033–1038.
93. Lim D, Mass WK. Reverse transcriptase-dependent synthesis of a covalently linked, branched DNA–RNA compound in *E. coli* B. *Cell* 1989;56:891–904.
94. Temin HM. Origin of retroviruses from cellular moveable genetic elements. *Cell* 1980;21:599–600.
95. Kohlstaedt LA, Wang J, Friedman JM, Rice PA, Steitz TA. Crystal structure at 3.5 A resolution of HIV-1 reverse transcriptase complexed with an inhibitor. *Science* 1992;256:1783–1790.
96. Lake JA. Tracing origins with molecular sequences: metazoan and eukaryotic beginnings. *Trends Biochem Sci* 1991;16:46–50.
97. Field KG, Olsen GJ, Lane DJ, Giovannoni SJ, Ghiselin MT, Raff EC, Pace NR, Raff RA. Molecular phylogeny of the animal kingdom. *Science* 1988;239:748–753.
98. Gogarten JP, Kibak H, Dittrich P, Taiz L, Bowman EJ, Bowman BJ, Manolson MF, Poole RJ, Date T, Oshima T, Konishi J, Denda K, Yoshida M. Evolution of the vacuolar $H^+$-ATPase: implications for the origin of eukaryotes. *Proc Natl Acad Sci USA* 1989;86:6661–6665.
99. Puhler G, Leffers H, Gropp F, Palm P, Klenk H-P, Lottspeich F, Garrett RA, Zillig W. Archae-

bacterial DNA-dependent RNA polymerase testify to the evolution of the eukaryotic nuclear genome. *Proc Natl Acad Sci USA* 1989;86:4569–4573.

100. Daniels SB, Peterson KR, Strausbaugh LD, Kidwell MG, Chovnick A. Evidence for horizontal transmission of the P transposable element between species. *Genetics* 1990;124:339–355.

101. Konieczny A, Voytas DF, Cummings MP, Ausubel FM. A superfamily of *Arabidopsis thaliana* retrotransposons. *Genetics* 1991;127:801–809.

102. Mizrokhi LJ, Mazo AM. Evidence for horizontal transmission of the mobile element jockey between distant *Drosophila* species. *Proc Natl Acad Sci USA* 1990;87:9216–9220.

103. Maruyama K, Hartl DL. Evidence for interspecific transfer of the transposable element mariner between *Drosophila* and *Zaprionus*. *J Mol Evol* 1991;33:514–524.

104. Toh H, Kikuno R, Hayashida H, Miyata T, Kugimiya W, Inouye S, Yuki S, Saigo K. Close structural resemblance between putative polymerase of a *Drosophila* transposable genetic element 17.6 and the *pol* gene product of Moloney murine leukemia virus. *EMBO J* 1985;4:1267–1272.

105. Webster TA, Patarca R, Lathrop RH, Smith TF. Potential structural motifs for reverse transcriptase. *Mol Biol Evol* 1989;6:317–320.

106. Poch O, Sauvaget I, Delarue M, Tordo N. Identification of four conserved motifs among the RNA-dependent polymerase encoding elements. *EMBO J* 1989;8:3867–3874.

107. Delarue M, Poch O, Tordo N, Moras D, Argos P. An attempt to unify the structure of polymerases. *Protein Eng* 1990;3:461–467.

108. Feng DF, Johnson MS, Doolittle RFJ. Aligning amino acid sequences: comparison of commonly used methods. *J Mol Evol* 1985;21:112–125.

109. Larder BA, Kemp SD, Purifoy DJM. Infectious potential of human immunodeficiency virus type I reverse transcriptase mutants with altered inhibitor sensitivity. *Proc Natl Acad Sci USA* 1989; 86:4803.

110. Saitou N, Nei M. The neighbor-joining method: a new method for reconstructing phylogenetic trees. *Mol Biol Evol* 1987;4:406–425.

111. Swofford DL. PAUP: phylogenetic analysis using parsimony. *Illinois Natural History Survey.* Champaign: 1991.

112. Feisenstein J. Phylogenies from molecular sequences: inference and reliability. *Annu Rev Genet* 1988;22:521–565.

113. Feng D-F, Doolittle RF. Progressive sequence alignment as a prerequisite to correct phylogenetic trees. *J Mol Evol* 1987;25:351–360.

114. Loeb DD, Padgett RW, Hardies SC, Shehee WR, Comer MB, Edgell MH, Hutchison CA III. The sequence of a large L1Md element reveals tandemly repeated 5′ end and several features found in retrotransposons. *Mol Cell Biol* 1986;6:168–182.

115. Hattori M, Kuhara S, Takenaka O, Sakaki Y. L1 family of repetitive sequences in primates may be derived from a sequence encoding a reverse transcriptase-related protein. *Nature* 1986;321:625–627.

116. Flavell AJ, Smith DB. A Ty1-copia group retrotransposon sequence in a vertebrate. *Mol Gen Genet* 1992;233:322–326.

117. Voytas DF, Cummings MP, Konieczny A, Ausubel FM, Rodermel SR. Copia-like retrotransposons are ubiquitous among plants. *Proc. Natl. Acad. Sci. USA* 1992;89:7124–7128.

118. Flavell AJ, Smith DB, Kumar A. Extreme heterogeneity of Ty1-copia group retrotransposons in plants. *Mol Gen Genet* 1992;231:233–242.

119. Felder H, Herzceg A, de Chastonay Y, Aeby P, Tobler H, Muller F. *Nucleic Acids Res* 1993; [in press].

120. de Chastonay Y, Felder H, Link C, Aeby P, Tobler H, Muller F. Unusual features of the retroid element PAT from the nematode *Panagrellus redivivus*. *Nucleic Acids Res* 1992;20:1623–1628.

121. Xiong Y, Burke WD, Eickbush TH. Pao, a retrotransposon from *Bombyx mori* with a highly divergent reverse transcriptase domain and unusual long terminal repeats. *Nucleic Acids Res* 1993; 21:2117–2123.

122. Cappello J, Handelsman K, Lodish H. Sequence of *Dictyostelium* DIRS-1: an apparent retrotransposon with inverted terminal repeats and an internal circle junction sequence. *Cell* 1985; 43:105–115.

123. Nei M. Relative efficiencies of different tree-making methods for molecular data. In: Miyamoto MM, Cracraft J, eds. *Phylogenetic analysis of DNA sequences.* New York: Oxford University Press; 1991:90–128.

124. Gojobori T, Yokoyama S. Rates of evolution of the retroviral oncogene of Moloney murine sarcoma virus and of its cellular homologues. *Proc Natl Acad Sci USA* 1985;82:4198–4201.
125. Weiner AM, Maizels N. tRNA-like structures tag the 3′ ends of genomic RNA molecules for replication: implications for the origin of protein synthesis. *Proc Natl Acad Sci USA* 1987;84: 7383–7387.
126. Varmus HE. Reverse transcription in bacteria. *Cell* 1989;56:721–724.
127. Wintersberger U, Wintersberger E. RNA makes DNA: a speculative view of the evolution of DNA replication mechanisms. *Trends Genet* 1987;3:198–202.
128. Cech TR. A model for the RNA-catalyzed replication of RNA. *Proc Natl Acad. Sci USA* 1986;83:4360–4363.
129. Ishihama A, Nagata K. Viral RNA polymerases. *Crit Rev Biochem* 1988;23:27–76.
130. Johnson MS, Sali A, Blundell TL. Phylogenetic relationships from three-dimensional protein structures. *Methods Enzymol* 1990;183:670–690.
131. Saiki RK, Gelfand DH, Stoffel S, Scharf SJ, Higuchi R, Horn GT, Mullis KB, Erlich HA. Primer-directed enzymatic amplication of DNA with a thermostable DNA polymerase. *Science* 1988; 239:487–491.
132. Shippen-Lentz D, Blackburn EH. Functional evidence for an RNA template in telomerase. *Science* 1990;247:546–552.
133. Lundblad V, Szostak JW. A mutant with a defect in telomere elongation leads to senescence in yeast. *Cell* 1989;57:633–643.
134. Lundblad V, Blackburn EH. RNA-dependent polymerase motifs in Est1: tentative identification of a protein component of an essential yeast telomerase. *Cell* 1990;60:529–530.
135. Rogers JH. How were introns inserted into nuclear genes? *Trends Genet* 1989;5:213–216.
136. Cavalier-Smith T. Intron phylogeny: a new hypothesis. *Trends Genet* 1991;7:145–148.
137. Camirand A, Brisson N. The complete nucleotide sequence of the Tst1 retrotransposon of potato. *Nucleic Acids Res* 1990;18:4929.
138. Rothnie HM, McCurrach KJ, Glover LA, Hardman N. Retrotransposon-like nature of Tp1 elements: implications for the organization of highly repetitive, hypermethylated DNA in the genome of *Physarum polycephalum*. *Nucleic Acids Res* 1991;19:279–286.
139. Springer MS, Davidson EH, Britten RJ. Retroviral-like element in a marine invertebrate. *Proc Natl Acad Sci USA* 1991;88:8401–8404.
140. Levin HL, Weaver DC, Boeke JD. Two related families of retrotransposons from *Schizosaccharomyces pombe*. *Mol Cell Biol* 1990;10:6791–6798.
141. O'Hare K, Alley MRK, Cullingford TE, Driver A, Sanderson MJ. DNA sequence of the Doc retroposon in the *white-one* mutant of *Drosophila melanogaster* and of secondary insertions in the phenotypically altered derivatives *white-honey* and *white-eosin*. *Mol Gen Genet* 1991;225:17–24.

# PART III

## Driving Forces in Evolution I: Processes Generating Genetic Diversity

The Evolutionary Biology of Viruses,
edited by Stephen S. Morse.
Raven Press, Ltd., New York © 1994.

# 8

# Mutation Rates and Rapid Evolution of RNA Viruses

Esteban Domingo and *John J. Holland

*Centro de Biología Molecular (CSIC-UAM), Universidad Autónoma de Madrid,
28049 Madrid, Spain; *Department of Biology and Institute for Molecular Genetics,
University of California, San Diego, La Jolla, California 92093*

## UNIQUE FEATURES OF RNA GENETICS

### Genetic Variation and Error Catastrophe

The RNA viruses constitute the most abundant group of parasites of animals and plants; it is estimated that more than 70% of the viruses that infect differentiated organisms are RNA viruses. They share with their hosts a great potential for variation, but they differ from their hosts in the extent and tolerance for such modifications. The following observations document the great variation potential of the eukaryotic chromosome: (a) Frequent recombination and transposition events—reverse transcription has played an important role in shaping the eukaryotic genome (1); (b) accumulation of mutations in pseudogenes, the inactive counterparts of functional genes (2); (c) hypervariability at defined immunoglobulin gene segments, a mechanism for the generation of immunoglobulin diversity (3,4).

It has been proposed that episodes of DNA hypermutability have contributed to accelerated evolution during certain time periods (5,6). Also, extensive DNA variation occurs in transformed cells, thereby contributing to the adaptation (invasiveness) of tumor cells (7). However, important constraints must be operating to limit the variation of DNA in normal cells. Studies by Eigen and Schuster and by others (8–10) have shown that high complexity of a genetic program necessitates a corresponding high accuracy of copying for it to be transmitted to offspring as a meaningful program. For a given genomic complexity, there is a critical-copying fidelity below which information can no longer be maintained. At this error threshold, any decrease in fidelity results in the so-called error catastrophe, a transition that can be equated with total loss ("melting") of information (10). In error catastrophe, the nucleotide sequences of viral RNA would become essentially random (8–10). In this view, it is expected that enzymes provide proofreading–repair activities for correcting misincorporated nucleotides in DNA evolved concomitantly with the increase in size of the DNA needed for cellular organization (8). Point mutation rates for cellular DNA have been estimated at $10^{-8}$ to $10^{-11}$ substitutions per nucleotide per round of replication (11–13). When proofreading–repair activities are suppressed (as in certain mutant forms of DNA polymerase or during the copying of homopolymeric templates), mutation rates approaching $10^{-4}$ substitutions per nucleotide per round of copying are attained (8,12,13). As documented in the next section, mutation rates averaging $10^{-3}$ to $10^{-5}$ substitutions per nucleotide copied are operating during replication or reverse transcription of viral RNA. No evidence of efficient proofreading–repair activities acting during RNA-dependent RNA or DNA synthesis has been obtained (14,15). The genomic size of RNA viruses (limited to the range of 3,000 to 30,000 nucleotides, albeit with highly compact information) is compatible with low-copying fidelities, without frequent incursions into error catastrophe (8–10). Recent experiments on the effect of chemical mutagenesis on mutation frequencies of poliovirus and vesicular stomatitis virus suggest that these viruses replicate very near the threshold for error catastrophe (16). It was not possible to increase mutation frequencies at defined single-base sites more than two- to threefold, without loss of virus infectivity.

High mutation rates, coupled with extremely rapid replication cycles, are features that distinguish RNA genetics from cellular DNA genetics. The DNA viruses, mainly those with limited genetic complexity, could also share with RNA viruses high-mutation rates. This would require, however, that the DNA polymerases they use for replication lacked proofreading–repair activities or that such activities were suppressed during infection (14). The possibility that at least some DNA viruses may behave as RNA viruses in their population structure and potential for rapid evolution has been previously suggested (17,18), and it must be considered an open question (14,17).

## Fate of Newly Arising Variants

A direct consequence of the high-mutation rates operating during RNA replication is that mutant genomes are continuously being generated (19,20). The implications of this fact, which itself is a hallmark of RNA genetics, are generally not fully comprehended. It means that in any infected cell (that may amplify one RNA molecule into $10^3$ or even $10^5$ progeny molecules) many single and multiple mutants arise. Normally, most or all of them will be maintained at low frequencies (or even eliminated entirely) by negative selection (19,21–23). This is the expected fate of most mutants if the virus is highly adapted to the environment (19). (This applies basically to viruses passaged many times in cell culture under defined conditions.) In contrast, if the virus replicates in an environment different from that in which its parental genomes were used to replicate (for example, upon adaptation of a natural isolate to cell culture), the probability of newly arising variants to become dominant will be higher. Rapid change may also be observed when unfit parental genomes (for example, a cDNA copy of a viral genome harboring deleterious mutations) initiate the infection (24,25). When a virus is observed to be unchanged, it is not because mutations have not occurred, but rather, because mutant progeny have been maintained at low levels, and the average nucleotide sequence of the viral population has remained invariant (19,26). This endows RNA genomes with extreme adaptability, since those mutant genomes, which on many occasions are doomed to extinction in one particular environment, may be the colonizers of a new ecological niche and become the effectors of virus survival. This continuous generation of variants underlies the quasispecies structure (extreme genetic heterogeneity) of RNA virus populations and the concept of population equilibrium (8,10,19,27–32) discussed in the following sections.

## MUTATION AND VARIATION OF RNA VIRUSES

### Mutation Rates and Frequencies

Because of the multiple implications of high mutation rates that were emphasized in the previous section, it is important to review the experimental evidence for such rates during RNA virus and retrovirus replication. There was considerable early evidence for rapid phenotypic variation in plant and animal RNA viruses (reviews in 30,32). However, it has been in only the last two decades that detailed analyses of nucleotide sequences of RNA genomes have unveiled the real magnitude of their variation.

We refer to *mutation rate* as the proportion of misincorporation events during nucleic acid synthesis, expressed as substitutions per nucleotide per round of template copying. In the literature of population genetics, mutation rate often refers to mutations that become incorporated into an evolving genome per unit time, expressed as substitutions per nucleotide (or per genome) per year (or generation). We use the term *rate of fixation* (or preferably *accumulation*, in view of the fleeting nature of genomic variants) of mutations to describe this second parameter. The *mutation* (or *mutant*) *frequency* measures the proportion of mutants (or mutant residues) in a population, and is expressed as substitutions per nucleotide. Mutant frequency can also refer to a phenotypic trait; for example, the frequency of RNA viruses resistant to neutralization by a monoclonal antibody (MAb) or to inhibition by an antiviral agent. These distinctions are very important, and they have not been rigorously followed in the literature of RNA virus evolution, thereby creating confusion on very important issues. Contrary to the mutation rate, which describes a biochemical event (independent of the biological properties of parental and mutant genomes), the mutation (or mutant) frequency is a population number heavily influenced by the competitive ability of the mutant genomes relative to their parental wild types (8,14,19,21–23,29–32).

Table 1 summarizes values of mutation rates and frequencies for several RNA viruses and retroviruses (23,33–57). It includes a few MAb-resistant (MAR) mutant

**TABLE 1.** *Point mutation rates and frequencies, and MAR mutant frequencies for RNA viruses*

| Virus | Procedure | Value[a] | Ref. |
|---|---|---|---|
| Phage Qβ | Reversion of extra-cistronic mutant | $10^{-4}$ | 33 |
| Vesicular stomatitis virus | Chemical determination of error oligonucleo-tides[b] | $10^{-3}$–$10^{-4}$ | 34,35 |
| Vesicular stomatitis virus | Reversion of amber mutant | $10^{-3}$–$10^{-4}$ | 36 |
| Vesicular stomatitis virus | MAR mutant frequency[c] | $10^{-4}$ | 37 |
| Influenza virus A | Repetitive nucleotide sequencing; plaque progeny | $10^{-5}$ | 38 |
| Poliovirus | Repetitive nucleotide sequencing | $<2 \times 10^{-6}$ | 38 |
| Poliovirus | Reversion of amber mutant | $2 \times 10^{-6}$ | 39 |
| Poliovirus | Reversion of guanidine dependence | $2 \times 10^{-4}$ | 40 |
| Poliovirus | Chemical determination of error oligonucleo-tides[b] | $3 \times 10^{-3}$–$5 \times 10^{-3}$ | 41 |
| Poliovirus | Specific U → C transition | $2 \times 10^{-5}$ | 25 |
| Poliovirus | MAR mutant frequency[c] | $10^{-4}$–$10^{-5}$ | 42–44 |
| Mengovirus | MAR mutant frequency[c] | $3 \times 10^{-3}$–$5 \times 10^{-5}$ | 45 |
| Foot-and-mouth disease virus | MAR mutant frequency[c] | $10^{-5}$ | 23 |
| Rhinovirus | MAR mutant frequency[c] | $10^{-4}$–$10^{-5}$ | 46 |
| Sindbis virus | Reversion of *ts* mutant | $<10^{-6}$ | 47 |
| Sindbis virus | Apparent reversion of *ts* mutants | $10^{-2}$–$10^{-5}$ ($<10^{-8}$ for some mutants) | 48,49 |
| Sindbis virus | MAR mutant frequency[c] | $10^{-3}$–$10^{-5}$ | 50 |
| Rous sarcoma virus | Selection of point mutant upon passage | $10^{-4}$ | 51,52 |
| Rous sarcoma virus | Denaturing electrophoresis; plaque progeny | $10^{-4}$ | 53 |
| Spleen necrosis virus | Retroviral vectors with marker | $5 \times 10^{-3}$–$10^{-5}$ | 54,55 |
| Murine leukemia virus (AKR 2A) | T1 fingerprinting and repetitive nucleotide sequencing | $2 \times 10^{-5}$ | 56 |
| Moloney murine leukemia virus | Ambar reversion on retroviral vector | $2 \times 10^{-6}$ | 57 |

[a]Expressed generally as substitutions per nucleotide site. However, in some cases, one specific base change was measured and in others any possible mutation at one or several sites was scored (see specific references for a precise interpretation of each value). MAR mutant frequency is the proportion of mutants resistant to neutralization by a MAb.

[b]Measured at many different sites of the viral genome.

[c]MAR mutant frequencies cannot be directly related to mutation rates or frequencies per site (see text).

frequencies, also measured for a large number of additional viruses not listed here (for review see 32,58). The MAR mutant frequencies depend on the number of amino acid replacements at the relevant epitope, that can lead to virus resistance to neutralization by the MAb. Also, such frequencies may be greatly underestimated by phenotypic masking of mutant genomes in wild-type envelopes (37,59) and, more modestly, by the lower fitness of the MAR mutants relative to the parental population from which they were isolated (23). Thus, low MAR mutant frequencies do not necessarily imply low mutation rates at the relevant sites (21–23,32,37,59). Table 1 includes results from chemical and genetic procedures that have been critically analyzed in several articles (10,14,17,20,29–32,60,61). Here, we will discuss only some general conclusions that may be derived from the values listed, in spite of the uncertainties involved in several determinations. For any virus analyzed, at least some values are in the range of $10^{-3}$ to $10^{-5}$ substitutions per site. Thus, the quasispecies structure must apply to all these viruses (14), as also supported by direct quantification of genetic heterogeneity (10,19–23,26–35,37,52, 60,61). Whenever mutation frequencies have been determined at several nucleotide sites—such as by application of the direct chemical method of Steinhauer and Holland (34) to vesicular stomatitis virus (34,35) or to poliovirus (41; see Table 1)—the results were very similar irrespective of the degree of conservation of the nucleotide site in different isolates of the virus. Thus, mutation and variation are entirely different concepts, a point rightly emphasized by Coffin (52). This is also demonstrated by the different extents of antigenic diversity of Picornaviridae, in spite of similar MAR mutant frequencies. Mengovirus, poliovirus, foot-and-mouth disease virus, and rhinoviruses include 1, 3, 7, and more than 100 described serotypes, respectively; yet, MAR mutant frequencies (determined using several different neutralizing MAbs) are very similar for all of them, ranging from $10^{-3}$ to $10^{-5}$, without any trend toward higher frequencies for rhinoviruses (see Table 1). Picornavirus variation has been recently reviewed (61).

Drake (62) has recently normalized several values included in Table 1 to mutation rates per genome per replication. He found that this rate, for lytic RNA viruses (Qβ, poliomyelitis, vesicular stomatitis, influenza A) is about 300-fold higher than for DNA-based microbes. Retroviruses, however, show average values about tenfold lower than lytic RNA viruses (62). It may be significant that treatment with azacitidine (5-azacytidine; AZC) could increase up to 13-fold the retroviral mutation rate (63), whereas several mutagenic agents could increase only two- to threefold poliovirus and vesicular stomatitis mutant frequencies without loss of virus viability (16). However, it cannot be excluded that this difference was affected by factors other than the spontaneous mutation rate, such as the target size (genome complexity) of the viral genomes undergoing chemical mutagenesis.

High mutation rates for RNA replicases and reverse transcriptases are also supported by many in vitro fidelity measurements with the purified enzymes (13,57, 64–68; review in 69). Misincorporation rates measured in vitro can be influenced by many environmental conditions and are difficult to relate to actual in vivo measurements, even though the agreement is often good. In vitro measurements have provided evidence that error rates are probably not uniform along an RNA genome and that some types of substitutions are more likely than others (13,14,32,57,64–69).

It has been observed many times that mutant genomes, randomly sampled from viral populations, show a decreased fitness relative to the parental genomes (19,21–23). In particular, repeated plaque-to-plaque passaging of RNA viruses led to a significant decrease in competitive fitness, as expected from the steady accumulation of slightly deleterious mutations (Muller's ratchet) (70–72). Those mutation frequencies that are based on quantification of the proportion of low-fitness clones will underestimate the true mutation rates. Since most newly arising mutants will tend to show lower fitness than their parental wild-type genomes, and most determinations have measured mutant frequencies, rather than mutation rates, it is likely that most values given in Table 1 represent underestimates of the true mutation rates.

## Quasispecies and Population Equilibrium

It is very rewarding to note that two independent fields of research led to the same conclusion: RNA genome populations could not be considered to be genetically defined populations. These fields of research were theoretical studies on replicons in early life (8–10) and experimental analyses of the individual genomes that compose viral populations (14,19,26,27,29–35,51,52,60,73). Eigen coined the term *quasispecies* to describe the distribution of related, nonidentical genomes that constitute replicon populations. To avoid misunderstandings, the concept conveyed by the word quasispecies must be considered unrelated to the concept of biological species (73), a controversial concept for higher organisms as well as for viruses, and a topic beyond the scope of this review. The theoretical formulations of Eigen, Schuster, and colleagues implied populations of infinite numbers of molecules in equilibrium (8–10). Here, the term *equilibrium* refers to selective equilibrium or selective rating of all mutants that may appear (29). That populations of RNA viruses are finite and generally far from equilibrium conditions in no way invalidates quasispecies as a suitable theoretical concept to understand the behavior of RNA viruses at the population level. Real RNA virus populations are quasispecies perturbed by many environmental changes as well as by random-sampling events, both types of perturbations being easily noticeable, but very poorly understood in their mode of action. The RNA virus quasispecies can suppress superior mutant progeny. A highly competitive mutant clone of vesicular stomatitis virus could overwhelm its parental population only when present above threshold level during dilute serial passage (74).

High mutation rates can, apparently paradoxically, result in an invariant average or consensus genomic sequence. This is not paradoxical provided equilibrium conditions prevail, and newly generated variants are maintained at low levels (8,10,14, 19,20,26–35). If there is disequilibrium (a situation often encountered during virus replication in vivo) a variant genome will overtake the previous average, leading to a new average or consensus population. Shifts in the conditions of equilibrium versus disequilibrium, rather than differences in mutation rates (at different genomic sites or for different RNA viruses) appear to be the main determinants of the observed variations. It is even expected that in a virus with a relatively low mutation rate (for example, a DNA virus with an active proofreading–repair system during its replication) a variant genome, once generated, could rapidly outcompete the parental genome. Again, variation must be considered to be quite different from mutation. If sufficiently low, mutation rates could become the rate-limiting step in long-term evolution. However, there is no evidence that any such limitation has been operating for RNA viruses.

## LONG-TERM EVOLUTION OF RNA VIRUSES

### Hypermutation, Recombination, Reassortment, and Defective Genomes in RNA Evolution

The genetic heterogeneity of RNA virus populations sets the scenario for adaptability, an essential feature for viral pathogenesis (adaptation within one organism and to certain target cells) as well as for long-term evolution (adaptation to many host organisms). It would be an oversimplification, however, to attribute to mutation alone the potential for long-term adaptation of viruses. Phenomena, such as hypermutation, biased hypermutation, RNA recombination, gene reassortment, and modulation of infectivity by defective genomes, all contribute to different extents (depending on the virus and its environment) to long-term evolution.

Biased hypermutation was first described in defective interfering (DI) RNAs of vesicular stomatitis virus (VSV) (75) and in highly mutated forms of measles virus found in cases of subacute sclerosing panencephalitis and measles inclusion body encephalitis (76). It is characterized by the overdominance of one type of replacement (for example, A→G within a stretch of a DI of VSV, or U→C in a defective matrix gene of measles virus in a brain). More recently, hypermutation and biased hypermutations have also been described for retroviruses (77,78). For human immunodeficiency virus (HIV), dislocation mutagenesis (slippage of the primer relative to the template for reverse transcription) has been proposed as the mechanism that generates molecules with many G→A transitions (78). Hypermutation may also be associated with error-prone subsets of polymerase molecules (77). A striking precise reversion of five substitutions in the poliovirus genome (25) might also be due to error-prone subsets of polymerase, but without mutational bias.

The occurrence of RNA *recombination* (the generation of covalently joined RNAs from two different parental molecules) has been described for many RNA viruses, including phage and animal as well as plant viruses (reviewed in 79,80). It may involve not only viral molecules, but also viral RNA and cellular RNA (81, 82). Zhang and Temin have used retroviral vectors to study the formation of chimeric viral–cellular RNA during a single cycle of retrovirus replication (83).

The rate of strand-switching mediated by a short region of sequence identity was about 0.1% to 1% of the rate of homologous retroviral recombination (83). The rescue of cellular RNA into viral genomes, which, until recently, seemed reserved to retroviruses, may actually be more frequent than thought, even with viral genomes unable to integrate into the cellular chromosome (82). Such chimeric RNAs often show altered virulence. Recombination may produce highly deviant forms of viable or defective RNA genomes and, in so doing, may provide more drastic evolutionary jumps than those mediated by point mutations. Recombination may also rescue high-fitness viral genomes from low-fitness parental viruses that harbor deleterious mutations. Even if many recombination events may be lethal, in rare cases recombination may lead to highly innovative exchange of functional modules (84). Elegant experiments to distinguish RNA breaking–rejoining from copy-choice by the replicase suggest the latter as the main mechanism of RNA recombination (85). The reader is referred to several recent articles (79,80,83–87) for a discussion of the molecular mechanisms involved, as well as for further discussion of biological implications of RNA recombination.

The occurrence of RNA *reassortment* (the exchange of genomic segments among different parents of a virus with a segmented genome) has played a very important role in the evolution of influenza viruses and in the emergence of new pathogenic strains for humans (reviews in 88,89). Segmented double-stranded RNA genomes also show high-frequency reassortment (90). Some reassortant rotaviruses can even include highly altered (rearranged) genomic segments (91).

In recent years, it has also become apparent that defective genomes are very frequent among the components of mutant spectra of RNA viruses and retroviruses (reviewed in 86,87). Initially, they were identified as incomplete virus particles that interfered with the replication of the corresponding standard, infectious virus (92). Defective RNAs may modulate (enhance or decrease) disease symptoms induced by RNA viruses (86,93,94). Application of the polymerase chain reaction (PCR) to the analysis of the individual genomes replicating in an infected host has revealed a surprisingly high proportion of nonfunctional, defective genomes (for example, containing premature stop codons in open-reading frames) in the viral quasispecies (87,95,96). The role that such defective genomes may play in human disease processes is not understood.

The dynamic nature of RNA virus populations is illustrated by the frequent appearance of infectious virus mutants resistant to interference by DI particles. In this manner, a coevolution of the standard viruses and their DI RNAs may be established that results in rapid evolution of the infectious agent (30–32,86,97). The DIs can act to promote recurring or continuous disequilibrium in RNA virus populations.

## Rates of Accumulation of Mutations in Nature

Rates of RNA virus evolution in nature, with a few exceptions, are not constant over long periods. This is not surprising in view of the several mechanisms that

**TABLE 2.** *Rate of accumulation of mutations (evolution rates) of RNA viruses*

| Virus | Genomic segment | Value[a] | Ref. |
|---|---|---|---|
| Influenza A virus | *NS* | $1.9 \times 10^{-3}$ | 89 |
| | *HA* (H3) | $6.7 \times 10^{-3}$ | 89 |
| | *HA* (H1)[b] | $5.8 \times 10^{-3}$–$1.7 \times 10^{-2}$ | 98 |
| | *NA* (N2) | $3.2 \times 10^{-3}$ | 89 |
| | *M1* | $0.8 \times 10^{-3}$ | 99 |
| | *M2* | $1.4 \times 10^{-3}$ | 99 |
| | *NP* | $1.6 \times 10^{-3}$–$2.2 \times 10^{-3}$ | 100,101 |
| | *NP*[b] | $3.5 \times 10^{-3}$–$2.4 \times 10^{-2}$ | 98 |
| Foot-and-mouth disease virus | *VP1*[c] | $9.0 \times 10^{-3}$–$7.4 \times 10^{-2}$ | 102 |
| | *VP1* | $6.5 \times 10^{-3}$ | 103 |
| | *VP1* | $1.4 \times 10^{-3}$ | 104 |
| Poliovirus | Average over genome | $6.9 \times 10^{-3}$–$1.4 \times 10^{-2}$ | 105 |
| Eastern equine encephalomyelitis virus | 26S structural gene | $1.4 \times 10^{-4}$ | 106 |
| Human immunodeficiency virus | *env* | $3.2 \times 10^{-3}$–$1.6 \times 10^{-2}$ | 107 |
| | *gag* | $3.7 \times 10^{-4}$–$1.8 \times 10^{-3}$ | 107 |
| Simian immunodeficiency virus | gp120[d] | $8.5 \times 10^{-3}$ | 108 |
| Equine infectious anemia virus | *env*[e] | $10^{-1}$–$10^{-2}$ | 109 |
| Cellular genes | | $10^{-8}$–$10^{-9}$ | 110,111 |

[a]Expressed as substitutions per nucleotide per year. Except where indicated ([b] to [e]) calculations are based on sequence comparisons of isolates during one or multiple epidemic outbreaks of the indicated virus.
[b]Immunodeficient child persistently infected with influenza virus.
[c]Cattle persistently infected with plaque-purified foot-and-mouth disease virus.
[d]Infections established with a molecular clone of simian immunodeficiency virus.
[e]Ponies chronically infected with a standard inoculum of equine infectious anemia virus.

generate diversity and that may modulate replication rates, as summarized in previous sections. However, average rates of accumulation of mutations estimated for different RNA viruses and retroviruses can be compared with average rates for cellular genes (Table 2).

Rates of evolution of RNA viruses are about $10^5$ to $10^6$ times higher than the rates estimated for eukaryotic cellular genes (30,89,98–111) (see Table 2). The values listed have necessarily been averaged over long periods, and may include periods of rapid evolution, alternating with others of relative stability (stasis) (99,103,104). It may be significant that when a rate of accumulation of mutations is measured in a single individual host persistently infected with a virus—as with influenza virus (98), foot-and-mouth disease virus (102), or lentiviruses (108,109)—values are higher than with the corresponding viruses measured on isolates from different individuals in an epidemic outbreak. The difference may reflect that in those persistently infected individuals (98,102,108,109), viral replication and selection are continuous, whereas, in an epidemic outbreak, they are not (103,104).

It may also be significant that some DNA viruses, such as hepadnaviruses, which include a reverse transcription step in their replication cycle (112), or parvoviruses, which do not, show rates of evolution in the range of $10^{-4}$ to $10^{-5}$ substitutions per nucleotide per year (18,113), not far from the values of some of the RNA viruses listed in Table 2. Whether this reflects high mutation rates during DNA virus replication cannot yet be determined from the available information. DNA recombination and rearrangements, or block additions and deletions, may be as critical as (or perhaps more than) point mutations in the generation of DNA virus diversity (114–116). This area deserves further exploration, because there is surprisingly little quantitative information concerning mutation rates for DNA animal viruses.

The dramatic difference in belonging to the rapidly evolving RNA world versus the relatively static DNA world was elegantly illustrated by Gojobori and Yokoyama (117), who showed that the rate of evolution of the v-*mos* gene from Maloney murine sarcoma virus ($1.31 \times 10^{-3}$ substitutions per nucleotide site per year) was a millionfold higher than the rate for its cellular counterpart, c-*mos*.

Quasispecies populations (understood as the ensemble of individual molecules that constitute the dynamic mutant spectrum of a virus population) are conceptually different from the classic concept of *genetic polymorphism* (genetic differences among individuals of one species), as applied to free-living cells and to sexually reproducing species. Nevertheless, differences between the average sequences of independent viral isolates could be (and often have been) considered equivalent to genetic polymorphisms in cellular organisms. Two opposed schools of thought have interpreted the origin and maintenance of genetic polymorphisms in different ways. The so-called selectionist versus neutralist controversy has recently entered the field of RNA virus evolution (for example, see refs. 20,52, and 118–121 for opposing points of view). We may ask whether the process of diversification of a viral quasispecies is driven by forces akin to those acting during the evolution of differentiated organisms.

## Neutralism Versus Selectionism in RNA Virus Evolution: Two Views of Genetic Variation

According to the *selectionist* view, genetic polymorphisms in higher organisms (122,123; reviews in 125–129) are maintained largely by selection. Different alleles are not phenotypically equivalent nor selectively neutral; rather, they represent a repertoire encoding nonidentical biological activities on which selection may act. Long-term evolution is shaped by selective forces acting on biologically distinguishable genomes.

In contrast, according to the *neutralist* view, most of the observed variation is phenotypically neutral or nearly neutral, and long-term evolution at the molecular level is mainly the result of random drift and random fixation of these neutral or nearly neutral alleles. Advantageous mutations are considered rare, and most mutants that are not selectively neutral are considered deleterious (129). Thus, they are eliminated from the population, contributing neither to polymorphism nor to evolution. This theory places an important weight on high mutational input in maintaining polymorphisms in natural populations (130). This premise is amply fulfilled by

RNA viruses, with mutation rates and frequencies and rates of accumulation of mutations exceeding by many orders of magnitude those of higher organisms (see Table 2 and preceding sections).

The selectionist versus neutralist controversy has dominated much of molecular evolution for two decades (118,125–133). Recently, RNA viruses have been recognized as excellent systems to study basic problems of molecular evolution. However, the connection between the quasispecies structure of individual viral populations and the process of long-term genetic diversification of viruses is not obvious. In particular, we must consider whether evolution is driven mainly by selective forces, by random sampling events, or by both.

### Positive Versus Negative Selection

One of the problems of interpretation arises in our view from imprecise distinction between positive Darwinian selection and negative selection (the one operating to eliminate deleterious variants and tending to perpetuate wild-type, prototype genomes). With RNA viruses' mutation frequencies averaging $10^{-3}$ to $10^{-4}$ substitutions per nucleotide (see Table 1), we must consider whether the suppression of all but a limited set of variants is, in fact, not equivalent to a positive-selection event. It has been argued that the most neutral variants dominate quasispecies (103,134), and even that random sampling of such variants may produce a transient molecular clock (103). However, relatively more neutral variants—those that are found when individual genomes are sampled at random from one population—may be far from being selectively equivalent (19,21–23), as emphasized by Coffin (52). Also, it must be stressed that neutrality is a relative term, necessarily referred to one biological environment. Different tissues within an infected individual may often provide disparate metabolic and physicochemical environments, thereby imparting different directions to virus evolution by the differential replication and survival of variants. In a different environment, a previously neutral variant may show low fitness, and another one, previously suppressed at undetectable levels, may become preponderant. Thus, selection is continuously shaping the quasispecies. According to the neutral theory, the rate of evolution per generation is equal to the rate of production of selectively neutral mutants per gamete per generation (118,129). Remarkably, in this formulation, the rate of evolution is independent of environmental conditions. In RNA virus evolution, much of the evidence points to a continuous, modulating role of the environment.

### Chance

That quasispecies are shaped by competitive selection does not necessarily imply that evolution also is. The contribution of chance in RNA virus evolution must be considered. In terms of the neutral theory, this would mean random drift among variants previously rated in the viral quasispecies. One mechanism of genetic drift that is probably a frequent occurrence is the random sampling of genomes during host-to-host transmission. Evidence that occasionally transmission of a virus involves one or a few infectious particles has been previously reviewed (29). Little is known of the random-sampling events that may take place within an infected individual (as in initiation of the viremic state in systemic infections, or in colonization of secondary replication sites, generally). Wain-Hobson has rightly pointed out that HIV variants may be amplified by the activation of latently infected lymphocytes by appropriate antigen, and that this amplification has nothing to do with fitness of the particular HIV variant (135). This is an example of a probabilistic event that does not deny the validity of quasispecies as a descriptor of RNA virus populations, and points out that quasispecies are often perturbed by a variety of chance events.

Both, within individuals and among individuals during transmission, sampling of viruses is independent of the virus population size. An individual may be infected by as many as $10^{12}$ particles, whereas a particular tissue may contain as few as $10^2$ or fewer (compare refs. 29 and 102 for experimental data). A randomly sampled particle may originate new rounds of genome amplification and new quasispecies distributions (103). Obviously, the situation is very different from that of sexually reproducing species in which the smaller the population size the more significant are the sampling errors (136). Nee and Maynard Smith (137) have suggested that transmission and colonization of a new host (they call this event *host-jumping*) might be a strategy to maintain the identity of a wild-type virus, even if the mutation rate within a host was close to (or temporarily above) the error threshold (8–10).

It could be argued that the foregoing description of RNA virus replication and evolution is a description of fluctuations among neutral or nearly neutral variants of the main prototypes. However, it has been selection (either negative or positive) that has shaped the main prototypes. Random sampling of the most neutral variants operates when selection has already constructed the raw material on which stochastic processes play their role. In the words of Kimura (118) in RNA virus evolution it is not exactly the "luckiest" that perpetuates itself, it is the "selected luckiest" that does.

### Synonymous Versus Nonsynonymous Replacements

The higher abundance of synonymous (silent) mutations versus nonsynonymous (amino acid altering) mutations has been used as an argument in support of the neutral theory (118,128,129,132). The rate of accumulation of synonymous substitutions is about fivefold higher than the rate for nonsynonymous substitutions, both for mammalian genes (138) and for influenza virus genes (139). It is often assumed that silent replacements have little, if any, phenotypic effect and, thus, are not subject to negative selection (reviewed in 128). However, there are several reports that show a preponderance of nonsynonymous substitutions during RNA virus or retrovirus evolution (98,102,108,140). In our view, the problem stems from considering all silent replacements as phenotypically neutral. Higher-order structures of

RNA often play functional roles. Examples are the various secondary structure domains at the internal ribosome entry site (or landing pad) of picornaviruses (141, 142) or the conservation of pseudoknots in the RNA of several plant and animal viruses (143,144). There is evidence that secondary structure constraints have imposed limitations for the evolution of RNA genetic elements (145,146). Thus, mutations that alter such higher-order structures—irrespective of their effect on the first, second, or third base position in codons—must be subjected to negative-selection. Structural and regulatory elements within the genomic RNA are as much a part of the phenotype as are viral protein-encoding segments. Negative-selection may act to eliminate synonymous as well as nonsynonymous substitutions. The dominant targets of negative-selection (RNA or protein domains) may depend on the environmental conditions being imposed on virus replication (type of cell, type of infection—acute or persistent, modulated by other RNAs—access of virus to antibodies and to other inhibitory molecules, and so on). The black box in RNA virus evolution is our understanding of selective constraints (52). Until we learn more about them, it will be difficult to interpret varying ratios of synonymous versus nonsynonymous replacements.

Given the available evidence, we suggest that both selection and random-sampling events probably contribute substantially in shaping the genetic composition of RNA virus populations. Currently, there is no evidence to suggest that one mechanism contributes disproportionately more than the other toward evolutionary diversification.

## SOME IMPLICATIONS OF QUASISPECIES POPULATION STRUCTURE AND RAPID RNA GENOME EVOLUTION

We briefly consider in the following discussion the implications of RNA virus population heterogeneity for viral pathogenesis, viral disease control, emergence of new pathogenic viruses, and for the general evolution of the biosphere, mainly the effect on differentiated organisms and on abundant microbial life forms.

The possible involvement of slow, persistent, and inapparent and latent virus infections in human autoimmune and degenerative disease has been discussed by Holland et al. (30), who emphasized the contribution that rare mutants from the mutant spectrum (present in any RNA virus population) could have in the establishment of disease. Some individuals may manifest "blindspots" or "immunocyte failure to control certain specific mutant virus types out of countless mutants arising during persistence" (30). Many degenerative diseases of humans (multiple sclerosis, presenile and senile dementia, amyotrophic lateral sclerosis, lupus erythematosus, Paget disease, Parkinson disease and others) remain highly enigmatic, and a possible viral etiology has neither been proved nor excluded. However, the association of the fatal brain disease subacute sclerosing panencephalitis (SSPE) with variant forms of measles virus has now been proved by epidemiological and virological evidence (76,147). There is still uncertainty concerning clearance of virus by the immune response and the survival of particular virus variants in each infected individual (14,30,95,135). Many reports, too numerous to be discussed here, have documented differences in biological properties among isolates of the same virus (reviewed in various chapters of ref. 148). Interestingly, the extent of a genetic lesion does not necessarily correlate with the magnitude of the phenotypic effect. Point mutations may have profound effects on virulence on cell tropism, whereas some block additions or deletions may have apparently unnoticeable consequences. This is highly relevant when considering the biological implications of quasi-species, since, in the error-prone replication of RNA viruses, point mutants are produced at high rates. A viral quasispecies is an extensive pool of deviant phenotypes, a number of which might exhibit detectable effects on pathogenesis were they to rise to dominance in the quasispecies populations.

We have recently reviewed, in detail, several complications of RNA virus heterogeneity for the control of viral disease by vaccination and by antiviral drug therapy (60,149). Only the main conclusions of our previous description are outlined here. Vaccines, either classic (whole virus) or synthetic, must include an equilibrated, broad repertoire of independent B- and T-cell epitopes. This is needed, not only to evoke an effective, protective response, but to avoid immunopathological sequelae of vaccination. Also, such a complete response will minimize the selection of variant viruses resistant to the induced response. Synthetic vaccines have potential danger for presenting the immune system with too limited a number of epitopes. Obviously, live-attenuated vaccines should not be administered to immunocompromised patients, nor should vaccines consisting of a retrovirus capable of inserting into the human chromosome, at least not without careful consideration of the risks involved. Excluding such obvious examples, as a general rule, whole-virus vaccines (attenuated or inactivated) are probably preferable to engineered empty capsids (which may be antigenically distinct from complete virions), and the latter, in turn, are preferable to complex multiepitopic proteins or peptides. For obvious reasons, vaccines that are based on a single protein or peptide are the most likely to be ineffective or to pose postvaccination problems. These are not just theoretical considerations, since vaccine-escape mutants of several viruses have been described (60, 149–152). The RNA virus populations are antigenically heterogeneous, and one or a few amino acid substitutions may trigger large antigenic shifts (153,154). The literature of recent years contains many reports of the success of synthetic vaccines in laboratory and field trials. Most of these results have not yet undergone widespread, epidemiologically more significant application. Ineffective vaccines might, in fact, promote virus variation; accordingly, new vaccines should be used with great care.

Parallel arguments apply to antiviral drug therapy (60,149). Combination therapy with multiple drugs, directed to independent viral loci, should be used whenever possible. To minimize the chance of selecting drug-resistant virus mutants, administration of a single drug (with a single target specificity) should be avoided. Computer-assisted modeling of antiviral compounds is problematic when applied to RNA virus quasispecies. Many invariant residues on an RNA virus sequence (i.e., strictly conserved when comparing sequences of many natural isolates of the virus) may, nevertheless, be mutated in minority genomes of the quasispecies. Thus, a drug designed to inhibit virus multiplication by interaction with such residues may, in fact, select for variant viruses resistant to the drug. It is also quite obvious from the preceding discussions that attempts to control chronic viral infections with an MAb should be avoided. Even if, in some cases, the lowering of virus levels may aid in clearing the infection by the immune response, the risk of selecting variants might often be high; damage could follow for the individual as well as for the community at large. We have previously suggested some general guidelines for the implementation of international vaccination programs and for the update of vaccines and antiviral agents (149). The reader is referred to that article for additional quantitative arguments and further discussion of this topic.

There is a great deal of interest and concern in the emergence of "new" viral pathogens. Not only acquired immunodeficiency syndrome (AIDS), but a number of new human and animal diseases have been associated with viruses, in particular, RNA viruses and retroviruses (30,60,155–158) (see also Chapter 16 by Morse). The great adaptability of viral quasispecies facilitates virus exploitation of new ecological niches. This must be favored by changes in human social habits, such as the great expansion of urban populations and migrations that increase contacts among many individuals. Population growth and evolving social organization increase livestock trade, again favoring opportunities for viruses to find potentially new hosts. In addition, human modification of the environment, such as deforestation, introduction of various pollutants (which alter microenvironments in specific ways), and global climatic changes, among others, may affect survival of viruses and their vectors. These influences could cause population disequilibrium and accelerated virus evolution. Large evolutionary jumps owing to RNA recombination (see preceding sections), combined with the fine-tuning provided by point mutations, have probably contributed to the emergence of new viral pathogens more frequently than originally thought. As pointed out by Temin, HIV and the AIDS epidemic "are different, but not unique" (20).

The evolution of unicellular as well as multicellular organisms is influenced by their interactions with rapidly evolving RNA genetic elements (1,5–7,30). A number of cell culture systems persistently infected with viruses have exhibited coevolution of cells and resident virus (159–163). Thus, cellular genetic makeup can be modified, not only by viruses able to integrate into the cellular chromosome, but also by viruses for which there is no evidence of integration. In these circumstances, viruses can act as selective forces to mediate differential growth of genetically altered cell types. de la Torre and colleagues identified at least six distinct cell phenotypes among stable cell clones (free of detectable virus) that they derived from a BHK-21 cell line persistently infected with foot-and-mouth disease virus (164). Even though the generation and selection of new cell genotypes is likely facilitated by the partial transformation state of many cultured cell lines—with their intrinsic genetic instability (7)—it is tempting to speculate that very early during the evolution of life on earth interactions of cells with viruses represented a driving force for cellular modification and differentiation. The stimuli for cellular modifications must have been very potent in view of the likely quasispecies organization of primitive replicons (8,9), irrespective of their having originated independently of the initial cells or from early cells.

Death and debilitation caused by infectious diseases clearly are modulators of populations of differentiated organisms (30). Rabies regulates fox populations, with 3- to 5-year oscillations in population density and disease prevalence. These effects are obviously not confined to RNA viruses. Myxoma virus (a *Leporipoxvirus*) was introduced into Australia in 1950 to help control a rabbit pest. The viruses that initially devastated the rabbit population became increasingly attenuated. Concomitantly, rabbits genetically resistant to these attenuated forms of myxoma virus were selected; thereby, coevolution was observed between a complex virus and its animal host organism.

Thus, RNA replicons, both in the form of complete, autonomous infectious virus or defective, subviral genomes, historically may well have exerted important influences in the evolution of the DNA-based biosphere. Today the emergence of new viral pathogens may merely represent a continuation of this ongoing modulating power, and it is to be expected that humans, in future years, will continue to be among the host populations subjected to viral modulation.

## SUMMARY, CONCLUSIONS, AND OUTLOOK

Mutation rates and frequencies for RNA viruses are generally in the range of $10^{-3}$ to $10^{-5}$ substitutions per nucleotide site per round of copying. These values are close to the maximum compatible with maintaining a meaningful genetic information, as suggested by theoretical and experimental studies. Genetic variability is the basis of adaptability and, as such, it is experienced by all known forms of life. However, it is an oversimplification to deny important, unique features of RNA genetics. In higher organisms, once a developmental plan has been expressed and the individual reaches adulthood, most normal biochemical activities are carried out following an essentially fixed program. It is only after prolonged periods that (through reshuffling of chromosomes and new introductions of genetic alterations) new genetic compositions can be tested in terms of their adaptive value. Thus, many years (millions of years in nature) are needed for substantial modifications of these complex life forms. The mechanisms involved in their evolution (selection, random sampling events) must necessarily be inferred indirectly by comparing organisms (at the morphological and molecular level) that have diverged over eons.

In contrast, RNA genetic elements and, in particular, riboviruses and retroviruses, are elemental life forms able to replicate their genomes severalfold in minutes. Viruses can survive in the form of virions undergoing no change, but as such they are inert complex macromolecules exerting no biological activities. Error-prone replication is inherent to the very existence of these molecular parasites. To replicate, they must divert cellular machinery to their activities, often causing a range of effects on their host organisms (from inapparent to severely pathological). DNA viruses may share many features of RNA genetic elements, but determination of mutation rates and frequencies and clonal analyses of population heterogeneities have been too limited for DNA viruses to permit meaningful comparisons.

The RNA virus mutants arise continuously, and are repeatedly subjected to competitive rating. It is this competitive rating, and not the absence of mutation events, that determines rapid evolution versus constancy of the consensus nucleotide sequence. We term population equilibrium as the state in which there is constancy of the consensus (average) sequence in a replicating RNA virus population. Equilibrium will tend to prevail when the initial virus genome is well adapted to the environment in which it continues further amplification. We suspect that, in the near future, there will be several additional viruses for which the distinction between absence of mutation and the rating of variants in equilibrium will be established. The latter is the most likely means of genome conservation in the absence of proofreading or repair mechanisms, the only known means to achieve error rates lower than $10^{-7}$ per nucleotide site. Conservation of average genomic sequences often hides extensive sequence heterogeneity among the individual genomes that compose the population. It is very likely that more detailed analyses of mutant spectra in viral quasispecies will be soon available in view of the progress of techniques for genomic isolation and for nucleotide sequencing.

Quasispecies can be clearly distinguished from genetic polymorphisms in cellular organisms. At most, polymorphisms might be equated with differences among the average sequences of independent viral isolates. Such differences are the result of diversification of viral quasispecies. The long-term evolution of RNA viruses is necessarily a complex process dominated by probabilities and uncertainties. Mutation, hypermutation, recombination, genomic rearrangements and reassortment, all alter RNA virus populations in the course of their replication. Then, on this exceptionally diverse raw material for evolutionary change, negative- as well as positive-selection, random-sampling events, and many additional influences must act. Among the latter, survival of genomes by phenotypic masking and spread of accompanying (nonselected) mutations by "driver" (selected) mutations (on the same or related genomes) may contribute to the diversification of mutant forms. Numerous mutants present as a minority in quasispecies may be unable to manifest their potential until they are selected by an appropriate environment. This is a form of periodic selection, first described for cellular microorganisms. The distinction between negative-selection and positive-Darwinian–selection does not seem obvious when considering the replication of viral quasispecies. In our view, the continuous modulating action of negative-selection together with episodes of positive-selection, and random drift of sequences must all play significant roles in RNA virus evolution.

Finally, in this review, we have discussed several implications of extreme RNA genome heterogeneity for viral pathogenesis, control of viral disease, emergence of new viral pathogens, and for the evolution of their host cells and organisms. Many of these implications are not immediately obvious because we are more familiar (and more comfortable) with classic DNA genetics, involving relatively fixed structures and information. We still lack full understanding of the nuances of RNA genetics, but a combination of further theoretical and experimental studies should provide better insights into RNA viruses and help in the rational design of methods to control the diseases they cause.

## ACKNOWLEDGMENTS

We are indebted to many colleagues for valuable discussions and new information before publication. Work in Madrid supported by DGICYT grant n. PB91-0051-C02-01, EC and Fundación Ramón Areces, and in La Jolla by NIH grant AI 14627.

## REFERENCES

1. Baltimore D. Retroviruses and retrotransposons: the role of reverse transcription in shaping the eukaryotic genome. *Cell* 1985;40:481–482.
2. Li WH, Wu CI, Luo CC. Nonrandomness of point mutation as reflected in nucleotide substitutions in pseudogenes and its evolutionary implications. *J Mol Evol* 1984;21:58–71.
3. Wabl M, Burrows PD, von Gabain A, Steinberg C. Hypermutation at the immunoglobulin heavy chain locus in a pre-B cell line. *Proc Natl Acad Sci USA* 1985;82:479–483.
4. Berek C, Milstein C. The dynamic nature of the antibody repertoire. *Immunol Rev* 1988;105:5–25.
5. Erwin DH, Valentine JW. "Hopeful monsters," transposons and Metazoan radiation. *Proc Natl Acad Sci USA* 1984;81:5482–5483.
6. Agur Z, Kerszberg M. The emergence of phenotypic novelties through progressive genetic change. *Am Nat* 1987;129:862–875.
7. Nicolson GL. Tumor cell instability, diversification, and progression to the metastatic phenotype: from oncogene to oncofetal expression. *Cancer Res* 1987;47:1473–1487.
8. Eigen M, Schuster P. *The hypercycle. A principle of natural self-organization.* Berlin: Springer-Verlag; 1979.
9. Swetina J, Schuster P. Self-replication with errors. A model for polynucleotide replication. *Biophys Chem* 1982;16:329–345.
10. Eigen M, Biebricher C. Sequence space and quasispecies distribution. In: Domingo E, Holland JJ, Ahlquist P, eds. *RNA genetics*, vol 3. Boca Raton: CRC Press; 1988:211–245.
11. Drake JW. Spontaneous mutation. *Nature* 1969;221:1128–1132.
12. Loeb LA, Kunkel TA. Fidelity of DNA synthesis. *Annu Rev Biochem* 1982;51:429–457.
13. Fry M, Loeb LA. *Animal cell DNA polymerases.* Boca Raton: CRC Press; 1986.
14. Holland JJ, de la Torre JC, Steinhauer DA. RNA virus populations as quasispecies. *Curr Top Microbiol Immunol* 1992;176:1–20.
15. Steinhauer DA, Domingo E, Holland JJ. Lack of evidence for proofreading mechanisms associated with an RNA virus polymerase. *Gene* 1992;122:281–288.
16. Holland JJ, Domingo E, de la Torre JC, Steinhauer DA. Mutation frequencies at defined single codon sites in vesicular stomatitis virus and poliovirus can be increased only slightly by chemical mutagenesis. *J Virol* 1990;64:3960–3962.
17. Smith DB, Inglis SC. The mutation rate and variability of eukaryotic viruses: an analytical review. *J Gen Virol* 1987;68:2729–2740.
18. Parrish CR, Aquadro CF. Strassheim ML, Evermann JF, Sgro JY, Mohammed HO. Rapid antigenic-type replacement and DNA sequence evolution of canine parvovirus. *J Virol* 1991;65:6544–6552.
19. Domingo E, Sabo D, Taniguchi T, Weissmann C. Nucleotide sequence heterogeneity of an RNA phage population. *Cell* 1978;13:735–744.
20. Temin H. Is HIV unique or merely different? *J AIDS* 1989;2:1–9.

21. Holland JJ, de la Torre JC, Clarke DK, Duarte E. Quantitation of relative fitness and great adaptability of clonal populations of RNA viruses. *J Virol* 1991;65:2960–2967.
22. Gonzalez MJ, Saiz JC, Laor O, Moore DM. Antigenic stability of foot-and-mouth disease virus variants on serial passage in cell culture. *J Virol* 1991;65:3949–3953.
23. Martínez MA, Carrillo C, González-Candelas F, Moya A, Domingo E, Sobrino F. Fitness alteration of foot-and-mouth disease virus mutants: measurement of adaptability of viral quasispecies. *J Virol* 1991;65:3954–3957.
24. Zibert A, Maass G, Strebel K, Falk MM, Beck E. Infectious foot-and-mouth disease virus derived from a cloned full-length cDNA. *J Virol* 1990;64:2467–2473.
25. de la Torre JC, Giachetti C, Semler BL, Holland JJ. High frequency of single base transitions, and extreme frequency of precise, multiple-base reversion mutations in poliovirus. *Proc Natl Acad Sci USA* 1992;89:2531–2535.
26. Steinhauer DA, de la Torre JC, Meier E, Holland JJ. Extreme heterogeneity in populations of vesicular stomatitis virus. *J Virol* 1989;63:2072–2080.
27. Domingo E, Dávila M, Ortín J. Nucleotide sequence heterogeneity of the RNA from a natural population of foot-and-mouth disease virus. *Gene* 1980;11:333–346.
28. Ortín J, Nájera R, López-Galíndez C, Dávila M, Domingo E. Genetic variability of Hong Kong (H3N2) influenza viruses. Spontaneous mutations and their location in the viral genome. *Gene* 1980;11:319–331.
29. Domingo E, Martínez-Salas E, Sobrino F, et al. The quasispecies (extremely heterogeneous) nature of viral RNA genome populations: biological relevance—a review. *Gene* 1985;40:1–8.
30. Holland JJ, Spindler K, Horodyski F, Grabau E, Nichol S, VandePol S. Rapid evolution of RNA genomes. *Science* 1982;215:1577–1585.
31. Steinhauer DA, Holland JJ. Rapid evolution of RNA viruses. *Annu Rev Microbiol* 1987;41:409–433.
32. Domingo E, Holland JJ. High error rates, population equilibrium, and evolution of RNA replication systems. In: Domingo E, Holland JJ, Ahlquist P, eds. *RNA genetics*, vol 3. Boca Raton: CRC Press; 1988:3–36.
33. Batschelet E, Domingo E, Weissmann C. The proportion of revertant and mutant phage in a growing population as a function of mutation and growth rate. *Gene* 1976;1:27–32.
34. Steinhauer DA, Holland JJ. Direct method for quantification of extreme polymerase error frequencies at selected single base sites in viral RNA. *J Virol* 1986;57:219–228.
35. Steinhauer DA, de la Torre JC, Holland JJ. High nucleotide substitution error frequencies in clonal pools of vesicular stomatitis virus. *J Virol* 1989;63:2063–2071.
36. White BT, McGeoch DJ. Isolation and characterization of conditional lethal amber nonsense mutants of vesicular stomatitis virus. *J Gen Virol* 1987;68:3033–3044.
37. Holland JJ, de la Torre JC, Steinhauer DA, Clarke D, Duarte E, Domingo E. Virus mutation frequencies can be greatly underestimated by monoclonal antibody neutralization of virions. *J Virol* 1989;63:5030–5036.
38. Parvin JD, Moscona A, Pan WT, Leider JM, Palese P. Measurement of the mutation rates of animal viruses: influenza A virus and poliovirus type 1. *J Virol* 1986;59:377–383.
39. Sedivy JM, Capone JP, Raj Bhandary UL, Sharp PA. An inducible mammalian amber suppressor: propagation of a poliovirus mutant. *Cell* 1987;50:379–389.
40. de la Torre JC, Wimmer E, Holland JJ. Very high frequency of reversion of guanidine resistance in clonal pools of guanidine-dependent type I poliovirus. *J Virol* 1990;64:664–671.
41. Ward CD, Flanegan JB. Determination of the poliovirus RNA polymerase error frequency at eight sites in the viral genome. *J Virol* 1992;66:3784–3793.
42. Emini EA, Jameson BA, Lewis AJ, Larsen GR, Wimmer E. Poliovirus neutralization epitopes: analysis and localization with neutralizing monoclonal antibodies. *J Virol* 1982;43:997–1005.
43. Minor PD, Schild GC, Bootman J, et al. Location and primary structure of a major antigenic site for poliovirus neutralization. *Nature* 1983;301:674–679.
44. Minor PD, Ferguson M, Evans DMA, Almond JW, Icenogle JP. Antigenic structure of polioviruses 1, 2 and 3. *J Gen Virol* 1986;67:1283–1291.
45. Boege U, Kobasa D, Onodera S, Parks GD. Palmenberg AC, Scraba DG. Characterization of Mengo virus neutralization epitopes. *Virology* 1991;181:1–13.
46. Sherry B, Mosser AG, Colonno RJ, Rueckert RR. Use of monoclonal antibodies to identify four neutralization immunogenes on a common cold picornavirus, human rhinovirus 14. *J Virol* 1986;57:246–257.
47. Durbin RK, Stollar V. Sequence analysis of the E2 gene of a hyperglycosylated, host restricted

mutant of Sindbis virus and estimation of mutation rate from frequency of revertants. *Virology* 1986;154:135–143.

48. Hahn YS, Strauss EG, Strauss JH. Mapping of RNA temperature-sensitive mutants of Sindbis virus: assignment of complementation groups A, B, and G to nonstructural proteins. *J Virol* 1989; 63:3142–3150.

49. Hahn CS, Rice CM, Strauss EG, Lenches EM, Strauss JH. Sindbis virus *ts* 103 has a mutation in glycoprotein E2 that leads to defective assembly of virions. *J Virol* 1989;63:3459–3465.

50. Stec DS, Waddell A, Schmaljon CS, Cole GA, Schmaljohn AL. Antibody-selected variation and reversion in Sindbis virus neutralization epitopes. *J Virol* 1986;57:715–720.

51. Coffin JM, Tsichlis PN, Barker CS, Voynow S. Variation in avian retrovirus genomes. *Ann NY Acad Sci* 1980;354:410–425.

52. Coffin JM. Genetic variation in retroviruses. In: Kurstak E, Marusyk RG, Murphy FA, Van Regenmortel MHV, eds. *Applied virology research*, vol 2. New York: Plenum Publishing; 1990:11–33.

53. Leider JM, Palese P, Smith FI. Determination of the mutation rate of a retrovirus. *J Virol* 1988;62:3084–3091.

54. Dougherty JP, Temin HM. High mutation rate of a spleen necrosis virus-based retrovirus vector. *Mol Cell Biol* 1987;6:4387–4395.

55. Dougherty JP, Temin HM. Determination of the rate of base-pair substitution and insertion mutation in retrovirus replication. *J Virol* 1988;62:2817–2822.

56. Monk RJ, Malik FG, Stokesberry D, Evans LH. Direct determination of the point mutation rate of a murine retrovirus. *J Virol* 1992;66:3683–3689.

57. Varela-Echavarria A, Garvey N, Preston BD, Dougherty JP. Comparison of Moloney murine leukemia virus mutation rate with the fidelity of its reverse transcriptase in vitro. *J Biol Chem* 1992;267:24681–24688.

58. Yewdell JW, Gerhard W. Antigenic characterization of viruses by monoclonal antibodies. *Annu Rev Microbiol* 1981;35:185–206.

59. Valcarcel J, Ortín J. Phenotypic hiding: the carry-over of mutations in RNA viruses as shown by detection of *mar* mutants in influenza virus. *J Virol* 1989;63:4107–4109.

60. Domingo E. RNA virus evolution and the control of viral disease. *Prog Drug Res* 1989;33:93–133.

61. Domingo E, Escarmís C. Martínez-Salas E, Martín Hernández AM, Mateu MG, Martínez MA. Picornavirus variation. In: Carrasco L, Sonenberg N, Wimmer E, eds. *Regulation of gene expression in animal viruses*. New York: Plenum Publishing; 1993:255–281.

62. Drake JW. Rates of spontaneous mutation among RNA viruses. *Proc Natl Acad Sci USA* 1993;90:4171–4175.

63. Pathak VK, Temin HM. 5-Azacytidine and RNA secondary structure increase the retrovirus mutation rate. *J Virol* 1992;66:3093–3100.

64. Ward CD, Stokes MAM, Flanegan JB. Direct measurement of the poliovirus RNA polymerase error frequency in vitro. *J Virol* 1988;62:558–562.

65. Roberts JD, Bebenek K, Kunkel TA. The accuracy of reverse transcriptase from HIV-1. *Science* 1988;242:1171–1173.

66. Preston BD, Poiesz BJ, Loeb LA. Fidelity of HIV-1 reverse transcriptase. *Science* 1988;242:1168–1171.

67. Roberts JD, Preston BD, Johnston LA, Soni A, Loeb LA, Kunkel T. Fidelity of two retroviral reverse transcriptases during DNA-dependent DNA synthesis in vitro. *Mol Cell Biol* 1989;9:469–476.

68. Ricchetti M, Buc H. Reverse transcriptases and genomic variability: the accuracy of DNA replication is enzyme specific and sequence dependent. *EMBO J* 1990;9:1583–1593.

69. Williams KJ, Loeb LA. Retroviral reverse transcriptases: Error frequencies and mutagenesis. *Curr Top Microbiol Immunol* 1992;176:165–180.

70. Chao L. Fitness of RNA virus decreased by Muller's ratchet. *Nature* 1990;348:454–455.

71. Duarte E, Clarke D. Moya A, Domingo E, Holland JJ. Rapid fitness losses in mammalian RNA virus clones due to Muller's ratchet. *Proc Natl Acad Sci USA* 1992;89:6015–6019.

72. Clarke DK, Duarte EA, Moya A. Elena SF, Domingo E, Hollad JJ. Genetic bottlenecks and population passages cause profound fitness differences in RNA viruses. *J Virol* 1993;67:222–228.

73. Domingo E, Holland JJ, Biebricher C, Eigen M. Quasispecies: the concept and the word. In: Gibbs A, Calisher C, García-Arenal F, eds. *Molecular evolution of viruses*. Cambridge: Cambridge University Press; 1993: [in press].

74. de la Torre JC, Holland JJ. RNA virus quasispecies can suppress vastly superior mutant progeny. *J Virol* 1990;64:6278–6281.

75. O'Hara PJ, Nichol ST, Horodyski FM, Holland JJ. Vesicular stomatitis virus defective interfering particles can contain extensive genomic sequence rearrangements and base substitutions. *Cell* 1984;36:915–924.

76. Cattaneo R, Schmid A, Eschle D, Baczko K, ter Meulen V, Billeter M. Biased hypermutation and other genetic changes in defective measles virus in human brain infections. *Cell* 1988;55:255–265.

77. Pathak VK, Temin HM. Broad spectrum of in vivo forward mutations, hypermutations, and mutational hotspots in a retroviral shuttlevector after a single replication cycle: substitutions, frameshifts, and hypermutations. *Proc Natl Acad Sci USA* 1990;87:6019–6023.

78. Vartanian JP, Meyerhans A, Asjö B, Wain-Hobson S. Selection, recombination and G→A hypermutation of human immunodeficiency virus type 1 genomes. *J Virol* 1991;65:1779–1788.

79. King A. Genetic recombination in positive strand RNA viruses. In: Domingo E, Holland JJ, Ahlquist P, eds. *RNA genetics*, vol 2. Boca Raton: CRC Press; 1988:149–165.

80. Lai MMC. Genetic recombination in RNA viruses. *Curr Top Microbiol Immunol* 1992;176:21–32.

81. Monroe SS, Schlesinger S. RNAs from two independently isolated defective interfering particles of Sindbis virus contain cellular tRNA sequences at their 5′ ends. *Proc Natl Acad Sci USA* 1983; 80:3279–3283.

82. Meyers G, Rumenapf T, Thiel HJ. Ubiquitin in a togavirus. *Nature* 1989;341:491.

83. Zhang J, Temin HM. Rate and mechanism of nonhomologous recombination during a single cycle of retroviral replication. *Science* 1993;259:234–238.

84. Zimmern D. Evolution of RNA viruses. In: Domingo E, Holland JJ, Ahlquist P, eds. *RNA genetics*, vol 2. Boca Raton: CRC Press; 1988:211–240.

85. Kirkegaard K, Baltimore D. The mechanism of RNA recombination in poliovirus. *Cell* 1986; 47:433–443.

86. Holland JJ. Defective viral genomes. In: Fields BN, Knipe DM, Chanock RM, Hirsch MS, Melnick JL, Monath TP, Roizman B, eds. *Virology*, vol 1. New York: Raven Press; 1990;151–165.

87. Roux L, Simon AE, Holland JJ. Effect of defective interfering viruses on virus replication and pathogenesis in vitro and in vivo. *Ad Virus Res* 1991;40:181–211.

88. Webster RG, Laver WG, Air GM, Schild GC. Molecular mechanisms of variation in influenza viruses. *Nature* 1982;296:115–121.

89. Smith FI, Palese P. Influenza viruses: high rate of mutation and evolution. In: Domingo E, Holland JJ, Ahlquist P, eds. *RNA genetics*, vol 3. Boca Raton; CRC Press; 1988:123–135.

90. Hundley F, Biryahwalo B, Gow M, Desselberger U. Genome rearrangement of bovine rotavirus after serial passage at high multiplicity of infection. *Virology* 1985;143:88–103.

91. Allen AM, Desselberger U. Reassortment of human rotaviruses carrying rearranged genomes with bovine rotavirus. *J Gen Virol* 1985;66:2703–2714.

92. Huang AS, Baltimore D. Defective viral particles and viral disease processes. *Nature* 1970;226: 325–327.

93. Kaper JM, Collmer CW. Modulation of viral plant diseases by secondary RNA agents. In: Domingo E, Holland JJ, Ahlquist P, eds. *RNA genetics*, vol 3. Boca Raton: CRC Press; 1988:171–194.

94. Huang AS. Modulation of viral disease processes by defective interfering particles. In: Domingo E, Holland JJ, Ahlquist P, eds. *RNA genetics*, vol 3. Boca Raton: CRC Press; 1988:195–208.

95. Meyerhans A, Cheynier R, Albert J, Seth M, Kwok S, Sninsky J, Morfeldt-Manson L, Asjö B, Wain Hobson S. Temporal fluctuations in HIV quasispecies in vivo are not reflected by sequential HIV isolations. *Cell* 1989;58:901–910.

96. Martell M, Esteban JI, Quer J, Genescà J, Weiner A, Esteban R, Guardia J, Gómez J. Hepatitis C virus (HCV) circulates as a population of different but closely related genomes: quasispecies nature of HCV-genome distribution. *J Virol* 1992;66:3225–3229.

97. Horodyski F, Holland JJ. Properties of DI particle-resistant mutants of vesicular stomatitis virus isolated from persistent infection and from undiluted passages. *Cell* 1983;33:801–810.

98. Rocha E, Cox NJ, Black RA, Harmon MW, Harrison CJ, Kendal AP. Antigenic and genetic variation in influenza A (H1N1) virus isolates recovered from a persistently infected immunodeficient child. *J Virol* 1991;65:2340–2350.

99. Ito T, Gorman OT, Kawaoka Y, Bean WJ, Webster RG. Evolutionary analysis of the influenza A virus *M* gene with comparison of the M1 and M2 proteins. *J Virol* 1991;65:5491–5498.

100. Altmuller A, Fitch WM, Scholtissek C. Biological and genetic evolution of the nucleoprotein gene of human influenza A viruses. *J Gen Virol* 1989;70:2111–2119.

101. Gorman OT, Bean WJ, Kawaoka Y, Webster RG. Evolution of the nucleoprotein gene of influenza A virus. *J Virol* 1990;64:1487–1497.

102. Gebauer F, de la Torre JC, Gomes I, Mateu MG, Barahona H, Tiraboschi B, Bergmann I, Augé de Mello P, Domingo E. Rapid selection of genetic and antigenic variants of foot-and-mouth disease virus during persistence in cattle. *J Virol* 1988;62:2041–2049.
103. Villaverde A, Martínez MA, Sobrino F, Dopazo J, Moya A, Domingo E. Fixation of mutations at the *VP1* gene of foot-and-mouth disease virus. Can quasispecies define a transient molecular clock? *Gene* 1991;103:147–153.
104. Martínez MA, Dopazo J, Hernández J, Mateu MG, Sobrino F, Domingo E, Knowles NJ. Evolution of the capsid protein genes of foot-and-mouth disease virus over six decades. Antigenic variation without accumulation of amino acid substitutions over six decades. *J Virol* 1992;66:3557–3565.
105. Kew OM, Nottay BK, Rico-Hesse R, Pallansch M. Molecular epidemiology of wild poliovirus transmission. In: Kurstak E, Marusyk RG, Murphy FA, Van Regenmortel MHV, eds. *Applied virology research*, vol 2. New York: Plenum Publishing; 1990:199–221.
106. Weaver S, Scot TW, Rico-Hesse R. Molecular evolution of eastern equine encephalomyelitis virus in North America. *Virology* 1991;182:774–784.
107. Hahn BH, Shaw GM, Taylor ME, Redfield RR, Markham PD, Salahuddin SZ, Wong-Staal F, Gallo RC, Parks ES, Parks WP. Genetic variation in HTLV-III/LAV over time in patients with AIDS or at risk for AIDS. *Science* 1986;232:1548–1553.
108. Burns DPW, Desrosiers RC. Selection of genetic variants of simian immunodeficiency virus in persistently infected rhesus monkeys. *J Virol* 1991;65:1843–1854.
109. Clements JE, Gdovin SL, Montelaro RC, Narayan O. Antigenic variation in lentiviral diseases. *Annu Rev Immunol* 1988;6:139–159.
110. Britten RJ. Rates of DNA sequence evolution differ between taxonomic groups. *Science* 1986;231:1393–1398.
111. Weissmann C, Weber H. The interferon genes. *Prog Nucleic Acids Res Molec Biol* 1986;33:251–301.
112. Bosch V, Kuhn C, Schaller H. Hepatitis B virus replication. In: Domingo E, Holland JJ, Ahlquist P, eds. *RNA genetics*. vol 2. Boca Raton: CRC Press; 1988:43–58.
113. Orito E, Mizokami M, Ina Y, Moriyama EN, Kameshima N, Yamamoto M, Gojobori T. Host-independent evolution and a genetic classification of the hepadnavirus family based on nucleotide sequences. *Proc Natl Acad Sci USA* 1989;86:7059–7062.
114. Crawford AM, Zelazny B. Evolution in *Oryctes Baculovirus*: rate and types of genomic change. *Virology* 1990;174:294–298.
115. Kool M, Voncken JW, Van Lier FLJ, Tramper J, Vlak JM. Detection and analysis of *Autographa californica* nuclear polyhedrosis virus mutants with defective interfering properties. *Virology* 1991;183:739–746.
116. Yañez R, Moya A, Viñuela E, Domingo E. Repetitive nucleotide sequencing of a dispensable DNA segment in a clonal population of African swine fever virus. *Virus Res* 1991;20:265–272.
117. Gojobori T, Yokoyama S. Rates of evolution of the retroviral oncogene of Maloney murine sarcoma virus and of its cellular homologues. *Proc Natl Acad Sci USA* 1985;82:4198–4201.
118. Kimura M. The neutral theory of molecular evolution and the world view of the neutralists. *Genome* 1989;31:24–31.
119. Air GM, Gibbs AJ, Laver WG, Webster RG. Evolutionary changes in influenza B are not primarily governed by antibody selection. *Proc Natl Acad Sci USA* 1990;87:3884–3888.
120. Gojobori T, Moriyama EN, Kimura M. Molecular clock of viral evolution and the neutral theory. *Proc Natl Acad Sci USA* 1990;87:10015–10018.
121. Fitch WM, Leister JME, Li X, Palese P. Positive Darwinian evolution in human influenza A viruses. *Proc Natl Acad Sci USA* 1991;88:4270–4274.
122. Hubby JL, Lewontin RC. A molecular approach to the study of genetic heterozygosity in natural populations. I. The number of alleles at different loci in *Drosophila pseudoobscura*. *Genetics* 1966;54:577–594.
123. Lewontin RC, Hubby JL. A molecular approach to the study of genetic heterozygosity in natural populations. II. Amount of variation and degree of heterozygosity in natural populations of *Drosophila pseudoobscura*. *Genetics* 1966;54:595–609.
124. Harris H. Enzyme polymorphisms in man. *Proc R Soc Lond [B]* 1966;164:298–310.
125. Lewontin RC. *The genetic basis of evolutionary change*. New York: Columbia University Press; 1974.
126. Lewontin RC. Population genetics. *Annu Rev Genet* 1985;19:81–102.

127. Ayala FJ. *Molecular evolution*. Sunderland, Mass: Sinauer Associates; 1976.
128. Nei M. *Molecular evolutionary genetics*. New York: Columbia University Press; 1987.
129. Kimura M. *The neutral theory of molecular evolution*. Cambridge: Cambridge University Press; 1983.
130. Ohta T. Mutational pressure as the main cause of molecular evolution and polymorphism. *Nature* 1974;252:351–354.
131. Kimura M. Evolutionary rate at the molecular level. *Nature* 1968;217:624–626.
132. King JL, Jukes TH. Non-Darwinian evolution. *Science* 1969;164:788–798.
133. Crow JF. Population genetics history: a personal view. *Annu Rev Genet* 1987;21:1–22.
134. Domingo E, Escarmis C, Martínez MA, Martínez-Salas E, Mateu MG. Foot-and-mouth disease virus populations are quasispecies. *Curr Top Microbiol Immunol* 1992;176:33–47.
135. Wain-Hobson S. Human immunodeficiency virus type 1 quasispecies in vivo and ex vivo. *Curr Top Microbiol Immunol* 1992;176:181–193.
136. Dobzhansky T, Ayala FJ, Stebbins GL, Valentine JW. *Evolution*. San Francisco: WH Freeman & Co.; 1977.
137. Nee S, Maynard Smith J. The evolutionary biology of molecular parasites. *Parasitology* 1990; 100:S5–S18.
138. Li WH, Wu CI, Luo CC. A new method for estimating synonymous and nonsynonymous rates of nucleotide substitution considering the relative likelihood of nucleotide and codon changes. *Mol Biol Evol* 1985;2:150–174.
139. Hayashida HH, Toh H, Kikuno R, Miyata T. Evolution of influenza virus genes. *Mol Biol Evol* 1985;2:289–303.
140. Diéz J, Dávila M, Escarmis C, Mateu MG, Dominguez J, Pérez JJ, Giralt E, Melero JA, Domingo E. Unique amino acid substitutions in the capsid proteins of foot-and-mouth disease virus from a persistent infection in cell culture. *J Virol* 1990;60:5519–5528.
141. Pilipenko EV, Blinov VM, Romanova LI, Sinyakov AN, Maslova SV, Agol VI. Conserved structural domains in the 5'-untranslated region of picornaviral genomes: an analysis of the segment controlling translation and neurovirulence. *Virology* 1989;168:201–209.
142. Pilipenko EV, Blinov VM, Chernov BK, Dmitrieva TM, Agol VI. Conservation of the secondary structure elements of the 5'-untranslated region of cardio- and aphthovirus RNAs. *Nucleic Acids Res* 1989;17:5701–5711.
143. Pleij CWA, Rietveld K, Bosch L. A new principle of RNA folding based on pseudoknotting. *Nucleic Acids Res* 1985;13:1717–1731.
144. van Belkum A, Abrahams JP, Pleij CWA, Bosch L. Five pseudoknots are present at the 204 nucleotides long 3' noncoding region of tobacco mosaic virus RNA. *Nucleic Acid Res* 1985; 13:7673–7686.
145. Yamamoto K, Yoshikura H. Relation between genomic and capsid structures in RNA viruses. *Nucleic Acids Res* 1986;14:389–396.
146. Fraile A, García-Arenal F. Secondary structure as a constraint on the evolution of plant viral satellite RNA. *J Mol Biol* 1991;221:1065–1069.
147. Cattaneo R, Billeter MA. Mutations and A/I hypermutations in measles virus persistent infections. *Curr Top Microbiol Immunol* 1992;176:63–74.
148. Domingo E, Holland JJ, Ahlquist P, eds. *RNA genetics*. Boca Raton: CRC Press; 1988.
149. Domingo E, Holland JJ. Complications of RNA heterogeneity for the engineering of virus vaccines and antiviral agents. In: Setlow JK, ed. *Genetic engineering, principles and methods*, vol 14. New York: Plenum Press; 1992:13–31.
150. Domingo E, Mateu MG. Martínez MA, Dopazo J, Moya A, Sobrino F. Genetic variability and antigenic diversity of foot-and-mouth disease virus. In: Kurstak E, Marusyk RG, Murphy FA, Van Regenmortel MHV, eds. *Applied virology research*, vol 2. New York: Plenum Publishing; 1990:233–266.
151. Martínez MA, Carrillo C, Plana J, Mascarella R, Bergada J, Palma EL, Domingo E, Sobrino F. Genetic and immunogenic variations among closely related isolates of foot-and-mouth disease virus. *Gene* 1988;62:75–84.
152. Carman WF, Zanetti AR, Karayiannis P, Waters J, Manzillo G. Tanzi E, Zuckerman AJ, Thomas H. Vaccine-induced escape mutants of hepatitis B virus. *Lancet* 1990;336:325–329.
153. Mateu MG, Martínez MA, Rocha E, Andreu D, Parejo J, Giralt E, Sobrino F, Domingo E. Implications of a quasispecies genome structure: effect of frequent, naturally occurring amino acid substitutions on the antigenicity of foot-and-mouth disease virus. *Proc Natl Acad Sci USA* 1989; 86:5883–5887.

154. Martínez MA, Hernández J, Piccone ME, Palma EL, Domingo E, Knowles NJ, Mateu MG. Two mechanisms of antigenic diversification of foot-and-mouth disease virus. *Virology* 1991;184:695–706.

155. Lorber B. Changing patterns of infectious diseases. *Am J Med* 1988;84:569–578.

156. Morse SS, Schluederberg A. Emerging viruses: the evolution of viruses and viral diseases. *J Infect Dis* 1990;162:1–7.

157. Morse SS. Emerging viruses: defining the rules for viral traffic. *Perspect Biol Med* 1991;34:87–107.

158. Kilbourne ED. New viruses and new disease: mutation, evolution and ecology. *Curr Opin Immunol* 1991;3:518–524.

159. Takemoto KK, Habel K. Virus–cell relationship in a carrier culture of HeLa cells and Coxsackie A9 virus. *Virology* 1959;7:28–44.

160. Ahmed R, Canning WM, Kaffman RS, Sharpe AH, Hallum JV, Fields BN. Role of the host cell in persistent viral infection: coevolution of L cells and reovirus during persistent infection. *Cell* 1981;25:325–332.

161. Ron D, Tal, J. Coevolution of cells and virus as a mechanism for the persistence of lymphotropic minute virus of mice in L cells. *J Virol* 1985;55:424–430.

162. de la Torre JC, Martínez-Salas E, Díez J, Villaverde A, Gebauer F, Rocha E, Dávila M, Domingo E. Coevolution of cells and viruses in a persistent infection of foot-and-mouth disease virus in cell culture. *J Virol* 1988;62:2050–2058.

163. Borzakian S, Couderc T, Barbier Y, Attal G, Pelletier I, Colbère-Garapin F. Persistent poliovirus infection: establishment and maintenance involve distinct mechanisms. *Virology* 1992;186:398–408.

164. de la Torre JC, Martínez-Salas E, Díez J, Domingo E. Extensive cell heterogeneity during persistent infection with foot-and-mouth disease virus. *J Virol* 1989;63:59–63.

The Evolutionary Biology of Viruses,
edited by Stephen S. Morse.
Raven Press, Ltd., New York © 1994.

# 9

# Is Antigenic Variation of HIV Important for AIDS and What Might Be Expected in the Future?

## Simon Wain-Hobson

*Unité de Rétrovirologie Moléculaire, Institut Pasteur, 75724 Paris, France*

I will presume that the reader of this volume has dipped into other chapters such that the notion of a viral quasispecies (1–4), perhaps novel to many, should now be firmly ensconced in short-term memory cells, at least. What we will discover is that the human immunodeficiency virus (HIV) is rather special. The issues common to all RNA viruses find themselves amplified. It is clear the HIV will be the harbinger in time—probably in my lifetime—of novel human diseases.

To help understand better the rather special position of retroviruses, and particularly HIV, among its fellow RNA viruses, it might be useful to explain briefly how retroviruses replicate as well as how HIV interacts with the immune system—that is to say, how we think it interacts with the immune system.

## POTTED RETROVIROLOGY

Retroviruses represent a diverse group of viruses. They are to be found in mammals, birds, reptiles, amphibians, and fish. Disease associations range from the inexistent, to cancers and AIDS, as well as neurological, pulmonary, and arthritic diseases. On the basis of biological and morphological criteria, retroviruses were classified into three groups, the oncoviruses, spumaviruses, and the lentiviruses. Among the oncoviruses are avian, murine, and human leukemia viruses. The lentiviruses are incredibly varied in their genetic structures and include the human and simian immunodeficiency viruses. Contemporary genetic data have extensively revised this classification. Thus, the oncoviruses encompass three to four distinct groups, some of which are not even oncogenic. Nonetheless, the original terms onco-, spuma-, and lentivirus are still extensively used.

The current inventory of retroviruses is unlikely to be complete. In one of the more fascinating of recent papers, a novel retrovirus was described from a dermal sarcoma of walleye fish (5). The provirus proved to be 13.2 kb, longer than most proviruses yet described, although some spumaviruses can attain nearly 13 kb (6).

All retroviruses are distinguished by their ability to metamorphose nucleic acids in the cytoplasm. That is to say, the single-stranded ( + ) virion RNA passes first to a single-stranded DNA ( − ) molecule and, finally, to a double-stranded (ds)DNA ( ± ) molecule. Such genetic magic is the work of the virally encoded enzyme, reverse transcriptase (RTase). Although this enzyme is uniquely a DNA polymerase, it may copy both RNA and DNA templates. The enzyme also manifests an RNase H activity, which acts on the neosynthesized DNA–RNA hybrid, degrading RNA some 14 to 18 bases following DNA polymerization (7).

Such an RNase H activity associated with a DNA polymerase is unique to RTases. Yet these enzymes, unlike virtually all other DNA polymerases, lack a 3′ exonuclease activity. A corollary is the inability to proofread and excise misincorporated bases. If such a mismatch occurs during DNA ( − ) strand synthesis, it will then rapidly become immortalized 14 to 18 bases later, as the RNase H activity of the RTase degrades the RNA template. On translocation of the double-stranded DNA molecule to the nucleus, any errors made during DNA ( + ) strand synthesis may be partly resolved. Within the nucleus, DNA mismatches may be corrected by cellular DNA repair enzymes, with varying degrees of efficiency (8). Consequently, there is profound asymmetry: any and all errors made during ( − ) DNA strand synthesis are maintained, whereas only a fraction of those occurring during ( + ) DNA strand synthesis are kept.

Within the nucleus, another singular event occurs—integration of the dsDNA viral genome into the host cell chromosome. Integration is specific for the viral sequences, yet random for chromosomal location. The integrated DNA form of the retroviral genome is called the *provirus*.

Although lack of a 3′ exonuclease activity is a property of all RNA viral systems, integration is unique to the retroviruses. Two consequences of this latter phenomenon are of interest to our tale. First, integration provides a formidable reservoir for the virus in the face of the immune system, ensuring infection for life. Second, variants accumulate with time.

Given such a replication strategy, it is evident that, in genetic terms, retroviruses must be described in terms of viral quasispecies.

## AND HIV?

As it turns out, HIV, indeed all the primate lentiviruses, are rather special. These viruses infect CD4-bearing helper T cells (T4 cells) by way of the CD4 glycoprotein at the cell surface, which acts as a viral receptor (9,10). Infection of macrophages by antibody–virus complexes through the Fc receptor has also been described (10–12). In peripheral blood, HIV proviruses are found mainly in the T4 memory cells, as opposed to T4 naive cell, population (13,14). Virtually all T4 memory cells in the body are small, resting lymphocytes. On contact with peptide antigens in the context of histocompatibility antigen (HLA) molecules presented by antigen-presenting cells, they become activated, giving rise to ephemeral T lymphoblasts.

The HIV genome is transcriptionally silent in resting T4 lymphocytes. However, as soon as the T4 cell has become activated in response to antigen, a low level of transcription from the integrated HIV genome ensues. Initial transcription results in the production of several 1.8- to 2-kb mRNAs that may direct the synthesis of two viral proteins *tat* and *rev*. Once produced *tat* amplifies viral transcription by interaction with the TAR element, which maps to the extreme 5′ end of all nascent RNAs. This phenomenon—referred to as *transactivation*—results in augmented transcription by a factor of 100 at least. Apparently, *rev* becomes active beyond a certain threshold and ensures the export of full-length genomic RNA or partially spliced mRNAs to the cytoplasm. Although *rev* negatively regulates its own expression, it has little effect on *tat*, which may exist in two forms. The first is a 14-kDa product, derived from a *rev*-independent 1.8-kb transcript, whereas the second, a 12-kDa product, is encoded by a *rev*-dependent 4.5-kb mRNA. The upshot is that the very powerful transactivation of the HIV genome lacks any negative-feedback, ensuring a huge burst of HIV production from a single cell (15). It is impossible to quantify the burst size, but it is probably in the realm of many hundreds of virions per infected cell. A large burst size probably reflects the need to produce progeny in as short a time as possible, given the short lifespan of a T lymphoblast, which is about 2 to 3 days.

Many of these virions will be cleared by antibody; some will infect activated T4 cells and, by a single round of replication, produce novel variants. Many of these progeny will be functionally defective. Some estimates have the proportion of infectious–defective virions in human plasma as low as $10^{-5}$, although there is clearly a range from $5 \times 10^{-1}$ downward (16–19). The important point is that probably only a few virions survive immune surveillance and remain replication-competent. As long as this number is greater than 1, then there will be a net propagation of HIV within the host.

As if the ability of HIV to remain transcriptionally silent within resting T4 cells was an insufficient means of escaping from immune surveillance, HIV can infect antigen-presenting cells—monocytes, macrophages, Langerhans cells, Kupffer cells, and the like. It appears that these cells are particularly refractory to clearance. Viral expression within macrophages is essentially constitutive (20), whereas that in T4 cells is coupled to the lifestyle of this cell. In addition, HIV-infected cells in the brain, essentially of the microglial lineage, allow replication in an organ where there is relatively limited immunological surveillance.

In short, HIV is, by virtue of being a retrovirus, by being endowed with a very powerful transactivation system that is triggered by T-cell activation, as well as its tropism for CD4 lymphocytes and macrophages, a redoubtable pathogen. These features ensure abundant particle formation and an effective insurance policy from ultimate clearance by the immune system.

Such properties, all combined into one, mean that the amplitude of the quasi-species, a feature of all RNA viruses, is inevitably going to be greater for HIV than for any other RNA viruses.

## IS THE HIV REVERSE TRANSCRIPTASE PARTICULARLY
## ERROR PRONE?

The frequency of base substitution errors per base per replication cycle has been estimated for three retroviruses: Rous sarcoma virus (RSV) of chickens, Moloney murine leukemia virus (MoMLV), and spleen necrosis virus of ducks (SNV), the latter two being virtual cousins in the evolutionary parlance of retrovirologists. The values come out to be $1.4 \times 10^{-4}$, $2 \times 10^{-5}$, and $7 \times 10^{-6}$ for RSV, MoMLV, and SNV, respectively (21–23). Given that retroviral genomes are approximately 8 to 9-kb in length, it is clear that error rates of the order of 0.1 to 1 substitutions per genome per cycle ensue.

No comparable values have yet been published for HIV-1, or for any lentivirus, for that matter. However, there are several reasons for believing that the fidelity of the HIV RTase will match at least that of the avian leukosis viruses, of which RSV is the most celebrated. There are numerous papers detailing the in vitro frequencies of base misincorporation for three RTases: those of HIV-1, MoMLV, and the avian myeloblastosis virus (AMV, akin to RSV) enzyme (24–27). Here, it must be understood that base misincorporation is actually composed of two steps. First, the non-complementary base must be polymerized opposite the template, and second, the RTase must be able to elongate beyond the mismatch. The latter step is obvious, for were it not to be so, replication would be aborted, and a provirus would never be completed; the mismatch would effectively function as a chain terminator.

The in vitro studies make it clear that, although the frequencies of base misincorporation are roughly comparable when averaged over a sufficient number of sites, the ability of the different RTases to elongate beyond a mismatch varies considerably. Thus, the HIV-1 RTase can elongate beyond a mismatch without great difficulty (25,27). The AMV enzyme does so less easily, and the MoMLV enzyme even less. From this, it may be concluded that the point substitution rate per base per cycle for HIV-1 should equal, if not exceed, that of RSV.

There are two criticisms of these data. First, they concern studies that have used DNA templates. The most recent data from RNA templates is rather contradictory, and it is unclear whether DNA synthesis from an RNA template is any more, or less, error-prone than from a DNA template (28,29). The second criticism is that the assay conditions used are decidely unphysiological. It remains to be seen whether this has any influence on the published findings.

## HIV QUASISPECIES

There is now a wealth of HIV sequence data, probably more than for any other virus. There is a specific HIV–SIV sequence repository (30). There are sequential samples taken during seroconversion (31,32); from asymptomatic phase and through to frank AIDS (17,33–42); donor–receiver pairs (13,43); mother–child samples (44); comparisons of quasispecies from peripheral blood, brain, and spleen, as well as different fragments of the same spleen (45–48). In addition there are at least four studies on the follow-up of experimentally infected macaques or African green monkeys by their respective SIVs (49–51). Indeed, the volume of information on HIV isolates and samples from all over the world continues unabated (30).

By integrating all this data, the following picture emerges: except by blood transfusion or maternal–fetal routes, infection is probably initiated by a single replication competent particle. Immediately following infection a rapid expansion of HIV ensues, such that virions and excess viral proteins can be detected in the peripheral blood. The resulting quasispecies, derived from a single virion some 5 days postinfection, are remarkably homogeneous, with the consensus sequence resembling one of the sequences present in the donor (32). By 2 weeks several single point mutations can be found (31).

In the initial expansion, the complexity of HIV quasispecies resembles that of acute-phase influenza A infection (52), which is perhaps not surprising, given that its replication is not subject to proofreading. However, by the time of seroconversion, which is generally complete by 4 weeks postinfection, the complexity of HIV quasispecies is greater than that of influenza A (31). This is because a proportion of variants are accumulated and can undergo subsequent replication. Hence, complexity increases with time. A single report showed that, during a persistent influenza A infection, in which the patient suffered from severe combined immunodeficiency, the complexity of influenza A quasispecies was greater during the otherwise acute-phase infection of an immunocompetent individual (53). Thus, it would seem that the persistence of genomes within the host is linked to an increased complexity of the viral quasispecies.

## HIV Quasispecies Change in Space

Several studies have shown differences between HIV quasispecies derived from peripheral blood and those from organs (45,47,48). This was attributed to organ-specific HIV variants by some. This interpretation was contingent on the assumption that HIV sequences were homogeneous throughout the organ. This turned out to be false (46). Not only did HIV quasispecies differ between distinct regions of the same spleen, they differed from the distribution in peripheral blood. Given that 98% of lymphocytes are located in organs and tissues, as opposed to peripheral blood, it can no longer be assumed that the latter is at all representative of the HIV mutant

The different spatial distributions of HIV variants were suggested to be due to localized activation of HIV within the myriad of discrete white pulps of the spleen. The hypothesis was the following: blood-borne pathogens would be either phagocytosed by antigen-presenting cells or picked up by antigen-transporting cells, only to be filtered out by the spleenic white pulp. Peptide antigens would be presented by dendritic cells, whereas whole antigen would become associated with follicular dendritic cells. Small, circulating, yet resting T lymphocytes in the pe-

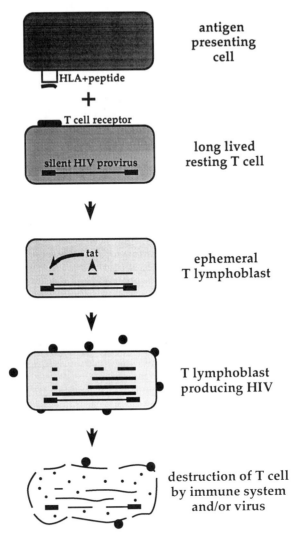

**FIG. 1.** Schema representing the amplification of an HIV genome by antigenic stimulation of a silently infected T cell. After being phagocytosed by macrophages, dendritic cells, or others, a pathogen is broken down to a collection of peptides, some of which are presented on the surface of the cell in the context of HLA class 1 molecules. Resting T lymphocytes bearing a specificity for the HLA–peptide complex will become activated (a T lymphoblast), including those CD4[+] cells harboring an HIV provirus, essentially memory cells. Initially, only a few RNA copies will be transcribed, all of which are spliced into small (1.8-2 kb) mRNAs coding for *tat* and *rev*. Production of *tat* has a positive effect on viral transcription, resulting in massive transcription of the provirus. After a threshold, *rev* ensures transport of full-length and partly spliced mRNA into the cytoplasm. Virion formation ensues. Cell destruction results either from the finite lifetime of the T lymphoblast, immune system clearance, or a direct cytopathic effect of the virus.

riphery, bearing T-cell receptors specific for these peptide antigens, would be arrested within the periarteriolar lymphocyte sheaths or germinal centers, become activated, and subsequently, pass into the periphery as ephemeral effector lymphocytes 2 to 3 days later. If any of the arrested T4 lymphocytes harbored a HIV provirus they would become activated, resulting in a local expansion of a few proviruses (Fig 1). In this scenario, amplification of a provirus would not be linked to any particular variant, but rather, because of the chance encounter of an antigen-presenting cell and a resting T cell bearing a cognate T-cell receptor. This inevitably gives rise to spatial inhomogeneity of HIV sequences.

On reflection, the link between virus replication and T4-cell activation is straightforward. It encompasses the fundamental aspect of the immune system, which is to expand only in response to antigen, otherwise remaining silent. The finding of HIV mainly in the T4 memory cell population supports this notion (14). Furthermore, the highly specialized HIV transactivation system can be so understood. Its *raison d'être* would be seen as a simple solution to the problem of producing huge amounts of virus, either before its emphemeral host cell dies or before the host cell is demolished by the extremely intense HIV-specific cytotoxic T-cell responses encountered in seropositive individuals. As virus replication is synonymous with error, the ensuing quasispecies come as no surprise. Finally, as we live in an antigenic world, it is axiomatic that there will be continual HIV replication.

Relative to HIV replication, there is one rider. As mentioned earlier, HIV infection of antigen-presenting cells would seem to result in constitutive viral expression (20). Thus, the infection of the two cell types results in overall viral expression. It is difficult to say which of the two is the more important. It may be a moot question— perhaps, both are important. Suffice to say that in lentiviral infections for which the predominant tropism is for monocytes–macrophages, the profound immunodepression, typical of AIDS, is much less marked (54).

## HIV Quasispecies Change in Time

Work with an experimental SIV infection, in which a virus derived from an infectious molecular clone was used, showed that the quasispecies become ever more complex. Although the average rate of fixation of base substitutions was about 1% per *env* sequence per year, similar to values established for several different RNA viral systems (1), this value actually belies a huge range of values, from 0% to 21% per *env* per year (49–51). Furthermore, such different mutant fixation rates occurred with the same animal. The underlying reasons for such startling differences remains unknown. However, they caution against overextrapolation of data and the application of any time scale to the radiation of lentiviral sequences.

On a shorter-time scale, HIV quasispecies in peripheral blood change in as little as 3 months and probably less (17). Perhaps a severe opportunistic infection may be capable of changing the HIV quasispecies in peripheral blood in a matter of days. Certainly, if the idea that antigenic stimulation is driving viral replication then, at a local level, quasispecies will most probably fluctuate on the same time scale as cellular immunological processes.

## A NUMBERS GAME

A combination of conventional polymerase chain reaction (PCR)-based studies have shown that the number of peripheral blood mononuclear cells (PBMCs) harboring HIV proviruses can vary from more than 1:100 to 1:50,000, depending on the disease stage (13,14,55,56). More recent data, based on in situ PCR methods, would suggest that the proviral load be even higher, as high as 10% of all PBMCs (57). In addition the proviral burden in the CD4$^+$ T lymphocytes of lymph nodes is a factor three- to tenfold greater than the frequency in peripheral blood (58–60,108).

In vivo, HIV is found essentially in CD4 lymphocytes, and then mainly within CD4$^+$ memory cells (14). Antigen-presenting cells of the monocyte–macrophage lineage, Langerhans and Kupffer cell, and microglial cells, also can be infected. There is debate whether or not follicular dendritic cells can become infected. Although there are reports of HIV infection of other nonleukocytic cell lines ex vivo (10), there is no good consensus on this point of whether these similar cell types may be infected in vivo.

An average adult has a peripheral lymphocyte count of about $2 \times 10^6$/ml. Given that the peripheral lymphocytes represent approximately 2% of the total pool, it is possible to estimate the total number of lymphocytes to be approximately $5 \times 10^{11}$. Assuming proviral loads of $10^{-1}$ to $10^{-5}$ for infected lymphocytes, at a minimum of a single proviral copy per cell, an HIV-infected patient will harbor between $5 \times 10^6$ to $5 \times 10^{10}$ infected lymphocytes, depending on the disease stage. This calculation ignores HIV in antigen-presenting cells of the monocyte–macrophage lineage, as well as cell-free virus, the titers of which can be particularly elevated during late-stage disease (19,55,56).

If the point mutation rate per genome per cycle for HIV is somewhere between 0.1 and 1, a range defined by those studies on RSV, MoMLV, and SNV (see foregoing), it is obvious that an HIV-infected individual will harbor something of the order of $5 \times 10^5$ to $5 \times 10^{10}$ genetically different variants. The consequences for therapy and prevention are all too clear.

Clearly, there is a tremendous sampling problem when describing HIV quasispecies. It is believable that, by sequencing 20 clones of 300 bp each from a mother and her child, it is possible to reliably draw conclusions? Given all that we know about the temporal and spatial distribution of HIV, it is, at least in my opinion, virtually impossible to do so. We will just have to ask our question in a different way.

## VIRAL PHENOTYPES AND DISEASE

Early studies indicated that primary viral isolates from asymptomatic carriers grew poorly in culture and could rarely be transmitted to established T-cell lines, whereas viruses from symptomatic individuals grew to very high titers and were readily propagated on T-cell lines (61). In addition, the latter isolates, unlike those from asymptomatic patients, frequently caused syncytia formation. This led to extensive and detailed genetic analyses (61,62), as well as those on the impact of syncytium-inducing (SI) isolates, as opposed to nonsyncytium inducing (NSI) isolates, in the progression to frank AIDS. These data always refer to isolates derived from cultures of immortalized cell lines, and of culturing results in the selective amplification of a few variants present (63). Certainly, an impressive correlation was noted between the rapid decline in CD4 cell count late in disease (64). Yet, if true, this would represent only part of the story, for the CD4 cell count declines slowly and inexorably following infection (65). It is not yet sufficiently clear what the contribution of the SI type of virus is to disease progression.

Several studies followed the evolution of the regulatory proteins, *tat* and *rev*, as well as the long-terminal repeat (LTR) over a 4-year period in vivo (17,33–35). Although there was continual evolution of the sequences, there was no evidence for the outgrowth of a variant endowed with a more efficient *tat*, *rev*, or LTR sequence (34). One or two variants showed hyperactive function. Yet, despite this, the genome harboring such a sequence did not dominate fellow genotypes.

## IMMUNE PRESSURE

### Antibody Escape?

The idea that HIV will always keep one escape mutant ahead of neutralizing antibody is most appealing, but does not stand up to analysis.

The reasons are the following:

1. Neutralizing titers are notoriously low. Passive transfer experiments with either concentrated high-titer human immunoglobulins selected for a certain HIV *env* V3 loop specificity, or else a human monoclonal antibody specific for the V3 loop, was able to protect against challenge (66). The V3 loop of the HIV-1 envelope glycoprotein gp120 elicits most neutralizing antibody (67). However, the titers were much greater than anything naturally observed. Therefore, one might question whether they constitute a strong selection pressure in the normal setting.

2. The virus can also spread by cell-to-cell contact, thereby further weakening the effect of neutralizing antibody; the replication of HIV in privileged organs as well as an apparent randomness of the virus, as a consequence of T-cell activation, weaken yet further the importance of neutralizing antibody.

3. The numbers of variant genomes present within an individual is huge. There is virtually every single and double nucleotide variant, with a decreasing proportion of triple and quadruple nucleotide variants. Given that most escape variants are due to single-base changes, there must be huge numbers of natural escape mutants of the infecting genotype. This requires multitude of suitably high-titer neutralizing antibodies to cope with all these distinct escape variants simul-

taneously. Given the low titers of neutralizing antibodies, even to the dominant genotype, it is not obvious how the immune system could contain so many variants.

4. Naturally arising antibody-escape mutants can occur in the absence of immune pressure. Work on foot-and-mouth disease virus (FMDV) has shown that many of the escape mutants that are observed in vivo can be found in in vitro passage of viral stocks (68). Thus, the emergence of an experimentally identifiable escape mutant is not evidence per se of antibody-selection pressure. Most recently, the same observation has been made for HIV (109).

5. The concept of antigenic sin should not be forgotten. This points out that the immune response to a variant of a known epitope will (a) cause a de novo response if the variant is immunogenic and (b) boost the response to the original epitope because there will be some partial recognition of the variant sequence by cells primed to the original epitope. Because the latter effectively behaves as a secondary immune response, its magnitude will be much greater than a primary response, even if the response is impaired somewhat by the poorer presentation. In HIV, for which there may be many variants of the original, infecting virus, the net result will be to boost the response to the primary genotype to the detriment of all other variants. This has already been documented for HIV (69).

6. Antigenic variation among the ungulate lentiviruses has been described for more than 20 years. Several reports on use of the equine infectious anemia virus (EIAV) system showed the febrile cycles and viral variation can occur before the development of neutralizing antibody (70). Antigenically distinct viral isolates appeared independent of the development of variant-specific neutralizing antibody (71). It was concluded that variant-specific antibody might be a secondary event, resulting from immune recognition of the preponderant virus type selected by nonimmunological factors. In visna virus infections, no correlation was noted between the neutralizing antibody titer to the inoculum virus and the occurrence of antigenic variants (72). Furthermore, antigenically distinct variants coexisted with the inoculum virus, indicating that emergence of the latter was not coupled to the elimination of the former (72,73).

Taken together it would seem that neutralizing antibody cannot be seen as a meaningful selection pressure.

## Cytotoxic T-Lymphocyte Escape

The cytotoxic T-lymphocyte (CTL) responses to HIV antigens are very intense, at least during the clinically asymptomatic phase, being on a par with those seen during acute-phase influenza or Epstein–Barr infections (74). Therefore, it is possible that they may constitute a strong selection pressure on the HIV quasispecies. There are now three reports that show no data in favor of this hypothesis (75–77). Basically, the experiments analyzed the fate of known CTL epitopes in vivo. Some

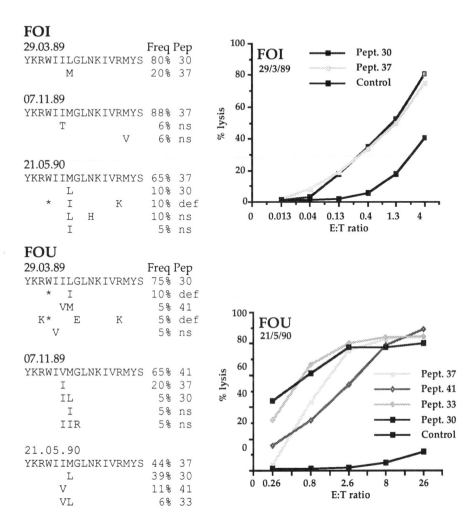

**FIG. 2.** Structure of HIV-1 *gag* quasispecies from two patients (FOU, FOI), all bearing the HLA-B27 antigen, over a 14-month period (75). Chromium-51 release of peptide-labeled P815(HLA-B27) targets as a function of the effector/target ratio. The peptides are identified by the code used on the left of the figure. The dates corresponding to the samples are given. All protein sequences are shown using the one-letter code which is: A, Ala; C, Cys; D, Asp; E, Glu; F, Phe; G, Gly; H, His; I, Ile; K, Lys; L, Leu; M, Met; N, Asn; P, Pro; Q, Gln; R, Arg; S, Ser; T, Thr; V, Val; W, Trp; and Y, Tyr. Those of the minor forms have been aligned relative to the major form. Only amino acid differences are noted. The frequency of each sequence is given on the right along with a peptide identifier code. An additional two residues at the COOH-terminus were sequenced as part of the amplified region, but are not shown. The sequences given correspond to the peptides synthesized. *Def* identifies a defective viral sequence owing to the presence of an in-phase stop codon (*): *ns* means that the corresponding peptide was not synthesized.

variant sequences were found, but nearly all remained restricted by the same HLA class I molecule (Fig. 2).

A few variants, however, were not recognized (76). These epitopes were clearly not recognized by patients' CTL. Could they be indicative of CTL-escape mutants? Two remarks can be made. First, there was no evidence, one way or the other, that these epitopes were encoded by replication competent proviruses. Second, an HIV-infected cell will present at least six HLA class I-restricting molecules, two A, two B, and two C. Each of these may present peptides. A single class I molecule may present in excess of 1 million different peptides essentially owing to recognition of the peptide backbone, rather than the peptide side chains, peptide anchor residues apart (78,79).

To complicate matters, it is known that a single peptide may elicit more than one T-cell clone (83). Thus, it is possible that, although a variant may escape recognition by one T-cell clone, it could be recognized by another.

Work on HIV has shown that there may be more than one peptide restricted by a given class I molecule per viral protein (80,81). Thus, for a true CTL-escape mutant to emerge, it would have to carry substitutions in virtually all of the viral epitopes recognized by the six restricting HLA class I molecules. In addition, it would have to escape a multitude of T-cell clones. As we have seen in the foregoing, the probability of a variant encoding such a large number of substitutions is probably greater than the inverse HIV proviral population size.

Thus, it seems unlikely that the intense CTL response during the asymptomatic phase constitutes a viable selection pressure on the HIV quasispecies.

### Immune System Overload?

This thesis supposes that, with the constant generation of sequence variation and peptide and conformational antigen motifs, more and more $CD4^+$ T-cells will be mobilized which will open themselves up for HIV infection and subsequent destruction (82). For the reasons given earlier, the impact of excessive variant production does not seen to be a problem for cellular immunity, given the enormous degeneracy of the peptide recognition and presentation system (77,83).

In addition, it is important to realize that only a fraction of all possible octameric and nonameric peptides will be presented by HLA molecules. Furthermore, although there may be huge numbers of variants, the problem of antigenic sin may ensure that not all are capable of eliciting an effective immune response.

In short, there appears no convincing evidence that HIV variation contributes to the development of immunosuppression.

### HOW FAR WILL HIV VARIATION GO? VERY FAR

Given the still relative novelty of HIV and the discovery of its incredibly complex quasispecies, this question is frequently asked. The answer, very far, may be derived from at least three lines of argument.

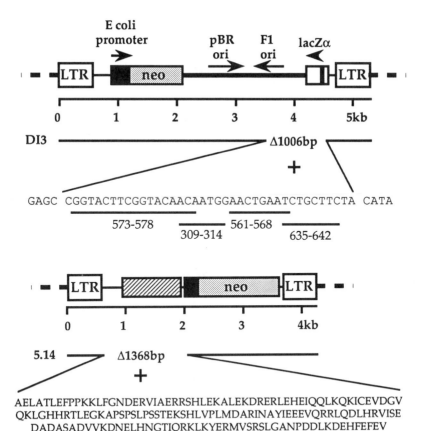

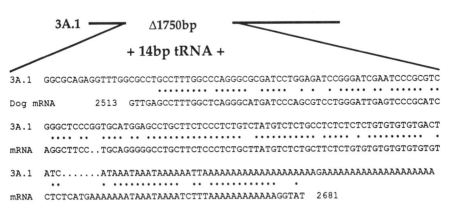

**FIG. 3.** Three examples of deletions with insertions during a single round of reverse transcription in the spleen necrosis virus system (87,88). The D13 mutant of the VP212 defective virus (97) carries a deletion of 1,006 bp, with an additional 39 bp that show homology to other parts of the VP212 genome (*underlined*). Mutant 5.14 of the JD216HyNeo-defective virus (88) carries a 1,368-bp deletion as well as an extra 489 bp of unknown origin. However, the sequence could code for 162 amino acids (shown in the one-letter code). Mutant 3A.1 (JD216HyNeo derivative) carries a larger deletion of 1,750 bp and an additional 697 bp, 505 of which arose from a partial duplication of the LTR, 14 bp of primer tRNA sequences, and 192 bp of nonviral origin. After screening the sequence banks, this sequence turned out to be 82% identical with canine blood-clotting factor IX mRNA sequence over a stretch of 161 bp (88).

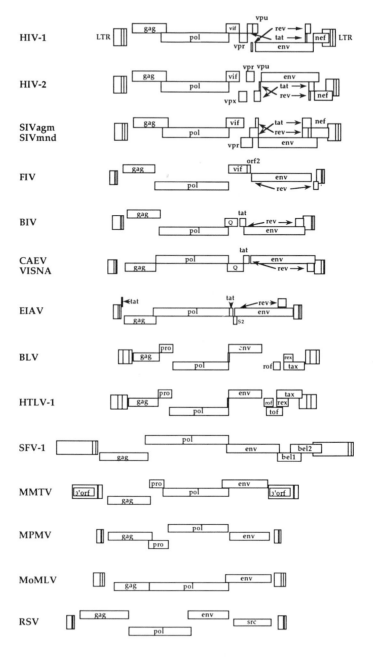

**FIG. 4.** Molecular organization of retroviral genomes. Open boxes correspond to open-reading frames (ORFs): *gag*, *pol*, and *env* encode polyprotein precursors. *pol* usually encodes the viral proteinase, reverse transcriptase–RNase H and integrase in that order. However, occasionally proteinase is encoded by a separate ORF (*pro*, in HTLV-1, BLV, MMTV, or MPMV) or part of the *gag* ORF (as in RSV). An additional dUTPase-coding sequence may be found between the

First, it must be realized the phenotypic differences necessarily result from small genetic differences. Given an intrinsic misincorporation rate of about $10^{-4}$ per base per cycle and a proviral size of approximately $10^{11}$, then if a novel genetic trait to required four to five substitutions, it would effectively prohibit its occurrence [i.e., $(10^{-4})^{4-5} << 1/10^{11}$ (i.e., 1 per proviral population)]. This suggests that small genetic changes must not be underestimated. Certainly, they have been implicated in changes of tropism (10, 84–86).

Although the point mutation rate is probably the most frequent source of genetic diversity, retroviruses are capable of generating a whole series of complex changes that, albeit a little less frequent, are, nonetheless, of potential import, given the proviral population sizes. These include duplications, deletions, transduction of host cell sequences (87,88), G→A hypermutation (20,63), insertional hypermutation, substitution with insertion, deletions with insertions (86,89), and complex strand displacement, some examples of which are given in Fig. 3. In addition, after formation of any heterozygous virion, efficient recombination ($f~0.2$/cycle) can add to the diversity (90). Although the steady accumulation of point mutations represent the familiar way of changing a viral sequence from A into Z by a series of increments, complex changes represent a sort of viral tunneling, whereby it is occasionally possible to go directly from A to Z.

Second, retroviruses have accumulated a vast array of genetic conformations, as a simple survey shows (Fig. 4). The figure represents a low-resolution description of the genetic organization of 16 retroviruses. However, many subtleties are still lost. Thus, to simply score retroviruses as possessing an envelope (env) protein sequence is as crude as saying that humans are primates. Both are true, yet the descriptions are hopelessly incomplete. Indeed, it is impossible to align the env sequences of many retroviruses, for example, HIV and RSV, or MMTV and HSRV. Even in a more restricted comparison, the env proteins of HIV-1 and

---

reverse transcriptase–RNase H and integrase domains for FIV, CAEV/Visna, EIAV, or sandwiched just upstream from the MMTV and MPMV proteinase ORF (107). The most 5' part of the *env* ORF of FIV, BIV, CAEV, Visna, and EIAV encodes the first exon of *rev*, which is assembled by differential splicing of genomic RNA. Likewise, the 5' region of the *env* ORF of the HTLV-1/BIV group corresponds to the first exon of the *tax* product. The HIV/SIV *tat*, HTLV-1/BLV *tax*, and SFV-1 *bel 1* products, all result in transactivation of proviral expression, even though there is no sequence homology. Likewise the lentiviral *rev* sequences and HTLV-1/BLV *rex* proteins are functionally equivalent. Yet again, there is no sequence homology. As mentioned in the text *env* sequences are exceedingly variable and should be collectively described only in general terms. The LTR (long-terminal repeat) sequences, which encode virtually all the major features necessary for transcription and reverse transcription, are highly variable in length and organization, and may even encode a viral superantigen (3' ORF in MMTV, ref. 59). The ORF *src* encodes a transduced host cell oncogene sequence. Abbreviations: SIVagm, SIV African green monkey virus; SIVmnd, SIV mandrill virus; FIV, feline immunodeficiency virus; BIV, bovine immunodeficiency virus; CAEV, caprine arthritis encephalitis virus; Visna, ovine maedi-visna virus; EIAV, equine infectious anemia virus; BLV, bovine leukemia virus; HTLV-1, human T-cell leukemia virus type 1; SFV-1, simian foamy virus type 1; MMTV, mouse mammary tumor virus; MPMV, Mason-Pfizer monkey virus; MoMLV, Moloney murine leukemia virus; RSV, Rous sarcoma virus.

HIV-2, although both conferring tropism for CD4-bearing cells, differ in their disulfide bridging and glycosylation, sequence variation and neutralization patterns.

If we continue the comparison of HIV-1 and HIV-2, there is a decided lack of reciprocity. The HIV-1 *tat* and *rev* proteins function equally well with the HIV-1 or HIV-2 cognate TAR and RRE sequences, respectively. However the HIV-2 proteins function less well with a HIV-1 target than their HIV-2 target (91). The HIV-1 RTase is sensitive to the nonnucleoside TIBO family of inhibitors, whereas the HIV-2 RTase is insensitive (92). Both the HIV-2 *tat* and *nef* protein sequences are larger than their HIV-1 counterparts. HIV-2 encodes a novel splice donor and acceptor sequences within the R/U5 region (93); and HIV-1 uniquely encodes a *vpu* protein, whereas HIV-2 encodes a *vpx* protein. In short, inspection of extant retroviruses shows an impressive array of different conformations.

Third, protein structure and function do not seriously restrict the degree of variation; rather, it reduces the rate of change. To illustrate this point, I would like to take the example of the HIV-1 proteinase. The proteinase, like those of all retroviruses, belongs to the aspartic proteinase family. Unlike their germline counterparts, they function as homodimers. As can be seen from Fig. 5, there is considerable sequence variation among all the retroviral proteinases. In fact, only seven residues are perfectly conserved. With the exception of the simian spumavirus (SFV) sequence, all proteinases keep approximately the same folding pattern in the order of β-sheet and α-helices. Indeed, the coordinates of the α-carbon atoms of the peptide backbones of the RSV and HIV-1 proteinases are virtually superimposable (94). Clearly, as long as the folding pattern is preserved, then, with time, most of the residues may be replaced.

Are there other examples? Despite having very little primary sequence in common, there is a remarkable similarity in the overall domain structure of the HIV-1 reverse transcriptase and that of the Klenow fragment of *Escherichia coli* DNA polymerase I (95). Again, when all the reverse transcriptase sequences are aligned, only five or six residues were perfectly conserved. Likewise, an alignment of all the retroviral integrase sequences revealed only ten perfectly conserved residues.

When it comes to the nucleocapsid structural proteins encoded by *gag*, their lack of enzymatic activity means that their primary amino acid sequences are even less constrained than those of *pol*. There are no perfectly conserved residues. One feature characteristic of the *gag* nucleic acid-binding protein was a highly basic sequence accompanied by a $Cys-X_2-Cys-X_4-His-X_4$ Cys motif. Although this motif is found duplicated in virtually all retroviral sequences, only a single copy may be found in the MLV-related viruses and none whatsoever in the spumavirus sequences, even though their COOH-terminal sequences are highly basic (6). Clearly, the spumaviruses have found another way to condense genomic RNA within the virion.

As a parallel, it is worth pointing out that the nucleocapsid structures of poliovirus, FMDV, rhinovirus, Mengo virus, and Theiler's disease virus, all of which are picornaviruses, have, apart from surface loops, a similar folding pattern for the VP1, VP2, and VP3 capsid proteins (96–100). Despite this, there is only 3% primary sequence identity (101).

```
        --a--      ---b----       -----c----  ---d--       --a'-
                   •                         • ••
HIV1    PQITLW---QRPLVTIRIG----------GQLKESLLDTGADDTVLEEMNLPGK----WKPKM-----
SIVcpz  PQITLW---QRPLIPVKVE----------GQLCEALLDTGADDTVIERIQLQGL----WKPKM-----
HIV2    PQFSLW---KRPVVTAHIE----------GQPVEVLLDTGADDSIVAGIELGNN----YSPKI-----
SIVmac  PQFSLW---RRPVVTAHIE----------GQPVEVLLDTGADDSIVTGIELGPH----YTPKI-----
SIVagm  -FELPLW---RRPIKTVYIE---------GVPIKALLDTGADDTIIKENDLQLSGP--WRPKI-----
SIVmnd  PEYSLS---RRPIEEVSVD----------GVTIRALLDTGADDTIFNERNIKLKGN--WQPKI-----
FIV     TTTTLE---KRPEILIFVN----------GYPIKFLLDTGADITILNRRDFQVKNSIENGRQN-----
VISNA   PYVVTE---APPKIEIKVG----------TRWKKLLVDTGADKTIVTSHD-MGI----PKGRII----
CAEV    -SYGITS---APPMVQVRIG----------SQQRNLLFDTGADRTIVRWHEGSGN----PAGRIK----
BIV     PSYIRLD--KQPFIKVFIG----------GRWVKGLVDTGADEVVLKNIHWDRIKG-YPGTPIK----
EIAV    -VTYNLE---KRPTTIVLIN----------DTPLNVLLDTGADTSVLTTAHYNRLKYRGRKYQGTG---
MMLV    QGGQGQDPPPEPRITLKVG----------GQPVTFLVDTGAQHSVLTQNPGPLS----DKSAW-----
FELV    QETQGQDPPPEPRITLRIG----------GQPVTFLVDTGAQHSVLTRPDGPLS----DRTAL-----
GALV    -QGSQGSDPLPQPRVTLTVE----------GTPIEFLVDTGAEHSVLTQPMGKVG----SRRTV-----
BAEV    -QGCQGSGAPPEPRLTLSVG----------GHPTTFLVDTGAQHSVLTKANGPLS----SRTSW-----
HSRV    -QMNPL--QLLQPLPGRIK----------GTKLLAHWDSGATITCIPESFLEDE----QPIKKTLIKT
SFV     -KMDPL--QLLQPLEAEIK----------GTKLKAHWDSGATITCVPEAFLEDE----RPIQTMLIKT
HTLV1   -PVIPLD-PARRPVIKAQVDTQTSHP-----TKIEALLDTGADMTVLPIALFSSN----TPLKNTS---
HTLV2   -PLIPLR-QQQQPILGVRISVMGQTP-----QPTQALLDTGADLTVIPQTLVPGP----VKLHDTL---
BLV     LSIPL--ARSRPSVAVYLSGPWLQPS---QNQALMLVDTGAENTVLPQNWLVRD----YPRIPAA---
MPMV    WVQPI--TCQKPSLTLWLD----------DKMFTGLIDTGADVTIIKLQDWPPN----WPITDTLYN-
MMTV    WVQEI--SDSRPMLHIYLN----------GRRFLGLLDTGADKTCIAGRDWPAN----WPIHQTESS-
RSV     LAMTME-HKDRPLVRVILTNTGSHPVKQRSVYITALLDSGADITIISEEDWPTD----WPVMEAANPQ
        --a--      -----b-----    -----c----   -d- --h-       --a'-- ..
```

```
        ---- -------b'------- ----c'----        -d'- ---h'-- --q--
                •                                •
HIV1    IGGIGGFIKVRQYDQIPVEICGHKAIGTVLVG--------PTPVNIIGRNLLTQIGCTLNF
SIVcpz  IGGIGGFIKVKQFDNVHIEIEGRKVVGTVLVG--------PTPVNIIGRNILTQLGCTLVF
HIV2    VGGIGGFINTKEYKNVEIEVLNKKVRATIMTG--------DTPINIFGRNILTALGMSLNL
SIVmac  VGGIGGFINTKEYKNVEIEVLGKRIKGTIMTG--------DTPINIFGRNILTALGMSLNL
SIVagm  IGGIGGGLNVKEYNDREVKIEDKILRGTILLG--------ATPINIIGRNLLAPAVPRLVMGQLSEKI
SIVmnd  IGGIGGNLRVKQYDNVYVEIRGKGTFGTVLIG--------PTPIDIIGRNIMEKLGGKLILAQLSDKI
FIV     MIGVGGGKRGTNYINVHLEIRDENYKTQCIFGNVCVLEDNSLIQPLLGRDNMIKFNIRLVMAQISDKI
VISNA   LQGIGGIIEGEKWEQVHLQYKDKMIKGTIVVL-------ATSPVEVLGRDNMRELGIGLIMANLEEKRI
CAEV    LQGIGGIVEGEKWNNVELEYKGETRKGTIVVL-------PQSPVEVLGRDNMARFGIKIIMANLFEKKRI
BIV     QIGVNGVNVAKRKTHVEWRFKDKTGIIDVLFS--------DTPVNLFGRSLLRSIVTCFTLLVHTEKIEPL
EIAV    IIGVGG-NVETFSTPVTIKKKGRHIKTRMLVA--------DIPVTILGRDILQDLGAKLVLAQLSKEIKFR-
MMLV    VQGATGGKRYRWTTDRKVHLATGKVTHSFLHV-------PDCPYPLLGRDLLTKLKAQIHFEGSG
FELV    VQGATGSKNYRWTTDRRVQLATGMVTHSFLTV-------PECPYPLLGRDLLTKLKAQIHFTGEG
GALV    VEGATGSKVYPWTTKRLLKIGHKQVTHSFLVI-------PECPAPLLGRDLLTKLKAQIQFSAEG
BAEV    VQGATGRKMHKWTNRRTVNLGQGMTIPKILVD-------TFDKWQILGRDVLSRLQASISIPEEV
HSRV    IHGEKQQNVYYVT-----------------------FKVKGRKVEAEVIASPYEYILLSPTDV
SFV     IHGEKQQDVYYLT-----------------------FKVQGRKVEAEVLASPYDYILLNPSDV
HTLV1   VLGAGGGQTQDHFKLTSLPVLIRLPFRTTPIVLTSCLV-DTKNNWAIIGRDALQQCQGVLYLPEAK
HTLV2   ILGASGQTNTQFKLLQTPLHIFLPFRRSPVILSSCLL-DTHNKWTIIGRDALQQCQGLLYLPDDP
BLV     VLGAGGVSRNRYNWLQGPLTLALKPEGPFITIPKILV-DTFDKWQILGRDVLSRLQASISIPEEV
MPMV    LRGIGQSNNPKQSSKYLTWRDKENNSGLIKPFV-----IPNLPVNLWGRDLLSQMK--IMMCSPNDIVTA-
MMTV    LQGLGMACGVARSSQPLRWQH-EDKSGIIHPFV-----IPTLPFTLWGRDIMKDIKVRLMTDSPDDSQDL
RSV     IHGIGGGIPMRKSRDMIELGVINRDGSLERPLLLFPA-VAMVRGSILGRDCLQGLGLRLTNL
        .flap..   -----b'------  ------c'-------   -d'- ---h'--- --q--
```

# RETROVIRAL PROTEASE SEQUENCES

**FIG. 5.** Alignment of retroviral proteinase sequences. Gaps (*dashed lines*) have been introduced into the sequence to maximize sequence homology. For a few sequences, the precise NH$_2$- and COOH-termini remain to be defined. The seven perfectly conserved residues are indicated by a dot (•). Regions *a–d*, *q*, *a'–d'*, and *q'* denote β-strands, and *h* and *h'* denote α-helices. *Flap* refers to those residues that surround the peptide distal to the catalytic site. (See Fig. 4 legend for virus abbreviations.)

Finally, there is a prejudice that the sequences of proteins involved in the regulation of viral transcription should evolve slowly. This is perhaps understandable in some ways, but just does not hold up under analysis. Thus, the sequences of the transactivator (*tat*) or mRNA transport (*rev*) proteins of the primate immunodeficiency vary enormously (Fig. 6). Comparison of the human and simian spumavirus sequences shows that the *bel 1* sequence, absolutely essential for proviral expres-

## tat

```
              .                            .    .  .  . ...   ..    .    .
HIV-1    MEPVDPRLEPWKH---------------------------GSQPKTACTN-CYCKKCCFHCQVCFMTKA
SIVcpz   MDPIDPDLEPWKH---------------------------GSQPRTVCTN-CYCKACCYHCIYCFTKKG
HIV-2    METHLKAPESSLESYNEPSSCTSEQGVTAQELAKQGEELLSQLHRPLEACTNSCYCKQCSFHCQLCFLKKG
SIVagm   MDKGEAEQIVSHQ----------------------DLSEDYQKPLQTCKNKCFCKKCCVHCQLQFLQKG
SIVmnd   MEPSGKEDHNCLP------------QDLGQEEIDYKQLLEEYYQPLQACENKCWCKKCCFHCMLCFHKKG
```

```
              ..   .                .
HIV-1    LGISYGRKKRRQRRR-AHQNSQTHQASLSKQPTSQSRGDPTGPKE
SIVcpz   LGISYGRKKRTTRRRTAPAGSKNNQDSIPKQPLSQSRGNKEGSEKSTKEVASKTEADQ
HIV-2    LGIWYARKSR--RRR-TPRKTKTHSSSASDKSISTRTGDSQPTKEQKTTETTMVTTCSLGR
SIVagm   LGVTYHAPRTR-RKK----IRSLNLAPLQHQSISTKWGRDGQTTPTSQEKVETTAGSN
SIVmnd   LGIRYHVY----RKR-GPGTNKKIPGGGEEAIRRAIDLCFFNRTCSRTHTANGQTTEKKKATA
```

## rev

```
              .              .              .    .  ...
HIV-1    MAGRSGDSDEE-LIRTVRLI-KLY-QSNPPPNPEGTR-----QARRNRRRRWRERQRQIHSISERIL
SIVcpz   MAGRSEPQDDARLLQAVKIIKILY-QSNPYPSPEGTR-----KARRNRRRRWRARQKQISEISGRVL
HIV-2    MHEKADGEE---LQERLRLIRLLH-QTNPYPHGPGYA-----SQRRNRRRRQRRWLRLVALADKLYT
SIVagm   MPLGPGE------RRFVRIW-LLY-STNTYPSGEGTA-----RQRRRARRRWRQQQDQIRVLVERLQ
SIVmnd   MSTGNVDQEL--IRRYLVVVKKLYEGLSTCAFSTGLAAEPIPQTARQRRRR----KQQLRTRRAQLR
```

```
HIV-1    STYLGRSAEPVPLQLPPLERLTLDCNEDCGTSGTQGVGSPQILVESPTVLESGTKE
SIVcpz   ATYLGRPPKPGDLELPELDKLSLQCVETTQDVGTSNTSQPQTATGETVPAGGNYSILGKGAKN
HIV-2    ---FPDPPTDSPLDRAIQDLQRLTIHELPDPPTDLPESNSNQGLAET
SIVagm   -------EQVYAVDRLADEAQHLAIQQLPDPPHSA
SIVmnd   --ELEGRILKQILDRGPDGLCQGVANLALAEKSESSN
```

## rev

```
                         .             .  ..                .  .
CAEV CO     --------------MDAGARYMRLTGKENWVEVTMDGEKERKREGFTAGQQDIQNSKYP
CAEV 63     --------------MDAGASYMRLTGEENWVEVTMDEEKERKGKDV---QQDIQNSKYP
VISNA 1514  MASKESKPSRTTWRDMEPPLRETWNQVLQELVKRQQQEEEQQGLVS-----GLQASKAD
VISNA SA    MASSKNMPSRITQKSMEPPLRETWQQVVQEMVMRKQRDEEEKQNLVT-----GLQASSGD
```

```
             . ...  .                            .         ...
CAEV CO     DIPTGHSHH---GNKSRRRRRKSGFWRWLRGIRQQRNKRKSDSTESLEPCLGALAELTLE
CAEV 63     NIPTGHSHH---GKKSRRRRRKSGFWRWLRGIRRQQNRPKSDSTESLEPCLGALAELTLE
VISNA 1514  QIYTGNSGDRSTGGIGGKTKKKRGWYKWLRKLRAREKNIPSQFYPDMESNMVGMENLTLE
VISNA SA    PIYTGNSSDRSTRGPGGKTKRRKGWFQWLRKLRAREKNIPSQFYPDMEGNCAGLENLTLG
```

```
             .                  .  .
CAEV CO     GAMEKGPAE----AARPSADDGNLDKWMAWRTPQK
CAEV 63     GAVEEEPAK----DAHPSSDNGNLDKWTAWRTPQK
VISNA 1514  TQLEDNALYNPATHIGDMAMDGRE---MAERTESAQKEKKKGGLSGQRTNAYPGK
VISNA SA    EGMEENPIYDSTAATNTANMDGRN---MAERT
```

**FIG. 6.** Alignment of lentiviral *tat* and *rev* protein sequences. Perfectly conserved residues are indicated by a *dot* (●). Virus abbreviations are given in Fig. 4.

sion, is more variable (6) that the envelope protein sequence. Other examples are easily found. Why such variation? Redundancy is unlikely, as RNA viruses are very economical. Could it be indicative of frequent transmission between different cell types or between species? It is difficult to say for the time being.

## AND THE HIV EPIDEMIC?

We have seen that the numbers of HIV proviruses in an individual may be approximately $10^{11}$, which must reflect the same number of reverse transcription steps. Yet, this is much like taking a still photograph of a crawling 11-month-old child. A few seconds later this child can be meters away. For HIV, we have effectively no idea of how much virus is being produced, shed, and cleared, nor the number of proviruses lost with CD4 cell destruction.

With the live-virus polio vaccination, it is known that some 3 to 7 days post-vaccination a child may shed something of the order of $10^3$ plaque-forming units (pfu) of poliovirus per gram of stools (102). For HIV, we have steady-state levels of the virus load, but little idea of the total virus production. Recent unpublished data would suggest that the plasma viremia loads have been underestimated by several orders of magnitude (19).

The importance of this relates to the frequency of rare, yet highly significant replication events. Thus, for the SNV system, it was estimated that the frequency of transduction of cellular DNA was approximately $10^{-3}$ to $10^{-4}$ per cycle (88). If the protein-coding capacity of the human genome is approximately 1%, then the probability of picking up some coding sequences might be about $10^{-5}$ to $10^{-6}$. For an AIDS patient harboring perhaps $10^{11}$ proviruses, there may be as many as $10^5$ recombinant HIV proviruses harboring human-coding sequences. Probably the vast majority of these will be functionally defective.

Yet this is not to forget the current AIDS epidemic. Various projections for the year 2000 put the number of HIV-seropositive individuals somewhere between 40 and 100 million. And time, of all things, does not stand still. With such a large pool of carriers, it seems inevitable that novel viral forms carrying transduced human DNA will arise. And some of these will carry transduced oncogenes. Already, an HIV provirus has been found integrated just 5' to the human *fps* oncogene in an aggressive CD4 lymphoma (103). A recent paper described the presence of chimeric HIV–human transcripts from clinical material (104), a prerequisite to gene transduction. Given that many transduced oncogenes were identified in the setting of experimental research laboratories (i.e., within a limited number of infections) and not during zoonotics, it would seem eminently possible that some sort of oncogenic derivatives of HIV will emerge.

More simpler rearrangements, such as deletions and rearrangements of existing HIV sequences, are to be expected. In this context, it is worth noting that the *vpx* open-reading frame of HIV-2 was derived from duplication of the *vpr* sequence (105). When it comes to small numbers of base substitutions, a profusion of novel variants is to be expected. From ex vivo work, it is known that HIV can infect established CD4$^-$ cell lines (Table 1) derived from hepatomas, osteosarcomas, neuroblastomas, gliomas, colon carcinomas, and fibroblasts (10). Again ex vivo work has identified an alternative receptor for HIV-1, a galactosylcerebroside (106). This is a remarkable finding, particularly as the dissociation constant for gp120 and this cerebroside is approximately $10^{-9}$ M. Other simple point mutations have been shown to broaden the host range of HIV-1 ex vivo (84–88).

Although the consequences of such findings can be of debatable importance of the time being, they must be indicative of the sorts of variants to be expected in the future. Probably, the variant forms will be clinically less terrifying than AIDS. However, it is possible that they, too, could develop into epidemic proportions.

**TABLE 1.** *Susceptibility of some nonhematopoietic cell lines to HIV*

| Susceptible cell types | CD4 dependence |
|---|---|
| HIV-1 | |
| Fibroblast | Dependent and independent |
| Osteosarcoma | Independent |
| Rhabdomyosarcoma | Independent |
| Neuroblastoma | Independent |
| Glioma | Independent |
| Hepatoma | Independent |
| Intestinal epithelium | Independent |
| Colorectal carcinoma | Dependent and independent |
| Lung epithelium | Dependent and independent |
| Teratocarcinoma | Dependent and independent |
| Trophoblast | Dependent |
| HIV-2 and SIV | |
| Fibroblast | Dependent and independent |
| Rhabdomyosarcoma | Independent |
| Teratocarcinoma | Dependent |

From Weiss, ref. 10.
CD4-dependence was determined by inhibition by soluble CD4 or anti-CD4 antibodies known to block HIV infection of T lymphocytes. Where cell types are labeled both CD4 dependent and independent, it refers to results from different laboratories or with different cell lines and sublines.

This belief in novel variants of HIV is not actually new. Holland adeptly put his finger on it when he wrote

> As human populations continue to grow exponentially, the number of ecological niches for human RNA virus evolution grows apace and new human virus outbreaks will likely increase apace. Most new human viruses will be unremarkable—that is they will generally resemble old ones. Inevitably, some will be quite remarkable, and quite undesirable. When discussing RNA virus evolution, to call an outbreak (such as AIDS) remarkable is merely to state that it is of lower probability than an unremarkable outbreak (1).

The RNA viruses are perpetual mutant factories. Retroviruses allow the accumulation of huge numbers of variants. AIDS reflects the first time a retrovirus has appeared in epidemic form. Given the huge numbers involved, proviral copy number, and individuals infected, it seems inevitable, and only a matter of time, that novel pathogenic variants of HIV will emerge.

One is reminded of JD Bernal's remark "whatever hasn't happened will happen and nobody will be safe from it." HIV has indeed escaped from Pandora's microbial box.

## ACKNOWLEDGMENTS

I would like to thank all my past and present collaborators. The literature is so vast that inevitably one can only select a few references from the many that merit citation. My apologies to colleagues for not being able to cite more.

This work was supported by grants from Institut Pasteur and l'Agence Nationale pour la Recherche sur le SIDA.

## REFERENCES

1. Holland JJ, ed. Genetic diversity of RNA viruses, *Curr Top Microbiol Immunol* 1992:176:1–20
2. Domingo E, Holland JJ, Ahlquist P, eds. *RNA genetics*. Boca Raton: CRC Press; 1988.
3. Eigen M, Biebricher CK. Sequence space and quasispecies distribution. In: Domingo E, Holland JJ, Ahlquist P, eds. *RNA genetics*. Boca Raton: CRC Press; 1988:211–245.
4. Wain-Hobson S. HIV quasispecies in vivo and in vitro. *Curr Top Microbiol Immunol* 1992;176: 181–193.
5. Martineau D, Bowser PR, Renshaw RR, Casey JW. Molecular characterization of a unique retrovirus associated with a fish tumour. *J Virol* 1992;66:596–599.
6. Kupiec JJ, Kay A, Hayat M, Ravier R, Périès J, Galibert F. Sequence analysis of the simian foamy virus type 1. *Gene* 1991;101:185–194.
7. Fu TB, Taylor J. When retroviral reverse transcriptases reach the end of their RNA templates. *J Virol* 1992;66:4271–4278.
8. Brown TC, Jiricny J. A specific mismatch repair event protects mammalian cells from loss of 5-methylcytosine. *Cell* 1987;50:945–950.
9. Dalgleish AG, Berverle PC, Clapham PR, Crawford DH, Greaves MF, Weiss RA. The CD4 (T4) antigen is an essential component of the receptor for the AIDS retrovirus. *Nature* 1984;312:763–767.
10. Weiss RA. Cellular receptors and viral glycoproteins involved in retrovirus entry. In: Levy JA, ed. *The retroviruses*, vol 2. New York: Plenum Press; 1993:1–108.
11. Homsey J, Meyer M, Tateno M, Klarkson S, Levy LA. The Fc and not CD4 receptor mediates antibody enhancement of HIV infection in human cells. *Science* 1989;244:1357–1360.
12. Takeda A, Tuazon CU, Ennis FA. Antibody-enhanced infection by HIV-1 via Fc receptor-mediated entry. *Science* 1988;242:580–583.
13. Schnittman SM, Psallidopoulos MC, Lane HC, et al. The reservoir for HIV-1 in human peripheral blood is a T cell that maintains expression of CD4. *Science* 1989;245:305–308.
14. Schnittman SM, Lane HC, Greenhouse J, Baselar M, Fauci AS. Preferential infection of CD4$^+$ memory T cells by human immunodeficiency virus type 1: evidence for a role in the selective T cell functional defects observed in infected individuals. *Proc Natl Acad Sci USA* 1990;87:6058–6062.
15. Cullen BR. Human immunodeficiency virus as a prototypic complex retrovirus. *J Virol* 1991; 65:1053–1056.
16. Brinchmann JE, Albert J, Vartdal F. Few infected CD4$^+$ T cells but a high proportion of replication competent provirus copies in asymptomatic human immunodeficiency virus type 1 infection. *J Virol* 1991;65:2019–2023.
17. Meyerhans AF, Cheynier R, Albert J, et al. Temporal fluctuations in HIV quasispecies in vivo are not reflected by sequential HIV isolations. *Cell* 1989;58:901–910.
18. Layne SP, Mergis MJ, Dembo M, et al. Factors underlying spontaneous inactivation and susceptibility to neutralization of human immunodeficiency virus. *Virology* 1992;189:695–714.
19. Piatak M, Saag MS, Yang LC, et al. High levels of HIV-1 in plasma during all stages of infection determined by competitive PCR. *Science* 1993;259:1749–1754.
20. Bachelerie F, Alcami J, Arenzana-Seisdedos F, Virelizier JL. HIV enhancer activity perpetuated by NF-κB induction on infection of monocytes. *Nature* 1991;350:709–712.
21. Pathak VK, Temin HM. Broad spectrum of in vitro forward mutations, hypermutations, and mutational hotspots in a retroviral shuttle vector after a single replication cycle: substitutions, frameshifts, and hypermutations. *Proc Natl Acad Sci USA* 1990;87:6019–6023.
22. Leider JM, Palese P, Smith FI. Determination of the mutation rate of a retrovirus. *J Virol* 1988;62:3084–3091.
23. Monk RJ, Malik FG, Stokesbury D, Evans LH. Direct determination of the point mutation rate of a murine retrovirus. *J Virol* 1992;66:3683–3689.
24. Preston BD, Poiesz BJ, Loeb LA. Fidelity of HIV-1 reverse transcriptase. *Science* 1988;242:1168–1171.

25. Ricchetti M, Buc H. Reverse transcriptases and genomic variability: the accuracy of DNA replication is enzyme specific and sequence dependent. *EMBO J* 1990;9:1583–1593.
26. Roberts JD, Bebenek K, Kunkel TA. The accuracy of reverse transcriptase from HIV-1. *Science* 1988;242:1171–1173.
27. Bebenek K, Abbotts J, Roberts JD, Wilson SH, Kunkel TA. Specificity and mechanism of error prone replication by human immunodeficiency virus 1 reverse transcriptase. *J Biol Chem* 1989; 264:16948–16956.
28. Yu H, Goodman MF. Comparison of HIV-1 and avian myeloblastosis virus reverse transcriptase fidelity on RNA and DNA templates. *J Biol Chem* 1992;267:10888–10896.
29. Ji J, Loeb LA. Fidelity of HIV-1 reverse transcriptase copying RNA in vitro. *Biochemistry* 1992; 31:954–958.
30. Myers G, Korber B, Smith RF, Berzofsky JA, Pavlakis GN. *Human retroviruses and AIDS*. Los Alamos: Los Alamos National Laboratory: 1992.
31. Pang S, Shlesinger Y, Daar ES, Moudgil T, Ho DD, Chen ISY. Rapid generation of sequence variation during primary HIV-1 infection. *AIDS* 1992;6:453–460.
32. Shaw GM, Pan P, Clark SJ, Hahn BH, Saag MS. Molecular and biologic transitions from acute to chronic HIV-1 infection. In: Girard M, Valette L, eds. *Retroviruses of human AIDS and related animal diseases*. Lyon: Fondation Marcel Merieux; 1991:73–77.
33. Delassus S, Cheynier R, Wain-Hobson S. Evolution of human immunodeficiency type 1 *nef* and long terminal repeat sequences over four years in vivo and in vitro. *J Virol* 1991;65:225–231.
34. Delassus S, Meyerhans A, Cheynier R, Wain-Hobson S. Absence of selection of HIV-1 variants in vivo based on transcription/transactivation during progression to AIDS. *Virology* 1992;188:811–818.
35. Pedroza Martins L, Chenciner N, Åsjö B, Meyerhans A, Wain-Hobson S. Independent fluctuation of HIV-1 *rev* and gp41 quasispecies in vivo. *J Virol* 1991;65:4202–4207.
36. Pedroza Martins L, Chenciner N, Wain-Hobson S. Complex intrapatient sequence variation in the V1 and V2 hypervariable regions of the HIV-1 gp120 envelope sequence. *Virology* 1992;191:837–845.
37. Simmonds P, Balfe P, Ludlam CA, Bishop JO, Leigh Brown A. Analysis of sequence diversity in hypervariable regions of the external glycoprotein of human immunodeficiency virus type 1. *J Virol* 1990;64:5840–5850.
38. Balfe P, Simmonds P, Ludlam CA, Bishop JO, Leigh Brown AJ. Concurrent evolution of human immunodeficiency virus type 1 in patients infected from the same source: rate of sequence change and low frequency of inactivating mutations. *J Virol* 1990;64:6221–6233.
39. Simmonds P, Balfe P, Peutherer JF, Ludlam CA, Bishop JO, Leigh Brown AJ. Human immunodeficiency virus-infected individuals contain provirus in small numbers of peripheral mononuclear cells and at low copy numbers. *J Virol* 1990;64:864–872.
40. Wolfs TFW, de Jong JJ, van den Berg H, Tijnagel JMGH, Krone WJA, Goudsmit J. Evolution of sequences encoding the principal neutralization epitope of human immunodeficiency virus 1 is host dependent, rapid and continuous. *Proc Natl Acad Sci USA* 1990;87:9938–9942.
41. Wolfs TFW, Zwart G, Bakker M, Valk M, Kuiken C, Goudsmit J. Naturally occurring mutations within HIV-1 V3 genomic RNA lead to antigenic variation dependent on a single aminoacid substitution. *Virology* 1991;185:195–205.
42. Kusumi K, Conway B, Cunningham S, et al. Human immunodeficiency virus type 1 envelope gene structure and diversity in vivo and after cocultivation in vitro. *J Virol* 1992;66:875–885.
43. Ou CY, Clesieiski CA, Myers G, et al. Molecular epidemiology of HIV transmission in a dental practice. *Science* 1992;256:1165–1171.
44. Wolinsky SM, Wike CM, Korber B, et al. Selective transmission of human immunodeficiency virus type-1 variants from mothers to infants. *Science* 1992;255:1134–1137.
45. Epstein LG, Kuiken C, Blumberg BM, et al. HIV-1 V3 domain variation in brain and spleen of children with AIDS: tissue specific evolution within host determined quasispecies. *Virology* 1991;180:583–590.
46. Delassus S, Cheynier R, Wain-Hobson S. Inhomogenous distribution of human immunodeficiency virus (HIV) type 1 genomes within an infected spleen. *J Virol* 1992;66:5642–5645.
47. Pang S, Vinters HV, Akashi T, O'Brien WA, Chen ISY. HIV-1 *env* sequence variation in brain tissue of patients with AIDS related neurologic disease. *J AIDS* 1991;4:1082–1092.
48. Steuler H, Storch-Hagenlocker B, Wildemann B. Distinct populations of human immunodeficiency virus type 1 in blood and cerebrospinal fluid. *AIDS Res Hum Retroviruses* 1992;8:53–59.

49. Burns DPW, Desrosiers RC. Selection of genetic variants of simian immunodeficiency virus in persistently infected rhesus monkeys. *J Virol* 1991;65:1843–1854.
50. Johnson PR, Hamm TE, Goldstein S, Kitov S, Hirsch VM. The genetic fate of molecularly cloned simian immunodeficiency virus in experimentally infected macaques. *Virology* 1991;185:217–228.
51. Overbaugh J, Rudensey LM, Papenhausen MD, Benveniste RE, Morton WR. Variation in simian immunodeficiency virus *env* is confined to V1 and V4 during progression to AIDS. *J Virol* 1991; 65:7025–7031.
52. Robertson JS, Nicolson C, Bootman JS, Major D, Robertson EW, Wood JM. Sequence analysis of the haemagglutinin (HA) of influenza A (H1N1) viruses in clinical material and comparison with the HA of laboratory derived virus. *J Gen Virol* 1991;72:2671–2677.
53. Rocha E, Cox N, Black RA, Harmon MW, Harrisson CJ, Kendall A. Antigenic and genetic variation in influenza A (H1N1) virus isolates recovered from a persistently infected immunodeficient child. *J Virol* 1991;65:2340–2350.
54. Narayan O, Clements JE. Lentiviruses. In: Fields BN, Knipe DM, eds. *Virology.* New York: Raven Press, 1990:1571–1589.
55. Ho DD, Moudgil T, Alam M. Quantitiation of human immunodeficiency virus type 1 in the blood of infected persons. *N Engl J Med* 1989;321:1621–1625.
56. Simmonds P. Variation in HIV virus load of individuals at different stages in infection: possible relationship with risk of transmission. *AIDS* 1990;4:S77–S83.
57. Bagasra O, Hauptman SP, Lischner HW, Sachs M, Pomerantz RJ. Detection of human immunodeficiency type 1 provirus in mononuclear cells by in situ polymerase chain reaction. *N Engl J Med* 1992;326:1385–1391.
58. Pantaleo G, Graziosi C, Butini L, et al. Lymphoid organs function as major reservoirs for human immunodeficiency virus. *Proc Natl Acad Sci USA* 1991;88:9838–9842.
59. Choi Y, Kappler JW, Marrack P. A superantigen encoded in the open reading frame of the 3′ long terminal repeat of mouse mammary tumour virus. *Nature* 1991;350:203–207.
60. Pantaleo G, Graziosi C, Demareset JF, et al. HIV infection is active and progressive in lymphoid tissue during the clinically latent stage of disease. *Nature* 1993;362:355–358.
61. Åsjö B, Morfeldt-Månson L, Albert J, et al. Replicative capacity of human immunodeficiency virus from patients with varying severity of HIV infection. *Lancet* 1986;2:660–662.
62. Tersmette M, De Goede REY, Al BJM, et al. Differential syncytium-inducing capacity of human immunodeficiency virus isolates: frequent detection of syncytium-inducing isolates in patients with acquired immunodeficiency syndrome (AIDS) and AIDS-related complex. *J Virol* 1988;62:2026–2032.
63. Vartanian JP, Meyerhans A, Åsjö B, Wain-Hobson S. Selection, recombination and G→A hypermutation of HIV-1 genomes. *J Virol* 1991;65:1779–1788.
64. Tersmette M, Labge JMA, DE Goede REY, et al. Association between biological properties of human immunodeficiency virus variants and risk for AIDS and AIDS mortality. *Lancet* 1989;1:983–985.
65. Phillips AN, Lee CA, Elford J, et al. Serial CD4 lymphocyte counts and development of AIDS. *Lancet* 1991;337:389–392.
66. Prince AM, Ressink H, Pascual D, et al. Prevention of HIV infection by passive immunization with HIV immunoglobulin. *AIDS Res Hum Retroviruses* 1991;7:971–973.
67. Emini EA, Schleif WA, Nunberg JH, et al. Prevention of HIV-1 infection of chimpanzees by gp120 V3 domain-specific monoclonal antibody. *Nature* 1992;355:728–732.
68. Diez J, Mateau MG, Domingo E. Selection of antigenic variants of foot-and-mouth-disease virus in the absence of antibodies, as revealed by an in situ assay. *J Gen Virol* 1989;70:3281–3289.
69. Zwart GH, Langedijk H, van der Hoek L, et al. Immunodominance and antigenic variation of the principal neutralization domain of HIV-1. *Virology* 1991;181:481–489.
70. Montelaro RC, Parekh B, Orrego A, Issel CJ. Antigenic variation during persistent infection by equine infectious anemia virus, a retrovirus. *J Biol Chem* 1984;259:10539–10544.
71. Carpenter S, Evans LH, Sevoian M, Chesebro B. Role of host immune response in selection of equine infectious anemia virus variants. *J Virol* 1987;61:3783–3789.
72. Lutley R, Petursson G, Palsson PA, Georgsson G, Klein J, Nathansson N. Antigenic drift in visna: virus variation during long-term infection of Icelandic sheep. *J Gen Virol* 1983;64:1433–1440.
73. Thormar H, Barshatsky MR, Arnesen K, Kozlowski PB. The emergence of antigenic variants is a rare event in long term visna virus infection in vivo. *J Gen Virol* 1983;64:1427–1432.

74. Hoffenbach A, Langlade-Demoyen P, Dadaglio G, et al. Unusually high frequencies of HIV-specific cytotoxic lymphocytes in humans. *J Immunol* 1989;142:452–462
75. Meyerhans A, Dadaglio G, Vartanian JP, et al. In vivo persistence of HIV-1 encoded HLA-B27-restricted cytotoxic T lymphocyte epitope despite specific in vitro reactivity. *Eur J Immunol* 1991;21:2637–2640.
76. Phillips RE, Rowland-Jones S, Nixon DF, et al. Human immunodeficiency virus genetic variation that can escape cytotoxic T cell recognition. *Nature* 1991;354:453–459.
77. Chen ZW, Shen L, Miller MD, Ghim SH, Hughes AL, Letvin NL. Cytotoxic T lymphocytes do not appear to select for mutations in an immunodominant epitope of simian immunodeficiency virus *gag*. *J Immunol* 1992;149:4060–4066.
78. Madden DR, Gorga JC, Strominger JL, Wiley DC. The structure of HLA-B27 reveals nonamer self-peptides bound in an extended conformation. *Nature* 1991;353:321–325.
79. Jardetsky TS, Lane WS, Robinson RA, Madden DR, Wiley DC. Identification of self peptides bound to purified HLA-B27. *Nature* 1991;353:326–329.
80. Dadaglio G, Leroux A, Langlade-Demoyen P, et al. Epitope recognition of conserved HIV envelope sequences by human cytotoxic T lymphocytes. *J Immunol* 1991;147:2302–2309.
81. Hadida F, Parrot A, Kieny MP, et al. Carboxy-terminal and central regions of human immunodeficiency virus-1 *nef* recognized by cytotoxic T lymphocytes from lymphoid organs. An in vitro limiting dilution analysis. *J Clin Invest* 1991;89:53–60.
82. Nowak MA, Anderson RM, McLean AR, Wolfs TW, Goudsmit J, May RM. Antigenic thresholds and the development of AIDS. *Science* 1991;254:963–969.
83. De Magistris MT, Alexander J, Coggeshall M, Altman A, Gaeta FCA, Grey HM, Sette A. Antigen analog-major histocompatibility complexes act as antagonists of the T cell receptor. *Cell* 1992; 68:625–634.
84. Takeuchi Y, Akutsu M, Murayama K, Shimuzu N, Hoshino H. Host-range mutant of human immunodeficiency type 1: modification of cell tropism by a single point mutation at the neutralization epitope in the *env* gene. *J Virol* 1991;65:1710–1718.
85. Cheng-Meyer C, Seto D, Tateno M, Levy JA. Biologic features of HIV-1 that correlate with virulence in the host. *Science* 1988;240:80–83.
86. O'Brien WA, Koyanagi Y, Namazie A, et al. HIV-1 tropism for mononuclear phagocytes can be determined by regions of gp120 outside the CD4-binding domain. *Nature* 1990;348:69–73.
87. Pathak VK, Temin HM. Broad spectrum of in vitro forward mutations, hypermutations, and mutational hotspots in a retroviral shuttle vector after a single replication cycle: deletions and deletions with insertions. *Proc Natl Acad Sci USA* 1990;87:6024–6028.
88. Pulsinelli GA, Temin HM. Characterization of large deletions occurring during a single round of retrovirus vector replication: novel deletion mechanism involving errors in strand transfer. *J Virol* 1991;65,4786–4797.
89. Pathak VK, Temin HM. 5-Azacytidine and RNA secondary structure increase the retrovirus mutation rate. *J Virol* 1992;66:3093–3100.
90. Hu WS, Temin HM. Genetic consequences of packaging two RNA genomes in one retroviral particle: pseudoploidy and high rate of genetic recombination. *Proc Natl Acad Sci USA* 1990; 87:1556–1560.
91. Guyader M, Emerman M, Sonigo P, Clavel F, Montagnier L, Alizon M. Genome organization and transactivation of the human immunodeficiency virus type 2. *Nature* 1987;326:662–669.
92. Merluzzi VJ, Hargrave KD, Labodia M, et al. Inhibition of HIV-1 replication be a nonnucleoside reverse transcriptase inhibitor. *Science* 1990;250:1411–1413.
93. Colombini S, Arya SK, Reitz MS, Jagodzinski L, Beaver B, Wong-Staal F. Structure of simian immunodeficiency virus regulatory genes. *Proc Natl Acad Sci USA* 1989;86:4813–4817.
94. Oroszlan S, Luftig RB. Retroviral proteinases. *Curr Top Microbiol Immunol* 1990;153–185.
95. Kohlstaedt LA, Wand J, Friedman JM, Rice PA, Steitz TA. 3.5Å resolution crystal structure of HIV-1 reverse transcriptase complexed with an inhibitor. *Science* 1992;256:1783–1790.
96. Hogle JM, et al. Three dimensional structure of poliovirus at 2.9Å resolution. *Science* 1985; 229:1358–1365.
97. Rossman MG, Arnold E, Erickson J, et al. Structure of a human common cold virus and functional relationship to other picornaviruses. *Nature* 1985;317:145–153.
98. Grant RA, Filman DJ, Fujinami RS, Icenogle JP, Hogle JM. Three dimensional structure of Theiler's virus. *Proc Natl Acad Sci USA* 1992;89:2061–2065.
99. Luo M, Vriend G, Kamer G, et al. The atomic structure of Mengo virus at 3.0Å resolution. *Science* 1987;235:182–191.

100. Acharya R, Fry E, Stuart D, Fox G, Rowlands D, Brown F. The three-dimensional structure of foot-and-mouth disease virus at 2.9Å resolution. *Nature* 1989;337:709–716.
101. Palmenberg AC. Sequence alignments of picornaviral capsid proteins. In: Semler BL, Ehrenfeld E, eds. *Molecular aspects of picornavirus infection and detection.* Washington, DC: American Society for Microbiology: 1989:211–241.
102. Horstmann DM, Opton EM, Klemperer R, Llado B, Vignec AJ. Viremia in infants vaccinated with oral poliovirus vaccine (Sabin). *Am J Hyg* 1964;79:47–63.
103. McGrath MS, Herndier BG, Reyes G, Shiramizu BT. HIV-associated T cell lymphoma: evidence that HIV may directly cause T cell transformation. (Abstract Q335.) *J Cell Biochem* Suppl 16E, 553.
104. Raineri I, Senn HP. HIV-1 promoter revealed by selective detection of chimeric provirus-host gene transcripts. *Nucleic Acids Res* 1992;20:6261–6266.
105. Tristem M, Marshall C, Karpas A, Petrik J, Hill F. Origin of *vpx* in lentiviruses. *Nature* 1990; 347:341–342.
106. Harouse JM, Bhat S, Spitalnik SL, et al. Inhibition of entry of HIV-1 into neural cell lines by antibodies against galactosyl ceramide. *Science* 1991;253:320–323.
107. Elder JH, Lerner DL, Hasselkus-Light CS, et al. Distinct subsets of retroviruses encode dUTPase. *J Virol* 1992;66:1791–1794.
108. Embretson J, Zupancic M, Ribas JL, et al. Massive covert infection of helper T lymphocytes and macrophages by HIV during the incubation period of AIDS. *Nature* 1993;362:359–362.
109. Sanchez-Palomino S, Rojas JM, Martinez MA, et al. Dilute passage promotes expression of genetic and phenotypic variants of human immunodeficiency virus type 1 in cell culture. *J Virol* 1993;67:2938–2943.

The Evolutionary Biology of Viruses,
edited by Stephen S. Morse.
Raven Press, Ltd., New York © 1994.

# 10

# The Future of Human Immunodeficiency Virus

Gerald Myers and Bette Korber

*HIV Sequence Database and Analysis Project, Theoretical Division,
Los Alamos National Laboratory, Los Alamos, New Mexico 87545*

> "All is flux."
> —Heraclitus

The phylogenetic processes of most viruses are slow compared with the rate at which their diseases spread. Even when the genetic variation of a pathogenic virus is unusually brisk, making vaccine modifications a perennial challenge, the dimensions of the disease typically remain unchanged (1). Human immunodeficiency virus (HIV), which causes acquired immunodeficiency syndrome (AIDS) in humans, is an exceptional case: as a result of a recent ecological breakthrough it appears to be engaged in a process of "fast-forward" evolution. This chapter will argue, on the basis of genetic indications, that the future of HIV will be characterized not only by continued rapid genetic variation, but also by a significant broadening of the clinical spectrum of AIDS.

Evolutionary biology is on safer ground when it inquires into the past, rather than the future. Although some unanswered questions remain about the "deep" history of the AIDS viruses, our understanding of their recent emergence has grown swiftly since the publication of the first HIV genetic sequences in 1985 (1–11; see also chap. 10). Although we will not systematically review that body of information, the first part of this chapter will consider certain aspects of the speciation and molecular epidemiology of HIV that we deem most predictive. What is known of its past suggests that HIV is in a singularly threatening state of evolutionary disequilibrium.

In the second part of the chapter, some quantitative features of HIV variation will be examined. First, the global rate of HIV variation will be discussed. Then we will turn to an extraordinary characteristic of all known primate immunodeficiency viruses: they are highly variable and also *complex*. (By complex we mean that they are able to regulate their own gene expression; 11). These would seem to be opposing evolutionary tendencies: Some viruses, such as the human T-cell lymphotropic retroviruses (HTLV), are complex, but manifest comparatively lower rates of genetic change; other RNA viruses mutate rapidly, but their near-total dependence on intricate aspects of host cell machinery places considerable constraint on their degrees of evolutionary freedom. The primate immunodeficiency viruses are degenerate in their complexity. They display inordinately high frequencies of nonsynonymous substitutions, not simply, as we might expect, in their envelope protein-coding sequences, but also in their regulatory and accessory genes. This motley evolutionary posture, coupled with the familiar collection of retroviral adaptive strategies–integration, legitimate and illegitimate recombination, transduction, pseudotyping, molecular mimicry and parallelism–has crucial pathogenic implications.

In the third part of the chapter, we will consider a few of the many different hypotheses about AIDS pathogenesis. We will then explore the association of different pathogenic characteristics with different genotypic patterns in both HIVs and simian immunodeficiency viruses (SIVs), focusing attention on the now well-studied HIV V3 loop, a small peptide in the viral coat protein that has an established association with HIV phenotype. On the basis of a *phenetic* analysis of HIV V3 loop variation, we will argue that the seemingly untrammeled diversification of HIV constitutes a significant factor in its pathogenic potential.

## THE RECENT EMERGENCE OF HIV

### Simian Origin of HIV: An Evolutionary Turning Point

There is strong evidence that HIVs and SIVs have naturally evolved from an unknown precusor within the highly diverse subfamily of lentiretroviruses, found also in ungulates and felids (12). The worldwide prevalence of these retroviruses suggests that they are old (13), but there is as yet no consensus about how old they are, especially in relation to the divergence of HIVs and SIVs. The European and South African visna viruses, which infect domesticated sheep, diverged from one another about 50 years ago, and one group of researchers has, on the basis of this fact, hypothesized a minimum "lookback" time for lentiviral radiation of merely 400 years (14). Eigen's calculations, in contrast, place the oldest common ancestor for known human and simian lentiviruses back at least 600 to 1,200 years (5), and it has been argued that even this seems too short a time for the dissemination of the felid lentiviruses (13). What is more, the African green monkey lentivirus (SIVagm), which may well be the distant predecessor of HIVs, is believed to have been present before the allopatric speciation of host guenons several thousand years ago; the four extant species of green monkeys dispersed throughout Africa harbor four distinguishable subtypes of SIV (15).

These reckonings—hundreds of years, thousands of years—are arguably short spans even on a microbial evolutionary scale. Yet, precisely because so much molecular genetic activity has gone on in the short evolutionary stretch of the lentiviruses, the deeper branches of the lentiviral tree are enveloped in darkness (9). This indeterminacy will not, however, be a consideration of any weight for our purposes, because our focus will be on the primate lentiviruses, particularly the recent emergence of HIVs from SIVs, to which we will now turn.

Three historical moments, each traceable to the 1970s, mark the onset of AIDS in the world. These three events are the epidemic of HIV-1 disease, the separate epidemic of HIV-2 disease, and the outbreak (epizootic) of AIDS-like disease in captive macaques, animals not found to be infected in the wild. Two earlier incidents of apparent HIV-1 infection have been documented retrospectively; the one referred to as the "Manchester sailor" case (16), involving a seropositive sample from 1959, is now undergoing molecular investigation. The preponderance of HIV-1 AIDS viewed from the dual standpoint of clinical and seroepidemiological records, however, came forward in Africa and the United States in the 1970s (2,10). What began as a handful of cases known anecdotally in the 1960s and 1970s has now resulted in the estimated infection of more than 10 million individuals: HIV-1 alone is expected to infect more than 40 million people by the year 2000 (17,18).

HIV-2 AIDS, which, until recently, has largely been associated with people of West Africa, does not have an equally well-documented seroepidemiologic record (10). However, the evidence of early cases of HIV-2 dissemination to India (19), Spain (20), Brazil (21), and even the near-epicentric Cameroon (22) is suggestive of a new and separate AIDS pandemic. The HIV-2 epidemic in India appears to be aggressive compared with the current low level of infection elsewhere outside of western Africa (19).

Finally, sooty mangabey viruses (SIVsmm) that are virtually indistinguishable from HIV-2s (see later) contributed to a devastating epizootic in the mid-1970s. The SIVsmm cross-infected stump-tailed macaques housed with the viruses' natural hosts, mangabeys from West Africa, first at the California Regional Primate Center in the 1960s, then subsequently at other United States primate centers (23,24).

This historical background, together with molecular evidence (discussed in a later section), has led AIDS researchers to conclude that three independent events of cross-species viral transmission and subsequent dissemination in the recipient species occurred at about the same time. Explanations for this striking improbability have focused on accidents and social conditions: (a) increased handling of animals associated with the sharp increase of animal exportation from Africa, (b) the widespread use of needles in association with human and captive monkey vaccination programs, and (c) urbanization of Africa (2,3,10,15,17,23). We may never know which of these contributing factors was most responsible for the origin of AIDS.

To illustrate the kind of molecular epidemiological evidence (reviewed in 9–11) that has supported the hypothesis of a recent simian origin of AIDS, one need only consider that it is not possible to tell from SIV/HIV-2 genetic sequences whether they were derived from an infected sooty mangabey or an infected human. We can add to this the fact that the range of sooty mangabeys in West Africa overlaps the epicenter of the HIV-2 epidemic (25,26). Moreover, several cases of apparent SIV infection of laboratory workers are now under investigation (27,28).

The close genetic relation of human and simian immunodeficiency viruses is apparent from the phylogenetic tree shown in Fig. 1. For the sake of clarity, only a very small number of known HIV and SIV sequences have been included in this tree analysis, which is based on relatively conservative homologous *pol* gene sequences (29). This tree is representative of the many sequences that have not been included,

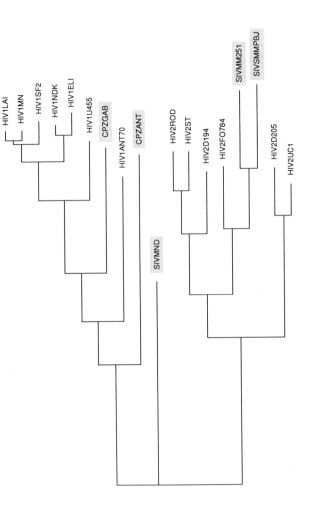

5 %

**FIG. 1.** Phylogenetic tree analysis comparing *pol* gene sequences of representative HIV-1, HIV-2, SIV, and CPZ samples. For the sake of clarity, only a very small number of known sequences have been included. Viral sequences derived from monkeys (SIV) and chimps (CPZ) have been highlighted. The HIV1ANT70 sequence may represent an animal viral sequence found on humans (see text). The tree was constructed by using the PAUP maximum parsimony algorithm. Of the 971 sites examined, 617 were varied. The lengths of the horizontal branches are proportional to single-base changes; these distances can be read as minimum percentage differences using the scale bar. The tree was rooted at the midpoint of the greatest patristic branch. Bootstrap analysis supported the tree topology as did tree analyses of other coding sequences.

and it is congruent with trees over other major viral-coding sequences. To emphasize some of the close relationships that are being encountered by AIDS researchers, SIV-derived sequences are highlighted in Fig. 1. Note that an "HIV-2" sample, designated F0784 in Fig. 1, taken from an asymptomatic Liberian individual (30), clusters with SIVs in the computer analysis. This is one of the most compelling pieces of evidence in support of the suspected mangabey virus origin of HIV-2. The close linkage of HIV-1 sequences with viral sequences isolated from wild-caught chimpanzees is the best clue yet for the origin of HIV-1s (31,32). It is apparent from Fig. 1 that a Cameroonian "HIV-1" isolate, denoted ANT70 (33), could be interpreted as a chimpanzee virus found in a human.

A highly diverse reservoir of immunodeficiency viruses have been catalogued in African green monkeys and their relatives (15,34,35). Although the genetic sequences determined thus far for these species do not directly account for the emergence of the two known types of HIV, recent phylogenetic studies suggest that the mangabey viruses found in West African animals are descendants of *sabaeus* monkey viruses also found in West Africa (Hahn B, personal communication). At least 13 species of feral African primates are known to be carrying HIV-related viruses (15,34); hence, additional zoonotic infections are clearly possible.

The epidemiological and epizootic moments implicit in Fig. 1 are historical and evolutionary turning points, suggestive of a discontinuity in the evolutionary flow. When there has been a breakthrough of this sort, Gould argues (as restated by Allan) that "punctuated equilibrium asserts that the rate of evolution is accelerated during times of environmental upheaval, resulting in the formation of new species" (15). This theory, which has traditionally been understood as *adaptive radiation*, has been described by Gould and colleagues in exact mathematical terms (36); referring to this process as *bottom-heavy cladism*, they predict a burgeoning of genetic forms during the period of evolutionary adjustment to a new host (here, humans). There is, in fact, evidence for rapid speciation of both HIV and SIV in captive animals concurrent with the AIDS pandemic.

### Rapid Speciation Indicates Disequilibrium

If one looks closely at the "leaves" of the phylogenetic tree shown in Fig. 1, one sees that there are at least three, distinct subtypes of HIV-2. It is not now known whether these subtypes arose through independent events of cross-species transmission, although that seems the likeliest possibility. Additional information about West African simian immunodeficiency viruses is needed to answer this question. The subtypes represented by HIV2ROD and HIV2D205 in Fig. 1 have an average difference of 23% in their coding sequences (37), and within each subtype distances of up to 18% are encountered (29). This degree of diversification indicates that the subtypes will probably evolve into distinct viral types, differing by at least 50% in their coding regions.

From newly determined HIV-1 *gag* and *env* gene sequences (29,38–40), five distinct sequence subtypes have now been identified in the major centers of the AIDS pandemic (Fig. 2;29). These HIV-1 subtypes appear to have diverged from one another as recently as 1960, that is to say, they may have proliferated in step with the pandemic. (The lookback calculations supporting this hypothesis are considered in the next section of this chapter.) Because the surveillance effort is still in its infancy, it is conceivable that additional prevalent viral sequence subtypes will be discovered in the next few years.

Many countries—Brazil, Thailand, Uganda, and India, to name a few—now have cocirculating strains of HIV-1. The fact that only a single sequence subtype of the five or more known subtypes has yet been reported in United States AIDS patients is the result, we suppose, of a very strong founder effect. Currently, no one subtype seems to be more successful or more virulent (but, see later section on variation and pathogenesis), nor do the subtypes appear to have host preferences. The measurements necessary for evaluation of such phenotypic viral polymorphism prove quite difficult, however, and subtle changes in transmissibility and the average incubation time for AIDS can be discerned only with data gathered over a period of many years. Although trafficking of extant HIV-1 subtypes will undoubtedly make a greater contribution than mutation to the virus's ecological variation in the coming decade, the future spectrum of HIV phenotypes, we will argue, will be shaped by mutation (including recombination) and selection.

### Transmission Contributes to Viral Diversification

Gould's theory of *bottom-heavy cladism* attempts to account for temporal asymmetries in the speciation process over long time periods. Although the existence of the five HIV-1 sequence subtypes shown in Fig. 2 is consistent with it, given the short time HIVs and SIVs have been observed, the HIV sequence subtypes are not sufficient to confirm the theory. Even so, the relationships of viruses shown in Fig. 2 are suggestive of a condition of disequilibrium that cannot be simply explained by a high mutation rate. Influenza A virus, for example, mutates about as rapidly as HIV-1, but its pattern of evolution is entirely different from that represented in Figs. 1 and 2: Flu-A trees are decidedly slender, whereas HIV-1 trees are bushy (11,41), which reminds us of Coffin's dictum that "mutation and variation are not the same" (42). This difference stems, for the most part, from the relative lack of intraspecific competition among HIV variants infecting resting cells. It is as if the niche for HIV within humans is more tolerant than that for influenza A. Until a steady-state evolutionary process is attained, HIV-1 variation might be epidemic-driven in a way that flu-A variation is not: specifically, the rate of HIV variation could be a nonlinear

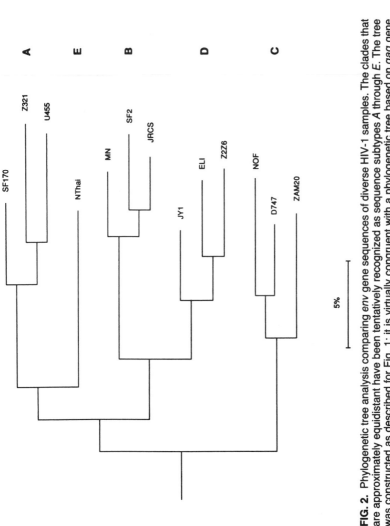

**FIG. 2.** Phylogenetic tree analysis comparing *env* gene sequences of diverse HIV-1 samples. The clades that are approximately equidistant have been tentatively recognized as sequence subtypes *A* through *E*. The tree was constructed as described for Fig. 1; it is virtually congruent with a phylogenetic tree based on *gag* gene sequences (29,38). At this still early time in the pandemic, subtype *A*, *C*, and *D* viruses are found preponderantly in Africa; subtype *B* viruses are found mostly in the United States and Europe; and subtype *E* viruses are characteristic of the outbreak in northern Thailand (but are also found in Central African Republic). The ANT70 sequence, included in Fig. 1, could represent a distinct HIV-1 subtype.

function of the annual AIDS incidence rate and the total number of infected individuals, irrespective of the hypothesized simian origin and any general concept of temporal asymmetry of viral adaptation. (The Luria–Delbruck equations, which describe a Darwinian process for rapidly mutating bacterial populations, turn out to be a useful starting point for exploration of this possibility.) The extent, if any, to which HIV-1 variation is epidemic-driven, should become apparent as a dilation of branch lengths in phylogenetic trees, perhaps by the year 2000.

Finally, the genetic proliferation of HIV in the coming decades can be considered from an ecological perspective. Ewald has argued that cultural characteristics will dictate, in very short times, the variability and virulence of a pathogen such as HIV (43,43a): "As frequencies of unprotected sexual contact with different partners increase, the [viral] benefits of increased durations of infectiousness should decline and virulence should increase." This prediction, which emphasizes variation through selection over mutability, can be corroborated or disproved in a few years, assuming that the critical timeframe of the HIV pandemic will be comparable with the outbreak and evolutionary resolution of the myxoma virus infection in Australian rabbits, a case thought to have some parallels with AIDS epidemicity. [Contrary to popular understanding, Ewald argues, the myxoma virus remains to this day significantly pathogenic in Australian rabbits.] Ewald proposes that "explaining the evolution of HIV thereore involves determination of how a benign pathogen [SIV] can evolve to high levels of virulence rather than how an extremely virulent pathogen [myxoma virus] can evolve to high levels of virulence" (43). The thrust of this argument is that human behavior—"sexual contact diversity" and intravenous drug use involving shared needles—will become an increasingly significant factor in the evolution and pathogenicity of HIV.

Whereas it has been argued by Temin that it is doubtful that HIV-1 will evolve into a worse pathogen, his argument is based on the assumption that the virus must simultaneously alter its infectiousness and its pathogenic potential, which he points out are "already optimized for spread and disease in the human population" (44). Temin goes on to suggest that HIV-1 may gradually evolve into a less pathogenic form. Ewald, on the other hand, theorizes that the pathogenic equilibrium could still shift upward as a result of human behaviors that provide a context for a decreased viral latency period. He also could imagine evolution toward a less pathogenic state, but only with the proviso of a more conservative human sexual behavior. Thus, the two outlooks differ in their assumptions about the major causal forces. HIV-2 disease, which already presents a greater clinical spectrum than HIV-1 disease while infecting far fewer individuals (see later section on variation and pathogenesis), provides an important context in which Ewald's causal hypothesis can be judged.

In this section we have reviewed the qualitative and semiquantitative evidence that supports the theory of a recent simian origin and rapid diversification of the HIVs. The probable implication of this viral history is that HIV will continue to be an atypically unstable and aggressive pathogen for some time. Before turning directly to questions about HIV pathogenicity, we will next review some relevant quantitative features of HIV molecular variation.

## MOLECULAR FACTORS CONTRIBUTING TO HIV VARIATION

### Rate of Variation

The theory of a recent simian origin and rapid speciation of HIV is supported by a determination of the average phenomenological rate of macroscopic variability. The possibility of determining such a quantity has been brought into question by Coffin, on the basis of widely observed intrapatient heterogeneity (42). The preceding discussion of HIV variation has emphasized global diversity, what can be called *macroscopic swarming*. Intrapatient heterogeneity, or *microscopic swarming*, is a second feature of diversification to be considered in any discussion of HIV evolution. In general, RNA viruses are highly mutable, but with RNA viruses such as foot-and-mouth-disease virus (FMDV), which infects cattle, and HIV, the variability is so great that there are populations of closely related viral genotypes within most infected individuals. These populations, or swarms, are termed *quasispecies* (4, 45,46; also see Chap. 10). Their relevance for progression to AIDS in HIV infections is a matter of some dispute (see section on variation and pathogenesis). Furthermore, it is unclear how selection comes into play when one individual infects another individual: since each transmission can be thought of as a recloning step that may or may not destabilize a quasispecies, Coffin has hypothesized that "cyclic patterns of variation" will arise and, therefore, that "there can be no such thing as a 'rate of variation'" (42). (Genetic bottleneck transmission, which depends on small innocula sizes, undoubtedly contributes to the indeterminancy emphasized by Coffin (47). While we agree with Coffin that empirical determinations of variation at a defined position in a coding sequence may prove intractable, this contention does not apply to macroscopic or global variation.

Molecular epidemiological investigations of mother–baby transmission, sexual transmission, and transmission from an infected dentist to his patients have revealed viral sequence similarities that were intermediate between those observed within a single individual and those observed between unlinked individuals (48–50). Furthermore, rate estimates of HIV-1 nucleotide substitution made by many investigators for many samples have been in surprisingly good agreement (reviewed in 11). On the basis of these findings, a continuous magnitude, representing an average phenomenological rate of macroscopic variability of HIV, may be determined. (Calculations based on this phenomenological rate presuppose a model akin to models of Brownian motion in which stochastic variables are handled as deterministic class-averages at the macroscopic level.)

To investigate this average quantity, we first determined frequencies of synonymous and nonsynonymous substitutions for a large number of HIV-1 *gag* and *env* gene sequences from a variety of infected individuals. Figure 3 shows a plot of the rates (52) of nonsynonymous ($d_n$) versus synonymous ($d_s$) nucleotide substitutions for 35 complete HIV-1 *env* gp120-coding sequences: $d_s$ is the corrected frequency of synonymous substitutions that could have occurred for each of the 595 possible pairwise sequence relations that have been recorded in Figure 3. This value

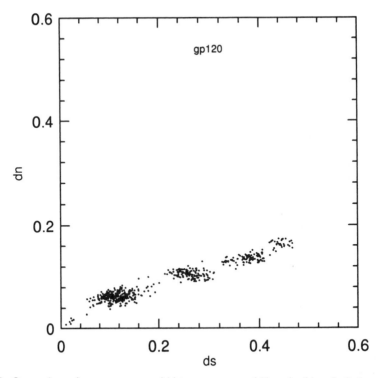

**FIG. 3.** Comparison of nonsynonymous ($d_n$) to synonymous ($d_s$) nucleotide substitutions in the HIV-1 *env* gp 120-coding sequences for 35 distinct samples. The Nei–Gojobori algorithm was employed (51). The slope, 0.4, is indicative of weak purifying selection. A similar analysis of HIV-1 *gag*-coding sequences (not shown) produces a slope of 0.2; for comparison, the influenza A hemagglutinin-coding sequence yields a slope of 0.3 (52). Only recently diverged *env* sequence relationships ($d_s$<0.5) are shown. The different sequence subtypes seen in Fig. 2 are responsible for the clustering effect. For relationships close to or greater than expected mutational saturation ($d_s$ = 0.75), the slope becomes less than 0.4.

is, accordingly, a fair indicator of the mutational distance separating relatively close sequences, such as those represented in the *env* gene tree in Fig. 2 ($d_s$<0.5). More distant relations involving the CPZ and ANT70 sequences (included in Fig. 1) have synonymous substitution values of about 0.75, the expected saturation state, and are not shown in Fig. 3.

The *gag* and *env* substitution frequencies ($d_s$) represented in these studies, taken together with similar rates determined by several investigators (reviewed in 11), support the hypothesis that the five recognized HIV-1 sequence subtypes (see Fig. 2) diverged from one another as recently as 1960, a plausible date for the onset of the AIDS pandemic. The reasoning is as follows: the average intertypic nucleotide substitution rate for HIV-1 *env* divergence is about 1.1% per year (11; and see Fig. 3), and the average intertypic nucleotide distance, the Hamming distance, shown in Fig. 2 is about 32%, suggesting a lookback time of about 30 years. The *gag* rate is just half the *env* rate (as is the ratio of $d_n$ to $d_s$ for *gag*-coding sequences) and the average intertypic distance is about 14% (29,38). With this same value for the *gag* fixation rate, Khan and colleagues calculated that the stump-tailed macaque virus apparently diverged from captive sooty mangabey and rhesus macaque viruses as recently as 1963, the approximate date of arrival of these animals into the colonies housing those other animals (23).

These estimates are consistent with the historical and seroepidemiologic records discussed in the earlier section on emergence of HIV. Hence, without evidence to the contrary, average rates of variation derived from large data sets suitably constituted from recently diverged viral sequences may be taken to have useful predictive value. However, the most important indicator for future HIV variation will be the nonsynonymous nucleotide substitution rate $(d_n)$, which is extraordinarily high. This is apparent from Fig. 3 in which we see first, that a linear relation of $d_n$ to $d_s$ is preserved for values of $d_s$ between 0.05 and 0.5, and second, that the slope in Fig. 3 (0.4) is relatively steep, greater, for example, than the equivalently determined slope for the influenza A hemagglutinin gene (0.3) (52). The ratio of $d_n$ to $d_s$ determined for 45 HIV-1 *gag*-coding sequences was approximately 0.2 (data not shown). The unusually high level of nonsynonymous substitution in HIV *env* sequences has been documented by several investigators, although with fewer sequences than were analyzed in Fig. 3 (9,53–56).

The ease with which amino acid replacements take place in HIVs is explained by their optimal mutation rate (57) and their unusual codon usage favoring nonconservative changes (11), but it is also a consequence of weak constraints. By one interpretation, the fact that synonymous substitutions predominate, supports Kimura's "neutral theory" (52). From another perspective, the relatively high ratio found in the HIV-1 *env* gene could be taken as evidence for "positive Darwinian selection" (58), or for "overdominant selection" (56,59). This latter possibility seems particularly plausible in the context of microscopic HIV evolution for which selection by an infected individual's immune response may favor change: extremely large ratios have been observed among microscopic swarms (48,54). Although we shall not attempt to weigh these alternatives in this chapter, the high ratios of nonsynonymous to synonymous substitution, which are indicative of weak purifying selection, have broad evolutionary implications.

## Breadth of Variability

Although we have discussed only *gag* and *env* gene variability, HIV proteins, in general, are subject to high frequencies of amino acid replacement. There is a common misperception about the genomic distribution of variation observed in HIV-1s and HIV-2s; namely, that it is limited, for the most part, to the hypervariable regions of the *env* gene. Most of the HIV regulatory and accessory proteins, as well as select portions of the *gag* polyprotein, are as variable as the envelope protein (9,11,29). These findings are evident in the comparisons of average information

**TABLE 1.** *Protein information densities expressed as amino acid equivalents*

| Viral sequences | gag | pol | tat | rev | env | nef |
|---|---|---|---|---|---|---|
| HIV-1/CIV (8) | 0.77 | 0.86 | 0.65 | 0.58 | 0.60 | 0.64 |
| HIV-2/SIV (7) | 0.84 | 0.90 | 0.55 | 0.70 | 0.73 | 0.55 |
| HIV-1/HIV-2/SIVs (15) | 0.56 | 0.62 | — | 0.30 | 0.35 | 0.34 |

The above data have been previously reported and discussed in refs. 9 and 11. The information density expresses the average number of invariant and chemically conserved residues at each position of the protein molecule. For example, there are 0.65 amino acid equivalents, on the average, at each position in the HIV-1/CPZ *tat* protein set under study. Comparisons within each row involve the same set of complete genomic sequences: 8 HIV-1/CPZs, 7 HIV-2/SIVs, and 15 combined HIV/SIVs. The values will decrease as more sequences are considered; in this discussion, the focus is on the ratios of protein variation. Thus HIV-1 *env* protein is actually no more variable than the HIV-1 *rev* protein (0.60 versus 0.58).

densities, expressed as amino acid equivalents, of homologous HIV proteins. Table 1, for example, provides a summary of *tat*, *rev*, and *nef* protein variability in comparison with the *gag*, *pol*, and *env* proteins; these three regulatory proteins are known to play crucial roles in the viral life cycle (60). The HIV-1 *rev* protein, for instance, is as variable as the HIV-1 *env* protein according to the measurements summarized in Table 1: on the average, HIV-1 *rev* protein manifests 0.58 amino acid equivalents at each residue position, versus 0.6 amino acid equivalents at each position in the *env* protein. The HIV-2 *tat*, *rev*, and *nef* proteins, all are as variable as their correlative *env* protein (0.55, 0.70, and 0.55, versus 0.73). The *rev* and *nef* values for the combined HIVs (0.30 and 0.34) are close to or less than the value for the combined *env* (0.35) and are also close to typical values encountered for eukaryotic proteins having very distant evolutionary relationships (11). In the previous analysis (see Fig. 3), we noted that HIV-1 *gag* gene variation was typical, but that *env* gene variation was not; the comparative data in Table 1 confirms the notion of weak purifying selection and extends it to several other HIV proteins. To the extent that HIV variation is more than merely a property of select (hypervariable) regions of the lentiviral envelope, balanced polymorphism becomes a major factor in HIV speciation.

Other molecular dimensions of HIV variability that undoubtedly contribute to its rich evolutionary potential are shown by the wide diversity of promoter–enhancer configurations and envelope glycosylation profiles: HIVs and SIVs may possess zero, one, or two NFϰB-binding sites in their long-terminal repeat (LTR) U3 regions (35,61,62) and may differ in their envelope modification at as many as 25 potential N-linked glycosylation sites (9,29,63,64).* Thus, the HIVs are both degenerate and complex. The combination of variability, degeneracy, existence as quasispecies, and complexity in HIVs allows many different lineages, and possibly

---

*The HIV-1 envelope gp120 molecule is one of the most heavily glycosylated proteins in nature; the hypervariability of the *env* gene appears to be strongly tied to mutational changes in the potential modification sites, NXS and NXT, where X is almost any amino acid.

different evolutionary scenarios, under a variety of selective forces. Complementation within the quasispecies (65), heterologous transactivation (11,60,61), transduction (66, also see Chapter 10), pseudotyping (67), and "bottleneck transmission" (47) are some of the compounding determinants of variability in these highly plastic pathogens. In view of these considerations, it is highly likely, as we will argue in the following, that different phenotypic lineages of HIV will emerge.

## HIV VARIATION AND PATHOGENESIS

In the first section of this paper we concluded with some certainty that the human AIDS-causing viruses have had a recent simian origin. From several points of view, this breakthrough was interpreted as a portent of viral genetic instability for some time to come. In the second section of the paper, some of the quantitative conditions of this instability were examined. We saw that the genetic volatility can rapidly scramble parts of the HIV *env* protein (as a result of a high ratio of nonsynonymous to synonymous substitutions; see Fig. 3) and that it involves the viral regulatory and accessory proteins as much as the viral coat (see Table 1). Accordingly, we can expect to see continued HIV speciation. What effect, if any, can we expect this to have on the global pathogenic potential of these viruses?

### Two Causal Views of HIV Pathogenesis

To attempt to summarize all of the current hypotheses about HIV pathogenesis is beyond our scope. (For a recent review of HIV immunopathogenesis, see ref. 68.) We shall limit our discussion to two classes of hypotheses that are most instructive in relation to the questions we have posed. One category of explanations for HIV pathogenesis emphasizes genetic variation in only general terms; it argues that HIV genetic variation needs to be only pervasive and nonspecific to produce pathogenic effects. The second class of hypotheses, supported by a now sizeable body of evidence, points to the occurrence of specific genetic determinants of the pathogenic phenotype. With the first class of hypotheses, the emphasis is on immunogenicity as the pathological force. The second class turns its focus toward direct cytolytic properties traceable to the HIV or SIV genotype. These two classes of explanations for HIV pathogenesis are not mutually exclusive: cell killing can be both direct and indirect.

Several examples of the first class of hypotheses may be mentioned. It has been proposed that viral variation could play a supportive role in a process of clonal antibody dominance: dominant B- and T-cell clones that have arisen in response to an initially homogeneous viral antigen ("original antigenic sin") might be stimulated by subsequent viral variants, irrespective of their precise makeup, resulting in an antibody "repertoire freeze" (69). A second possibility here is that of a "superantigen effect" (70). In a third example that involves viral variation only in a broad sense, HIV diversification is thought sufficient to exhaust the immune system: an essential asymmetry between antigen and antibody stems from the fact that all viral variants can threaten cells through interaction with the CD4 receptor, but each antibody clone is type-specific for a single variant (71,72). Many other mechanisms that fall under this category have been put forward, among which might be mentioned the hypothesis that AIDS is essentially an autoimmune disease (73). Common to all of these various theories is the concept that variation is significant only in a general way. As Haseltine puts it, "transmission, pathogenesis and incubation of the disease is similar regardless of the infecting strain" (74).

The foremost example of specific genetic determination of atypical AIDS pathogenesis is an acutely lethal SIV variant that arose after a single passage in a macaque. The parent virus in this case was benign in its natural host, the mangabey, but through mutating to the highly lethal form in the macaque—able to kill macaques in weeks rather than years—it acquired aggressive pathogenic properties when placed back into mangabeys (75,76). A less dramatic, but no less informative, example is encountered in the comparison of two HIV-2 variants found to be infectious in macaques. The HIV-2 variant with enhanced cytopathogenicity and replicative rate has significantly fewer envelope modification sites (the protein sites mentioned earlier) than the second, less cytopathic variant (77). Cases have been observed in which naturally occurring attenuated forms have yielded cytopathic variants (78) and, conversely, highly cytopathic forms have generated nonpathogenic viral progeny (79,80). We are reminded that AIDS neuropathogenesis is a distinct evolutionary concern when we learn about HIV-1 variants with different properties of gene expression in the mouse brain (81).

### Evidence For Different Pathogenic Phenotypes of HIV

Although the strongest argument for differing viral phenotypes is derived mostly from animal studies, evidence exists for different pathogenic phenotypes of HIV in the global setting. Early in the epidemic, a Swedish research team noted that persons harboring viruses that could be characterized as having "rapid-high" phenotypes in vitro were closer to clinical AIDS than persons whose viruses were "slow-low" in culture (82). Further investigation of these two HIV-1 states—one predominant in early infection, the other predominant in late infection—has linked them, first, to the capacity for not inducing syncytia (NSI) versus inducing syncytia (SI) in vitro and, second, to abilty to grow on macrophages (macrophage-tropic or monocytotropic) versus ability to grow on transformed T-cell lines (T-cell–tropic) (83–91). Thus, generally speaking, progression to HIV-1 AIDS has been more rapid in persons harboring the SI viral phenotype that are said to be T-cell-tropic (83,84). The correlation is not without exceptions—for example, an extremely cytopathic virus that is also macrophage-tropic has been reported (92); however, this has been the finding of many investigators.

With few qualifications, the functionally critical determinant of the SI phenotype has been traced to a region of the viral envelope known as the V3 loop or principal neutralizing determinant (PND), a 35-amino acid peptide located in the third hypervariable domain of the envelope protein. No single region of the HIV genome dictates the variability underlying HIV phenotypic diversity, however, the V3 loop has received the greatest attention because (a) it serves as an epitope for potent immunological responses (93,94); (b) it affects fusion and viral virulence (85); and (c) it is implicated as a primary determinant of HIV-1 cell tropism (86,87). A single amino acid substitution in this variable protein can have phenotypic ramifications (88,89). Positively charged amino acid residues, most often arginines, found in the flanking regions of the V3 loop have been associated with the SI, or the more cytopathic, phenotype (90).

The correlation of V3 peptide-coding sequences with pathogenesis has almost entirely been focused upon HIV-1 variants, as exemplified by the hundreds of HIV-1 V3 sequences deposited in the HIV sequence database (29). Samples of HIV-1 subtype D viruses, characteristic of many infections in Central Africa (see first section and Fig. 2), have shown a higher frequency of the SI phenotype than have subtype B viruses, which are typical of HIV-1 infections in the United States. However, because the spectrums of opportunistic diseases are very different in the United States and Africa, and because viral sampling has not been systematically controlled, it is not possible to conclude simply that the one viral subtype is intrinsically more pathogenic than the other viral subtype.

Even so, *phenetic* analyses of V3 loop protein sequences, which emphasize structural similarities, irrespective of the evolutionary pathway, suggest that uncannily similar V3 loop proteins can be encountered in otherwise dissimilar nucleotide sequence subtypes (95). This manifestation of evolutionary parallelism can be viewed as sequence convergence, although we prefer to interpret it as simply lack of divergence: certain V3 sequence lineages, suggestive of the monocytotropic forms of HIV-1, appear to be more stable than other V3 lineages, especially those of the T-cell–tropic variety. Thus, some Ugandan V3 loop protein sequences of one *cladistically* determined sequence subtype are found to tightly cluster with United States V3 loop sequences of another sequence subtype (95,96).

Although they have not been studied with the same intensity, HIV-2 V3 loop peptide sequences yield an intriguing pattern that has a suggestive correlation with the health status of the person from which the viral sequence was derived. As shown in Fig. 4, diverse HIV-2 viral sequences evaluated from a cladistic point of view (see Fig. 4B) can, nevertheless, share common V3 loop proteins evaluated from a phenetic point of view (see Fig. 4A). Correlations of HIV-2 V3 protein sequences with SI and non-SI phenotypes have not been attempted; however, the majority of infected persons with divergent forms of the V3 protein shown in Fig. 4A were symptomatic, whereas the persons with the relatively stable form of V3 were mostly asymptomatic (11 of the 13 tightly clustered V3 sequences in Fig. 4A were taken from asymptomatic blood donors in Guinea Bissau) (97). As we have previously noted, positively charged amino acids in the flanks of HIV-1 V3 loops have been associated with the more cytopathic phenotype (90); therefore, it is noteworthy to

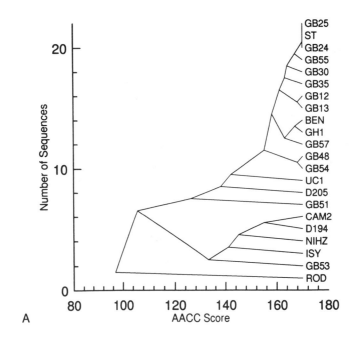

A

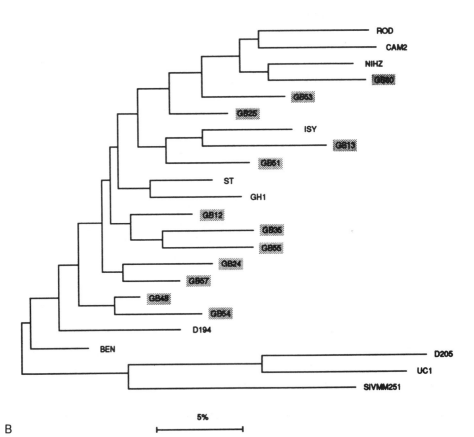

B

find an abundance of positive charges in the homologous regions of the divergent HIV-2 V3 loop sequences taken from the symptomatic individuals represented in the Fig. 4 analyses.

We think that there may be parallels between the apparently stable HIV-2 V3 lineage and the previously mentioned HIV-1 V3 loops that are similar at the protein level, despite being distant by cladistic analysis. Neither form possesses the positively charged flanking residues characteristic of more cytopathic HIV-1s. Moreover, highly conserved V3 loop protein sequences are encountered in SIVagm sequences and in two chimpanzee viral sequences, none of which appear to be pathogenic in their wild-caught hosts.

We tentatively conclude from these findings that viral diversification in and of itself is not a condition for pathogenicity. Diversification of the HIV V3-coding sequence appears to be a major determinant of, or a major effect deriving from, HIV pathogenicity. However, V3 variation cannot be the sole marker factor for progression to AIDS. Macaques infected with molecularly cloned SIV can progress to disease without any apparent dependence on the V3 loop (98,99). Although the focus of this discussion has been on molecular evolutionary parallelism, a large part of the story is the conspicuous lack of genetic stability, the diversification of the V3 peptide sequence in both HIV-1- and HIV-2-infected individuals who typically possess T-cell–tropic viruses and are symptomatic. Genetic flux, not conservation, is the predominant observation at this time.

## CONCLUDING REMARKS

The clinical spectrum of HIV-2 disease in the world is now wider than the spectrum for HIV-1 disease. Some of the viral samples represented in the analyses of Fig. 4 were taken from persons with fullblown AIDS; others were taken from persons with only neurological impairments; still others represent what may prove to be benign forms of HIV-2. Although HIV-2 affects far fewer people, it sets the minimal boundary conditions, so to speak, for the HIV-1 pathogenic potential. The recent observation that a group of long-term survivors of HIV-1 infection harbor closely related viruses derived from the same individual is especially significant (100). The possibility of emergence of less pathogenic forms of HIV-1 (Temin's expectation, discussed in the earlier section on emergence of HIV) is prefigured by the less pathogenic forms of HIV-2 in Senegal, for example (101). What remains to

FIG. 4. A comparison of HIV-2 V3 loop protein sequence diversification, represented by a phenogram in **A**, to nucleotide sequence diversification, represented by a cladogram in **B**. Figure 4B was generated from the V3 region-coding sequences analogous to the construction of the trees shown in Figs. 1 and 2. Figure 4A was generated from V3 loop amino acid sequences using an "amino acid class covering" algorithm that clusters sequences on the basis of protein chemistry and that minimizes cladistic bias (95). Sequences at the top of **A**, are nearly identical (AACC score of 175) indicative of highly conserved V3 loop protein sequences primarily among samples from asymptomatic seropositive individuals from Guinea Bissau. This tight-clustering pattern is not observed in the phylogenetic reconstructions based on nucleotide sequences (highlighted in B). The comparatively divergent V3 loop sequences in **A**, joined by similarity scores less than 140, were derived from symptomatic patients (as discussed in the text).

be determined is the extent to which HIV-1s and HIV-2s will take more, not less, pathogenic courses in some societies with high rates of viral transmission (Ewald's expectation, as discussed in the section on emergence of HIV).

One of the most frequently asked questions concerns the route of transmission of HIVs: How likely is it that HIV could be transmitted through airborne droplets? There is no simple answer to this question. Some nonprimate lentiviruses are thought by some to be transmitted through sputum and the air, and one, the equine infectious anemia virus, is transmitted by flies. Although there is no reason to think that the modes of HIV transmission will change in the coming decades, what gets transmitted could become an urgent question. Wain-Hobson raises the disturbing possibility that HIV gene capture, and attendant oncogenesis, becomes increasingly possible as the AIDS incidence rate rises (see Chap. 10).

This Heraclitean picture is not complete without some mention of the profound ecological disturbances caused by HIV. Of these, the resurgence of tuberculosis, the leading cause of death in the world, is clearly the most serious (17,102,103). Although there is controversy about the extent to which emerging tuberculosis, and the increase in the number of strains resistant to multiple antibiotics, is due to AIDS, no one to our knowledge denies the connection.

## ACKNOWLEDGMENTS

The HIV Sequence Database and Analysis Project, which has been the springboard for much of this inquiry, is funded by the AIDS Division of the National Institute of Allergy and Infectious Diseases through an interagency agreement with the US Department of Energy. Kersti MacInnes, Rhoshel Lenroot, April Gifford, and Miranda McEvilly assisted with the alignments and analyses. We heartily thank Lynda Myers for her editorial help. The CPZant and HIV1ANT70 sequences, analyzed in Fig. 1 were kindly provided before publication by Marleen Vanden Haesevelde and Innogenetics, Belgium, LAUR 93–475.

## REFERENCES

1. Levine AJ. *Viruses*. New York: Scientific American Library; 1992: Chap. 7.
2. Grmek MD. *History of AIDS: emergence and origin of a modern pandemic*. Princeton: Princeton University Press; 1990. Maulitz RC, Duffin J, translators.
3. Diamond J. The mysterious origin of AIDS. *Nat Hist* 1992;9:24–29.
4. Eigen M. The AIDS debate. *Naturwissenschaften* 1989;76:341–350.
5. Eigen M, Nieselt-Struve K. How old is the immunodeficiency virus? *AIDS* 1990;4(suppl 1):S85–S93.
6. Temin HM. Evolution of retroviruses and other retrotranscripts. In: Bolognesi D, ed. *Human retroviruses, cancer and AIDS: approach to prevention and therapy*. New York: Alan R Liss; 1988:1–28.
7. Gojobori T, Moriyama EN, Ina Y, Ikeo K, Miura T, Tsujimoto H, Hayami M, Yokoyama S. Evolutionary origin of human and simian immunodeficiency viruses. *Proc Natl Acad Sci USA* 1990;87:4108–4111.

8. Sharp PM, Li W-H. Understanding the origins of AIDS viruses. *Nature* 1988;336:315.
9. Myers G, MacInnes KA, Korber B. The emergence of simian/human immunodeficiency viruses. *AIDS Res Hum Retroviruses* 1992;8:373–386.
10. Myers G, MacInnes K, Myers L. Phylogenetic moments in the AIDS epidemic. In: Morse SS, ed. *Emerging viruses*. Oxford: Oxford University Press; 1993:chap 12.
11. Myers G, Pavlakis GN. Evolutionary potential of complex retroviruses. In: Levy JA, ed. *The retroviruses*, vol 1. New York: Plenum Press; 1992:51–104.
12. Coffin JM. Structure and classification of retroviruses. In: Levy JA, ed. *The retroviruses*, vol. 1. New York: Plenum Press;1992:19–50.
13. Olmstead RA, Langley R, Roelke ME, Goeken RM, Adger-Johnson D, Goff JP, Albert JP, Packer C, Laurenson MK, Caro TM, Scheepers L, Wildt DE, Bush M, Martenson JS, O'Brien SJ. Worldwide prevalence of lentivirus infection in wild feline species: epidemiologic and phylogenetic aspects. *J Virol* 1992;66:6008–6018.
14. Querat G, Audoly G, Sonigo P, Vigne R. Nucleotide sequence analysis of SA-OMVV, a visna-related ovine lentivirus: phylogenetic history of lentiviruses. *Virology* 1990;175:434–447.
15. Allan JS. Viral evolution and AIDS. *J NIH Res* 1992;4:51–54.
16. Grmek MD. *History of AIDS: emergence and origin of a modern pandemic*. Princeton: Princeton University Press; 1990:128. Maulitz RC, Duffin J, translators.
17. Lederberg J, Shope RE, Oaks SC Jr., eds. *Emerging infections. Microbial threats to health in the United States*. Washington, DC: National Academy Press; 1992.
18. Merson MH. Slowing the spread of HIV: agenda for the 1990s. *Science* 1993;260:1266–1268.
19. Pfutzner A, Dietrich U, von Eichrel Un, von Briesen H, Brede HD, Maniar JK, Rubsamen-Waigmann H. HIV-1 and HIV-2 infections in a high-risk population in Bombay, India: evidence for the spread of HIV-2 and presence of a divergent HIV-1 subtype. *J AIDS* 1992;5:972–977.
20. Estebanez P, Sarasqueta C, Rua-Figueroa M, et al. Absence of HIV-2 in Spanish groups at risk for HIV-1 infection. *AIDS Res Hum Retroviruses* 1992;8:423–425.
21. Pieniazek D, Peralta JM, Ferreira JA, et al. Identification of mixed HIV-1/HIV-2 infection in Brazil by polymerase chain reaction. *AIDS* 1991;5:1293–1299.
22. Zekeng L, Salla R, Kaptue L, et al. HIV-2 infection in Cameroon: no evidence of indigenous cases. *J AIDS* 1992;5:319–320.
23. Khan AS, Galvin TA, Lowenstein LJ, Jennings MB, Gardner MB, Buckler CE. A highly divergent simian immunodeficiency virus (SIV-stm) recovered from stored stump-tailed macaque tissues. *J Virol* 1991;65:7061–7065.
24. Novembre FJ, Hirsch VM, McClure HM, Fultz PN, Johnson PR. SIV from stump-tailed macaques: molecular characterization of a highly transmissible primate lentivirus. *Virology* 1992;186:783–787.
25. Hirsch VM, Olmsted RA, Murphey-Corb M, Purcell RH, Johnson PR. An African primate lentivirus (SIV) closely related to HIV-2. *Nature* 1989;339:389–391.
26. Marx PA, Li Y, Lerche NW, Sutjipto S, Gettie A, Yee JA, Brotman BH, Prince AM, Hanson A, Webster RG, Desrosiers RC. Isolation of a simian immunodeficiency virus related to human immunodeficiency virus type 2 from a West African pet sooty mangabey. *J Virol* 1991;65:4480–4485.
27. Centers for Disease Control. Seroconversion to simian immunodeficiency virus in two lab workers. *MMWR* 1992;41:678–681.
28. Centers for Disease Control. Anonymous survey for simian immunodeficiency virus (SIV) seropositivity in SIV-laboratory researchers—United States, 1992. *MMWR* 1992;41:814–815.
29. Myers G, Korber B, Smith RK, Berzofsky J, Pavlakis G, eds. *Human retroviruses and AIDS 1992*. Los Alamos: Theoretical Biology and Biophysics, Los Alamos National Laboratory; 1992.
30. Gao F, Yue L, White AT, Pappas PG, Barchue J, Hanson AP, Greene BM, Sharp PM, Shaw GM, Hahn BH. Human infection by genetically diverse $SIV_{SM}$-related HIV-2 in West Africa. *Nature* 1992;358:495–499.
31. Huet T, Cheynier R, Meyerhans A, Roelants G, Wain-Hobson S. Genetic organization of a chimpanzee lentivirus related to HIV-1. *Nature* 1990;345:356–359.
32. Peeters M, Fransen K, Delaporte E, Van den Haesevelde M, Gershey-Damet G-M, Kestens L, van der Groen G, Piot P. Isolation and characterization of a new chimpanzee lentivirus: simian immunodeficiency virus isolate CPZ-ANT from a wild-captured chimpanzee. *AIDS* 1992;6:447–451.
33. DeLeys R, Vanderborght B, Haesevelde MV, Heyndrick L, van Geel A, Wauters C, Bernaerts R, Saman E, Nijs P, Willems B, Taelman H, van der Groen G, Piot P, Tersmette T, Huisman JG, van Heuverswyn H. Isolation and partial characterization of an unusual human immunodeficiency retrovirus from two persons of West-Central African origin. *J Virol* 1990;64:1207–1216.

34. Johnson PR, Myers G, Hirsch VM. Genetic diversity and phylogeny of nonhuman primate lenti-viruses. In: Koff W, et al, eds. Annual review of AIDS research, vol 1. New York: Marcel Dekker;1991:47–62.
35. Hirsch VM, Dapolito GA, Goldstein S, McClure H, Emau P, Fultz PN, Isahakia M, Lenroot R, Myers G, and Johnson PR. A distinct African lentivirus from Sykes' monkey. *J Virol* 1993;67: 1517.
36. Gould SJ, Gilensky NL, German RZ. Asymmetry of lineages and the direction of evolutionary time. *Science* 1987;236:1437–1441.
37. Dietrich U, Adamski M, Kreutz R, Seipp A, Kuhnel H, Rubsamen-Waigmann H. A highly divergent HIV-2-related isolate. *Nature* 1989;342:948–950.
38. Louwagie J, McCutchan FE, Peeters M, Brennan TP, Sanders-Buell E, Eddy GA, van der Groen G, Fransen K, Guy-Dermat G-M, Deleys R, Burke DS. Comparison of *gag* genes from seventy international HIV-1 isolates provides evidence for multiple genetic subtypes. *AIDS* 1993;7:769–780.
39. Dietrich U, Grez M, von Briesen H, Panhans B, Geibendorfer M, Kuhnel H, Maniar J, Mahambre G, Becker WB, Becker MLB, Rubsamen-Waigmann H. HIV-1 strains from India are highly divergent from prototypic African and US/European strains but are linked to a South African isolate. *J AIDS* 1992;7:23–27.
40. Ou C-Y, Takebe Y, Luo C-C, Kalish M, Auwanit W, Bandea C, de la Torre N, Moore JL, Schochetman G, Yamazaki S, Gayle HD, Young NL, Weniger BG. Wide distribution of two subtypes of HIV-1 in Thailand. *AIDS Res Hum Retroviruses* 1992;8:1471–1472.
41. Buonagurio DA, Nakada S, Parvin JD, Krystal M, Palese P, Fitch WM. Evolution of human influenza A viruses over 50 years: rapid, uniform rate of change in NS gene. *Science* 1986;232: 980–982.
42. Coffin JM. Genetic diversity and evolution of retroviruses. *Curr Top Microbiol Immunol* 1992; 176:143–164.
43. Ewald PW. Transmission modes and the evolution of virulence. *Hum Nat* 1991;2:1–30.
43a. Ewald PW. The evolution of virulence. *Sci Am* 1993;268:86–93.
44. Temin HM. Is HIV unique or merely different? *J AIDS* 1989;2:1–9.
45. Steinhauer DA, Holland JJ. Rapid evolution of RNA viruses. *Annu Rev Microbiol* 1987;41:409–433.
46. Wain-Hobson S. HIV-1 quasispecies in vivo and ex vivo. *Curr Top Microbiol Immunol* 1992; 176:181–193.
47. Clarke DK, Duarte EA, Moya A, et al. Genetic bottlenecks and population passages cause profound fitness differences in RNA viruses. *J Virol* 1993;67:222–228.
48. Wolinsky SM, Wike CM, Korber B, Hutto C, Parks W, Rosenblum L, Kuntsman K, Furtado M, Munoz J. Selective transmission of human immunodeficiency virus type 1 (HIV-1) variants from mothers to infants. *Science* 1992;255:1134–1137.
49. Burger H, Weiser B, Flaherty K, Gulla J, Nguyen P-N, Gibbs RA. Evolution of human immunodeficiency virus type 1 nucleotide sequence diversity among close contacts. *Proc Natl Acad Sci USA* 1991;88:11236–11240.
50. Korber B, Myers G. Signature pattern analysis: a method for assessing viral sequence relatedness. *AIDS Res Hum Retroviruses* 1992;8:1549–1560.
51. Nei M, Gojobori T. Simple methods for estimating the numbers of synonymous and nonsynonymous nucleotide substitutions. *Mol Biol Evol* 1986;5:418–426.
52. Gojobori T, Moriyama EN, Kimura M. Molecular clock of viral evolution, and the neutral theory. *Proc Natl Acad Sci USA* 1990;87:10015–10018.
53. Leigh-Brown A, Monaghan P. Evolution of the structural proteins of human immunodeficiency virus: selective constraints on nucleotide substitution. *AIDS Res Hum Retroviruses* 1988;4:399–407.
54. Simmonds P, Balfe P, Ludlam CA, Bishop JO, Leigh-Brown AJ. Analysis of sequence diversity in hypervariable regions of the external glycoprotein of human immunodeficiency virus type 1. *J Virol* 1990;64:5840–5850.
55. Burns DPW, Desrosiers RC. Selection of genetic variants of SIV in persistently infected rhesus monkeys. *J Virol* 1991;65:1843–1854.
56. Shpaer EG, Mullins JI. Rates of amino acid change in the envelope protein correlate with pathogenicity of primate lentiviruses. *J Mol Evol* 1993;37:57–65.
57. Nowak M. HIV mutation rate. *Nature* 1990;347:522.

58. Fitch WM, Leiter JME, Li X, Palese P. Positive Darwinian evolution in human influenza A viruses. *Proc Natl Acad Sci USA* 1991;88:4270–4274.
59. Hughes AL, Nei M. Nucleotide substitution at major histocompatibility complex class II loci: evidence for overdominant selection. *Proc Natl Acad Sci USA* 1989;86:958–962.
60. Cullen BR. Human immunodeficiency virus as a prototypic complex retrovirus. *J Virol* 1991; 65:1053–1056.
61. Gaynor R. Cellular transcription factors involved in the regulation of HIV-1 gene expression. *AIDS* 1992;6:347–363.
62. Ross EK, Bucker-White AJ, Rabson AB, Englund G, Martin MA. Contribution of NF-ϰB and Sp1 binding motifs to the replicative capacity of human immunodeficiency virus type 1: distinct patterns of viral growth are determined by T-cell types. *J Virol* 1991;65:4350–4358.
63. Myers G, Lenroot L. HIV glycosylation: what does it portend? *AIDS Res Hum Retroviruses* 1992; 8:1459–1460.
64. Lee W-R, Syu W-J, Du B, Matsuda M, Tan S, Wolf A, Essex M, Lee T-H. Nonrandom distribution of gp 120*N*-linked glycosylation sites important for infectivity of human immunodeficiency virus type 1. *Proc Natl Acad Sci USA* 1992;89:2213–2217.
65. Li Y, Hui H, Burgess CJ, Price RW, Sharp PM, Hahn BH, Shaw GM. Complete nucleotide sequence, genome organization, and biological properties of human immunodeficiency virus type 1 in vivo: evidence for limited defectiveness and complementation. *J Virol* 1992;66:6587–6600.
66. Swain A, Coffin JM. Mechanism of transduction by retroviruses. *Science* 1992;255:841–845.
67. Le Guern M, Levy JA. Human immunodeficiency virus (HIV) type 1 can superinfect HIV-2-infected cells: pseudotype virions produced with expanded cellular host range. *Proc Natl Acad Sci USA* 1992;89:363–367.
68. Levy JA. Pathogenesis of human immunodeficiency virus infection. *Microbiol Rev* 1993;57:183–289.
69. Kohler H, Goudsmit J, Nara P. Clonal dominance: cause for a limited and failing immune response to HIV-1 infection and vaccination. *J AIDS* 1992;5:1158–1168.
70. Coffin JM. Superantigens and endogenous retroviruses: a confluence of puzzles. *Science* 1992;255: 411–413.
71. Nowak MA, Anderson RM, McLean AR, Wolfs TFW, Goudsmit J, May RM. Antigenic diversity thresholds and the development of AIDS. *Science* 1991;254:963–966.
72. Nowak MA. Variability of HIV infections. *J Theor Biol* 1992;155:1–20.
73. Dalgleish AG, Collizi V. Role of major histocompatibility complex recognition in the protection and immunopathogenesis of AIDS. *AIDS* 1992;6:523–525.
74. Haseltine WA. Human immunodeficiency virus (HIV) gene expression and function. In: Gallo RC, Jay G, eds. *The human retroviruses.* San Diego: Academic Press; 1991:69–106.
75. Dewhurst S, Embretson JE, Anderson DC, Mullins JI, Fultz PN. Sequence analysis and acute pathogenicity of molecularly cloned SIV-smm-pbj14. *Nature* 1990;345:636–640.
76. Martin MA. Fast-acting slow viruses. *Nature* 1990;345:572–573.
77. Misher L, Schmidt A, Steele J, Morton W, Hu S-L. In vivo passage of HIV-2 EHO in *Macaca nemestrina* results in persistent infection, disease progression and altered cellular tropism. Presented at the 10th Annual Symposium on Nonhuman Primate Models for AIDS, San Juan, Puerto Rico, 17–20 November, 1992.
78. Hoxie JA, Brass LF, Pletcher CH, Haggarty BS, Hahn BH. Cytopathic variants of an attenuated isolate of human immunodeficiency virus type 2 exhibit increased affinity for CD4. *J Virol* 1991; 65:5096–5101.
79. Courgnaud V, Laure F, Fultz PN, Montagnier L, Brechot C, Sonigo P. Genetic differences accounting for evolution and pathogenicity of simian immunodeficiency virus from a sooty mangabey monkey after cross-species transmission to a pig-tailed macaque. *J Virol* 1992;66:414–419.
80. Cheng-Mayer C, Shioda T, Levy JA. Host range, replicative, and cytopathic properties of human immunodeficiency virus type 1 are determined by very few amino acid changes in *tat* and gp120. *J Virol* 1991;65:6931–6941.
81. Corboy JR, Buzy JM, Zink MC, Clements JE. Expression directed from HIV long terminal repeats in the central nervous system of transgenic mice. *Science* 1992;258:1804–1808.
82. Åsjö B, Morfeldt-Månson L, Albert J, Biberfeld G, Karlsson A, Lidman K, Fenyo EM. Replicative capacity of human immunodeficiency virus from patients with varying severity of HIV infection. *Lancet* 1986;2:660–662.
83. Schuitemaker H, Koot M, Kootstra NA, Wouter Dercksen M, de Goede REY, van Steenwijk RP,

Lange JMA, Eeftink Schattenkerk JKM, Miedema F, Tersmette M. Biological phenotype of human immunodeficiency virus type 1 clones at different stages of infection: progression of disease is associated with a shift from monocytotropic to T-cell-tropic virus populations. *J Virol* 1992;66: 1354–1360.

84. Boucher CAB, Lange JMA, Miedema FF, Weverling GJ, Koot M, Mulder JW, Goudsmit J, Kellam P, Larder BA, Tersmette M. HIV-1 biological phenotype and the development of zidovudine resistance in relation to disease progression in asymptomatic individuals during treatment. *AIDS* 1992;6:1259–1264.

85. Page KA, Stearns SM, Littman DR. Analysis of mutations in the V3 domain of gp160 that affect fusion and infectivity. *J Virol* 1992;66:524–533.

86. Westervelt P, Gendelman HE, Ratner L. Identification of a determinant within the human immunodeficiency virus 1 surface envelope glycoprotein critical for productive infection of primary monocytes. *Proc Natl Acad Sci USA* 1991;88:3097–3101.

87. Hwang SS, Boyle TJ, Lyerly HK, Cullen BR. Identification of the envelope V3 loop as the primary determinant of cell tropism in HIV-1. *Science* 1991;253:71–74.

88. Wolfs TFW, Zwart G, Bakker M, Valk M, Kuiken CL, Goudsmit J. Naturally occurring mutations within HIV-1 V3 genomic RNA lead to antigenic variation dependent on a single amino acid substitution. *Virology* 1991;185:195–205.

89. Takeuchi Y, Akutsu M, Murayama K, Shimuzu N, Hoshino H. Host range mutant of human immunodeficiency virus type 1: modification of cell tropism by a single point mutation at the neutralization epitope in the *env* gene. *J Virol* 1991;65:1710–1718.

90. Fouchier RAM, Groenink M, Kootstra NA, Tersmette M, Huisman HG, Miedema HG, Schuitemaker H. Phenotype-associated sequence variation in the third variable domain of the human immunodeficiency virus type 1 gp120 molecule. *J Virol* 1992;66:3183–3187.

91. Tersmette M, Gruters RA, de Wolf F, de Goede REY, Lange JMA, Schellekens PTA, Goudsmit J, Huisman HG, Miedema F. Evidence for a role of virulent human immunodeficiency virus (HIV) variants in the pathogenesis of acquired immunodeficiency syndrome: studies on sequential isolates. *J Virol* 1989;63:2118–2125.

92. Collman R, Balliet JW, Gregory SA, Friedman H, Kolson DL, Nathanson N, Srinivasan A. An infectious molecular clone of an unusual macrophage-tropic and highly cytopathic strain of human immunodeficiency virus type-1. *J Virol* 1992;66:7517–7521.

93. Javaherian K, Langlois AJ, LaRosa GJ, Profy AT, Bolognesi DP, Herlihy WC, Putney SD, Matthews TJ. Broadly neutralizing antibodies elicited by the hypervariable neutralizing determinant of HIV-1. *Science* 1990;250:1590–1593.

94. Emini EA, Schleif WA, Nunberg JH, et al. Prevention of HIV-1 infection of chimpanzees by gp120 V3 domain-specific monoclonal antibody. *Nature* 1992;355:728–732.

95. Korber B, Myers G. Contrasting HIV phylogenetic relationships and V3 loop protein similarities. In: Girard M, ed. *Proceedings of the Septieme Colloque Des Cent Gordes*. Paris, France; Merieux Foundation; 1993:151–158.

96. Albert J, Franzan L, Jansson M, Scarlatti G, Kataaha PK, Katabira E, Mubiro F, Rydaker M, Rossi P, Pettersson U, Wigzell H. Ugandan HIV-1 V3 loop sequences closely related to the US/European consensus. *Virology* 1992;190:674–681.

97. Boeri E, Giri A, Lillo F, Ferrari G, Varnier OE, Ferro A, Sabbatani S, Saxinger WC, Franchini G. In vivo genetic variabilitiy of the human immunodeficiency virus type 2 V3 region. *J Virol* 1992;66:4546–4550.

98. Overbaugh J, Rudensy LM, Papenhausen MD, et al. Variation in simian immunodeficiency virus *env* is confined to V1 and V4 during progression to simian AIDS. *J Virol* 1991;65:7025–7031.

99. Johnson PR, Hamm JE, Goldstein S, et al. The genetic fate of molecularly cloned simian immunodeficiency virus in experimentally infected macaques. *Virol* 1991;185:217–228.

100. Learmont J, Tindall B, Evans L, Cunningham A, Cunningham P, Wells J, Penny R, Kaldor J, Cooper DA. Long-term symptomless HIV-1 infection in recipients of blood products from a single donor. *Lancet* 1992;340:863–867.

101. Romieu I, Marlink R, Kanki P, M'Boup S, Essex M. HIV-2 link to AIDS in West Africa. *J AIDS* 1990;3:220–230.

102. Bloom BR, Murray CJL. Tuberculosis: commentary on a reemergent killer. *Science* 1992;257: 1055–1062.

103. Bermejo A, Veeken H, Berra A. Tuberculosis incidence in developing countries with high prevalence of HIV infeciton. *AIDS* 1992;6:1203–1206.

*The Evolutionary Biology of Viruses*,
edited by Stephen S. Morse.
Raven Press, Ltd., New York © 1994.

# 11

# Evolution of Genetic Exchange in RNA Viruses

## Lin Chao

*Department of Zoology, University of Maryland, College Park, Maryland 20742*

Evolution provides the conceptual framework for understanding biological diversity. In this chapter, I will discuss, from an evolutionary viewpoint, the process of genetic exchange in RNA viruses. From that perspective, genetic exchange and sex are synonymous (1) and sex is surprisingly similar in RNA viruses and eukaryotes. In both types of organisms, the same hypotheses can account for the evolution of sex, and equivalent problems arise.

## SEX IN DNA VIRUSES

In all viruses, sexual reproduction occurs when two or more viruses coinfect the same host cell, and hybrid progeny are produced through genetic exchange between the coinfecting parent viruses.

In DNA viruses, genetic exchange is prompted by *recombination*, which is defined here as the generation of a new nucleotide strand containing information from two or more parental strands. Recombination in DNA viruses, just as in eukaryotes, is by a mechanism of breakage-and-reunion (2,3). Following a physical break in the parental DNA molecules, pieces from different parental molecules can be joined to form a recombinant. Breakage-and-reunion does not require replication of the parental DNA (4).

## SEX IN RNA VIRUSES

### Mechanisms of Genetic Exchange

Although sex and recombination are universal in DNA viruses, they are not in RNA viruses. Despite intensive searches, recombinants have never been identified in RNA bacteriophages and several eukaryotic RNA viruses [e.g., vesicular stomatitis (VSV) and Newcastle disease virus of fowl] (5,6). The ability of detecting recombinants is obviously limited by the reversion rate of the markers used in a cross, but recombinants are not observed in some phages, even when nonreverting markers are used (5). These results show that recombination is either absent or occurs at a level no greater than the mutation rate in these RNA viruses. As a result, genetic maps of RNA viruses that did not produce recombinants had to be determined by methods other than crosses (3,5). One method was the sequencing of RNA, and the first organism to have its genome completely sequenced was the RNA phage MS2 (7).

In many RNA viruses (e.g., the families Picornaviridae and Coronaviridae), recombination is common and genetic maps are frequently constructed through crosses (8–11). The recombinational mechanism, however, is believed to be copy-choice, rather than break-and-union. In a *copy-choice* model, recombinants result from template-switching by the replicase during synthesis of a new strand. Evidence in support of copy-choice is that, if RNA replication is inhibited, no recombinants are formed (12).

In other RNA viruses (e.g., the members of the families Reoviridae, Birnaviridae, Orthomyxoviridae, Bunyaviridae, Arenaviridae, Nodaviridae, and Cleviridae; 13,14), sex may be achieved by an alternative mechanism to recombination: genome segmentation. Viruses in the genus *Reovirus*, for example, have genomes that are fragmented into 10 to 12 distinct double-stranded RNA molecules or segments, which range from 680 to 4,500 nucleotides (n) in length (15). Such viruses are segmented, single-component because one virion generally carries a full complement of chromosomes (the entire genome).

Whenever two or more segmented, single-component viruses infect the same host cell, the resulting progeny consist of individuals randomly reassorted from segments descending from the coinfecting parents (16). Although reassortment is a coarser form of genetic exchange than recombination, it allows for reciprocal exchange between viral genomes. In a typical coinfection, recombinants are generated at a frequency of 10% to 20% (17–19).

Not all segmented RNA viruses are, however, single-component. Viruses in the groups Tobravirus, Cucumovirus, and Bromovirus are segmented and multicomponent (13). For example, in the three-component brome mosaic virus (BMV) the genome consists of three single-stranded RNA molecules of about 3,200, 2,900, and 2,100 nucleotides, and there are three types of viral particles (20). Each type of particle carries a different segment, and all three particle types must coinfect the same host cell for a successful infection.

The noticeable absence or rarity of recombination in some RNA viruses led Pressing and Reanney (21) and Chao (22) to suggest that reassortment evolved as an alternative to recombination for the purpose of promoting sex. Supporting this argument are the results that recombination between homologous segments is rare or nonexisting (16,17,23). In instances where recombination between segments is observed, recombinants are obtained only after intense selection over several generations (24–26). If recombination occurs in segmented viruses, it is much less than the rate of segment reassortment.

Thus, RNA viruses manifest a wide variety of "mating systems," each offering what may be interpreted to be different degrees of sex, outcrossing, and inbreeding. For example, reproduction in RNA viruses lacking both recombination and genome segmentation is effectively asexual. Reproduction by single-component RNA viruses that are either segmented or capable of copy-choice recombination can be either sexual or asexual. A lone single-component virus infecting a single host cell reproduces asexually, whereas coinfecting viruses can reproduce sexually. In multicomponent viruses, reproduction is almost always sexual, and results in outcrossing, because it requires coinfection. Multicomponent viruses and eukaryotes, therefore, are similar in that sex in both is an automatic consequence of reproduction. A difference is that in three-component viruses, an offspring may be the product of three parents.

## Sex and Selfish RNA Elements

Besides promoting genetic exchange, sex has a second and striking similarity in RNA viruses and eukaryotes. The need for sex requires a genome to expose itself to foreign genetic material. This introduces genic selection for selfish genetic elements (27). If an element increases its representation during the genetic exchange, it can gain an advantage, even if it is deleterious to individual genomes. In diploid sexual organisms, meiotically driven or segregation distortion alleles are an example of selfish elements. Because selfish elements highlight the similarities of sex in RNA viruses and eukaryotes and because they are later invoked to account for the evolution of segmentation, they are reviewed here.

Opportunity for genic selection (or selection on RNAs) arises in RNA viruses during coinfection (28,29). If an RNA has a higher replication or encapsidation rate inside a coinfected cell, it can increase in frequency within a population of viral RNAs. A rate advantage can be acquired by an RNA by several mechanisms. First, it can become smaller and replicate faster. Second, it may have extra sequences recognized by encapsidation or replication enzymes.

Selection on RNAs has produced selfish elements, also known as defective interfering (DI) RNAs, that exploit either or both of the foregoing strategies in segmented and nonsegmented RNA viruses (26). The advantage gained by such RNA elements is marked. While working with poliovirus, Cole and Baltimore (30) showed that, when particles containing standard RNAs and ones with DI RNAs coinfect the same cell, the progeny is enriched by about 5% for particles containing DI RNAs.

What prevents DI RNAs from increasing uncontrollably and driving viral RNAs to extinction? The likely answer is that selection intervenes from a higher level. In their evolution, DI RNAs have lost gene-coding regions and have come to rely on coinfecting viruses for gene functions. This parasitism has a negative effect on the coinfection group (the group of viruses coinfecting a cell). In the experiments by Cole and Baltimore, the total yield of polioviruses produced by an infected cell is inversely proportional to the frequency of DI RNAs in the coinfection group; hence, the name DI. Thus, although selection on RNAs favors DI RNAs, selection at the level of coinfection groups opposes them, much the same way that selection on individuals may stop deleterious meiotically driven alleles from becoming fixed (29). Selection on diploid sexual individuals and selection on coinfection groups can be viewed as analogous, the only difference being that ploidy in a coinfection group is variable and statistically distributed.

## EVOLUTION OF SEGMENTATION

Although the existence of genetic exchange and selfish RNAs highlights the resemblance of sex in RNA viruses and eukaryotes, RNA viruses may have evolved copy-choice and segmentation as alternatives to breakage-and-reunion recombination. If that is true, sex in RNA viruses and in eukaryotes is an independent evolution, and sex in RNA viruses becomes a particularly instructive model for the evolution of sex as a general phenomenon. The sections to follow take advantage of this fortuitous event and examine specifically the benefits that sex may offer RNA viruses. Because results from segmented viruses are currently more informative, the focus is on the evolution of segmentation. The advantage of segmentation must be evaluated in relation to traditional hypotheses for the evolution of sex; hence, the latter are reviewed first.

### General Models

The effect of sex in a population is to promote linkage equilibrium. Not surprisingly, most models for the evolution of sex ascribe an advantage to sex by assuming a disequilibrium. For example, if the only viruses in a population are of the genotypes *AB* and *ab*, a disequilibrium exists because *Ab* and *aB* are absent. If the missing genotypes are favored by selection, sex is advantageous because it can generate those genotypes from the existing ones. As Felsenstein (31) and Maynard Smith (32) point out, the disequilibria that are considered by the models can be classified by two criteria. First, the disequilibrium can be due to beneficial or deleterious mutations. Second, it can result from genetic drift or selection. Hereafter, selection for beneficial mutations is referred to as *positive-selection* and that against deleterious mutations as *purifying-selection*.

### Positive-Selection Models

The popular view that sex is advantageous because it generates genetic variation assumes that beneficial mutations favored by positive-selection are in linkage disequilibrium.

If a finite population is exposed to a new environment, certain favorable genetic combinations may not exist owing to genetic drift. Sex is advantageous because it can create the missing combinations and accelerate evolution (33,34). A finite population is required because, in an infinitely large population, all genetic combinations will be generated by mutations at the frequency expected from linkage equilibrium.

If populations are infinitely large, a disequilibrium may result if selection changes with time. For example, if the environment fluctuates temporally such that *AB* and *ab* genotypes are favored and then *Ab* and *aB* are favored, sex is again advantageous (35,36). Other forms of variable positive selection (e.g., spatially heterogeneous environments, changes in selection optima, and host–parasite coevolution) also provide an advantage to sex, but some are more restrictive (32).

All models of beneficial mutations assume, however, that the environment is variable, either spatially or temporally.

### Purifying-Selection Models

If mutations are primarily deleterious, the role of purifying-selection is to eliminate mutations and preserve the wild-type genotype. If mutation rates are in addition high, sex can also be advantageous. However, unlike models with beneficial mutations, models with deleterious mutations do not require a changing environment. Instead, they require that the genome changes by accumulating mutations and the linkage disequilibrium is between mutations and parts of the genomes that have not been mutated. Whether the disequilibrium is generated by genetic drift or selection depends on whether the effect of deleterious mutations is independent or synergistic. A mutation is *independent* if its effect is constant, regardless of how many other deleterious mutations are present in a genome. It is *synergistic* if its deleterious effect is negatively epistatic (i.e., larger in genomes with many mutations than in ones with fewer mutations).

If deleterious mutations are independent, the equilibrium mean fitness of an infinitely large, asexual and haploid population is $e^{-U}$ (37–39), where $U$ is the average number of mutations per genome replication (the genomic mutation rate). Populations at this equilibrium, or mutation–selection balance, contain a distribution of genomes with zero to many mutations. Haigh (38) has shown that if the average effect of a deleterious mutation is $(1-S)$ and $0<S<1$, the distribution is Poisson, and the frequency of genomes free of mutations (the mutation-free class) is $C(0) = e^{-U/S}$. Thus, for an RNA virus with a genome of $10^4$ nucleotides and a mutation rate of $10^{-4}$ errors (see following), $U = 1$ and if $S = 0.1$, $e^{-U/S} = 5 \times 10^{-5}$. If selection is weaker, $C(0)$ becomes even smaller. The small size of $C(0)$ is what makes sex advantageous, if populations are finite.

For example, in a finite population, the mutation-free class can be lost by chance (genetic drift). If lost, the one-mutation free class becomes the least-loaded class, but that, in turn, can be lost just as easily. With each loss, the mutational load increases. This directional increase first postulated by Muller (40) is often termed

*Muller's ratchet* (31). Although often considered only in relation to finite asexual populations, Muller's ratchet also operates in finite sexual populations. The difference is that in sexual populations the least-mutated class pertains to the smallest unit of assortment, which is the allele, and Muller's ratchet advances whenever the nonmutated allele is lost by drift.

If Muller's ratchet is operating, sex can be beneficial because it slows the ratchet and buys time for back and compensatory mutations to appear and correct deleterious mutations accumulated through the ratchet. There are two equally correct accounts of how the slowing is achieved. From the perspective of the genome, sex simply recreates through genetic exchange among more-mutated genomes any less-mutated class that may have been lost. From the perspective of the allele, sex increases the equilibrium frequency of the least-mutated class. For example, if $V$ is the allelic mutation rate, the equilibrium frequency in a haploid sexual population it is $e^{-V/S}$, which is greater than $e^{-U/S}$ because $V << U$. Although an allele may be difficult to define in eukaryotes, it is the segment in segmented viruses because of the low rate of recombination between segments (see foregoing). Thus, $V = U/10$ for a ten-segment virus and the expected frequency of the least-mutated class is $e^{-U/10S}$ (22).

Back mutations cannot stop Muller's ratchet, because backward rates are anticipated to be less than forward rates (38,41). Although any mutation can change the sequence of a gene (and perhaps harm it), only one back mutation can restore the original sequence. However, Muller's ratchet can be stopped by compensatory mutations, which are mutations that occur away from the deleterious mutations and correct for the effects of the latter without changing it back to its original sequence (42). But, whether compensatory mutations are sufficiently common relative to deleterious mutations to stop Muller's ratchet is an empirical issue.

A population in a mutation–selection balance according to Haigh's (38) model is similar to a "quasispecies" in the origin of life models by Eigen (43). A difference is that Haigh assumes that the probability of a mutation changing a genome from a class to any higher class is same for all classes. In other words, the probability of changing from $C(0)$ to $C(1)$ is the same as that of $C(2)$ to $C(3)$. Because Eigen's model corrects for genome size, the probabilities are not the same. For example, if genome size is $M$ nucleotides, there are $M$ targets for the change $C(0)$ to $C(1)$, but $M - 1$ for $C(1)$ to $C(2)$, because one nucleotide will already have been mutated. If $M$ is small and $U$ is high, the bias can make the equilibrium value of $C(0)$ much smaller than that in Haigh's model. For origin of life scenarios, Eigen's model is more appropriate because $M$ is likely small. However, for RNA viruses, $M$ is between $10^3$ and $10^4$ nucleotides (discussed later), in which case, the two models are approximately equivalent.

If deleterious mutations are synergistic, the equilibrium mean fitness of an infinitely large asexual population remains $e^{-U}$, but that of an equivalent sexual population is increased, in which event sex should evolve without genetic drift (22,44, 45). The stronger the synergism, the greater is the fitness advantage of sexual populations. Sex and synergism increase population fitness by increasing the efficiency of purifying selection. For example, if a genome with one mutation has a fitness of 0.9 and one with two mutations has a fitness *less* than $0.9 \times 0.9$, selection is more efficient because it rids two mutations in one genome more easily than two mutations in separate genomes. Selection creates the linkage disequilibrium because it disproportionately eliminates the more-mutated genomes. The role of sex is to re-create the latter.

## Population Selection

Because the foregoing models compare the fitness of sexual and asexual populations, they invoke population (or group) selection, which is often judged to be inappropriate for evolutionary models. The reason is that evolution is argued to result primarily from selection on individuals and not on populations. In this instance, a population approach is taken to facilitate the analysis, and it has its merits. As Felsenstein and Yokoyama (46) point out, individual and population selection may be the same in the evolution of sex because sexual and asexual populations cannot exchange genes. The unit of selection, therefore, is the population (or the lineage), and the evolution of sex is correctly a population problem. Where this argument fails is that it can explain only the presence or absence of sex, but not how much sex (e.g., adjustments in the rate of recombination). Modifier models are necessary to account for the latter (36,46–48). Fortunately, as Felsenstein (49) observes, conditions that increase the fitness of sexual populations in population models also favor the evolution of recombination in modifier models with individual selection.

Despite such objections, population models will be invoked throughout this chapter. The justification is that the discussion to follow addresses primarily the evolution of asexual and sexual reproduction in RNA viruses and not the degree of sexuality. If recombination is rare in segmented RNA viruses, genetic exchange is minimal between segmented and nonsegmented (asexual) viruses, in which case, Felsenstein and Yokoyama's arguments apply, and population arguments are appropriate.

## TESTING THE MODELS

Does a positive- or purifying-selection model, with or without drift, better account for the evolution of segmentation in RNA viruses? To discriminate between the models, the selective force and the source of the linkage disequilibrium must be identified. Although the results are not always clear-cut, studies on three RNA viruses, influenza A virus, $\phi 6$, and $Q\beta$, shed some light on the relevance of these models. These studies are reviewed next. The review will examine the advantages of segmentation, but not the costs. The evolution of segmentation ultimately requires that the advantages exceed the costs. For multicomponent viruses, a cost is failure to reproduce when coinfection groups lack one of the particle types (see foregoing). For single-component viruses, little is known. There may be a cost for however viral particles are assigned a full complement of segments. Alternatively, the cost may be that some viral particles do not receive a full complement. Regardless, because there are no male viruses, the total cost in single-component viruses is likely to be much less than the twofold cost of sex in some eukaryotes (41,50).

## Influenza A Virus

Influenza A virus (19,51–53) has a genome that is divided among eight segments. Humans, pigs, ducks, and possibly additional mammals and birds serve as its hosts. Escaping the immune response of the host is an important factor governing the spread of the virus. Two surface proteins, hemagglutinin (HA) and neuraminidase (NA), are the major antigens involved. Viruses with a new HA and NA subtypes spread through human populations every 10 to 30 years (a process called *antigenic shift*) and cause major pandemics. During the interpandemic years, minor changes (*antigenic drift*) can occur in the antigens and cause regional epidemics.

Antigenic drift results from single or few amino acid changes in the antigens. The generally high mutation rate of RNA viruses (discussed later) facilitates such changes. Antigenic drift should not be confused with genetic drift. Although both represent genetic changes in a population, genetic drift results from chance events and antigenic drift from antigenic selection.

Antigenic shift also results from antigenic selection, but unlike antigenic drift, it requires the introduction of a distinctly new antigen. For example, before 1968, the prevalent influenza A virus in the human population had an HA and NA serotype of H2N2. In 1968, an antigenic shift and its resulting pandemic occurred. The new virus, A/Hong Kong/68, had a new H3 but the old N2 subtype. When A/Hong Kong/1968 was dissected genetically (54), it possessed seven segments that were homologous to the earlier H2N2 virus. One of the seven encoded for the N2 neuraminidase. The H3 hemagglutinin was on the eighth segment, which came from some other virus, possibly an avian influenza A virus. These events suggest that viruses in nonhuman animal populations form a reservoir that can exchange segments with human viruses.

### *Positive-Selection*

Any consideration of positive-selection models for sex in influenza A virus is struck by the advantage of segmentation during antigenic selection. During the antigenic shift that produced A/Hong Kong/68, a linkage disequilibrium resulted from the absence of the H3 antigen in human H2N2 virus. Antigenic shift is clearly the product of positive-selection. However, closer examination reveals that the advantage may not be sufficient to account for the evolution of segmentation.

Although antigenic shift makes segmentation beneficial, it is infrequent. As indicated, pandemics occur every 10 to 30 years in human viruses. In laboratory culture, populations of influenza A virus can double in less than one hour (19,28).

Clearly, thousands of generations (if not more) must ensue between pandemics. Unless selection during pandemics is extraordinarily strong and the cost of segmentation is equally small, it is difficult to imagine that a trait evolved from selection acting only once in 1,000 or more generations. The strongest argument against positive-selection and antigenic shift comes from the observation that influenza B virus, a close relative of influenza A virus, undergoes antigenic drift, but not antigenic shift (19). Yet, influenza B virus has a genome that is segmented into eight segments.

Antigenic drift occurs more frequently than antigenic shift, but antigenic drift alone cannot generate a linkage disequilibrium because it requires mutational changes on only one segment. However, influenza A virus evolves so rapidly (55,56) that other changes must occur during interpandemic years. For example, in 1977 an H1N1 virus appeared, and its outbreaks were concentrated in primary and middle schools (19). A few years later, it began affecting the previously spared younger and older populations. Could genes determining age of host and antigenic drift have been in linkage disequilibrium? H1N1 virus cocirculates currently with H3N2 virus, and reassortants are occasionally detected. Are reassortants favored by positive-selection? They have not replaced their H1N1 and H3N2 parents, so they may be just part of the "recombinational load."

Unfortunately, the answers to the foregoing questions are not known. Positive-selection is always difficult to study because its confirmation requires the identification of an environmental cause. Assessing the models is additionally complicated because it is not known whether the observed disequilibria result from drift or selection. Thus, until more answers are obtained, the role of positive-selection, with or without drift, remains undetermined for the evolution of segmentation.

### Purifying-Selection

Unlike positive-selection, purifying-selection is easier to identify because it can operate in a constant environment and requires only a high-mutation rate and that mutations are, on the average, deleterious. RNA viruses easily satisfy the first requirement. Whereas the mutation rate for DNA is estimated to be $10^{-9}$ to $10^{-10}$ errors per nucleotide replication, it is $10^{-3}$ to $10^{-5}$ for RNA (6,22). Thus, if the average rate of $10^{-4}$ is representative, larger RNA viruses, which have a genome in excess of $10^4$ nucleotides, would have a genomic mutation rate $U$ greater than one error per genome replication. The deleterious nature of these mutations has been confirmed by several studies.

Such a study was done by Domingo and colleagues (57), who worked with Q$\beta$, a nonsegmented RNA phage with a genome size of 4,500 nucleotides. Because Q$\beta$ also lacks recombination, it is asexual. By T1 fingerprinting, which samples about 10% of the Q$\beta$ genome, Domingo and coworkers found that 15% of isolates from their populations had sequences that deviated from a wild-type sequence. The large heterogeneity was shown to be maintained by a selection–mutation balance. After

just 50 generations, a population founded by a single clone was just as heterogeneous as an older established population. When the fitness values of the deviants
were measured, they ranged from 0.8 to 0.9, whereas wild-type fitness was 1.

A second confirmation of purifying-selection in RNA viruses comes from molecular information. When Saitou and Nei (55) examined sequence data on four
proteins from natural isolates of influenza A virus, they found the characteristic
stamp of purifying-selection. The substitution rate at the third position of codons
was 2.3 times greater than that for the first and second positions. This outcome is
not inconsistent with the results of Fitch and associates (56), who identified rapid
sequence evolution by positive-selection on the *HA* gene of the virus. Although the
rate of nonsynonymous substitutions is greater than that of synonymous substitutions in antigenic sites of the *HA* gene (56), the substitution rate at third-codon
positions is greater than that at first and second positions in sites over the entire *HA*
gene (55). Thus, the predominant pattern of selection across the influenza A genome is purifying.

A third case in support of purifying-selection comes from φ6, which is an RNA
phage with a genome size of 13,379 nucleotides (58,59). To test the reality of
Muller's ratchet, Chao (60) subjected twenty φ6 lineages to intensified drift. This
was achieved by forcing each lineage through successive bottlenecks of one individual, which was always chosen randomly. After 40 cycles of population bottleneck
and expansion, the fitness of the lineages was measured relative to the original φ6
clone used to start the experiment. Mean fitness of the 20 lineages was 22% lower
than the original clone.

However, Muller's ratchet alone does not provide an advantage for segmentation.
For there to be such advantage, the φ6 lineages subjected to Muller's ratchet must
meet and exchange segments through coinfection. Additionally, the lineages must
harbor mutations that are on different segments (i.e., mutation pressure must create
a linkage disequilibrium). Chao and associates (61) have searched for a disequilibrium by crossing the lineages and then measuring the fitness of the hybrid reassortants. Although their sample size is limited, they found evidence for a disequilibrium because the hybrids had, on average, a higher fitness than their parent phage.

These studies with φ6 demonstrate four things. First, because the mean fitness of
the lineages after being subjected to bottlenecks was less than that of the original
clone, the average effect of the mutations accumulated by Muller's ratchet is deleterious. Second, mutation pressure is sufficiently strong to advance Muller's ratchet
in a segmented RNA virus. Third, compensatory mutations are not sufficiently
common to stop Muller's ratchet within the range of fitness values that were examined. Finally, Muller's ratchet creates a linkage disequilibria among the lineages
and, as a result, an advantage for segmentation.

The magnitude of the gain that sex provides against Muller's ratchet is illustrated
by comparing estimates of the frequency of the mutation-free class in Qβ and in a
hypothetical φ6 that is asexual. Following Haigh's model (see foregoing), let the
frequency of the mutation-free class in Qβ be $C(0,Q\beta) = e^{-U/S}$. Because the size
ratio of φ6 and Qβ genomes is about 3, the genomic mutation rate for φ6 is three

times greater and the mutation-free class for an asexual $\phi6$ is $C(0,\phi6) = e^{-3U/S}$. Assuming that $S$ is the same for both phages, $C(0,Q\beta)^3 = C(0,\phi6)$. If mutations are randomly distributed in the $Q\beta$ population examined by Domingo and colleagues (see foregoing), $C(0,Q\beta) = (1-p)^{1/t} = (1-0.15)^{1/.1} = 0.2$, where $t$ is the percentage of the $Q\beta$ genome sampled by T1 fingerprinting and $p$ is the frequency of deviants. Thus, $C(0,\phi6) = (0.2)^3 = 0.008$. If population size for these phages were $10^3$, for the purpose of illustration, the expected number $Q\beta$ viruses free of mutations is 200 individuals. For $\phi6$, the expected number is only eight. Clearly, $\phi6$ is much more sensitive to genetic drift and Muller's ratchet, and it has much more to gain by being segmented.

The importance of Muller's ratchet to viruses can be questioned because viral populations are often enormous. However, simply achieving large sizes does not mean that their populations are "effectively" large. In the $\phi6$ experiments, the populations were expanded to a size of $8 \times 10^9$ and yet Muller's ratchet was not stopped when the lineages were also bottlenecked to one individual. From a population genetics viewpoint, whenever there are bottlenecks, the population size is effectively much closer to the size of the bottleneck (62). If viral transmission in nature is often by a single propagule, Muller's ratchet may be active in wild populations.

## Comparative Results

Comparative evidence supports the hypothesis that a purifying-selection model may be responsible for the evolution of segmentation in RNA viruses. As Pressing and Reanney (21) observed: in general, segmented viruses tend to have larger genomes. RNA phages show the same pattern. All known RNA phages are asexual, except $\phi6$ which has a genome about three times larger than that of the other phages (5,57–59). This relation between genome size and segmentation is predicted by models of purifying-selection because $U$ should be higher in larger viruses. The appeal of these results is that they explain why some RNA viruses are asexual (see foregoing) and others are segmented. Assuming that there is a cost to segmentation, it is overcome only in larger viruses.

There are exceptions to the relation between genome size and segmentation. For example, some large RNA viruses, VSV and the New World alphaviruses, which have genomes of approximately 11,162, and 11,700 nucleotides, respectively, are likely asexual (nonsegmented and lacking high rates of recombination) (63; Weaver, S, personal communication). However, alphaviruses and VSV have an additional similarity; they or some of their subtypes can have a very low level of diversity within populations (64,65). Moreover, most alphaviruses have a remarkably low rate of evolution for an RNA virus. Compared with that of influenza A (55,56), the 1–$4 \times 10^{-4}$ nucleotides per year substitution rate of alphaviruses is about one order of magnitude smaller (65).

It is unlikely that these properties of alphaviruses and VSV are due to low mutation rates. Although mutation rates for alphaviruses have not been measured, they

have been shown to be high in VSV (66), which can evolve very rapidly in laboratory culture (6). Could the asexuality, low diversity, and slow evolution have a
common cause? They could if population size were often immensely large in these
viruses.

The arguments come from a combination of two purifying-selection models:
Muller's ratchet and Ohta's (67,68) slightly deleterious model of molecular evolution. By Muller's ratchet, VSV and alphaviruses would not need sex if their populations were immensely large, such that genetic drift were minimal. If that were
correct, the lack of diversity and low rate of evolution is in turn consistent with
Ohta's model, which asserts that most of molecular evolution is due to the random
fixation of deleterious mutations.

To appreciate Ohta's model, imagine if alphaviruses or VSV were instead viruses
with many small and isolated populations. Within each population, drift operates
and the deleterious mutations are fixed. The substitution rate is high because drift
operates continuously, and the diversity across populations is great because each
population fixes different alleles. On the other hand, if each virus were one large
population, drift is diminished and evolution is slowed down because few deleterious mutations are fixed. Instead, the mutations are quickly removed by selection,
and the population is homogeneous because every virus is approximately a wild-
type sequence. In alphaviruses, the movement of viremic hosts (birds or bats) may
provide sufficient migration to unify isolated viral populations into one large population (65). Thus, these characteristics of alphaviruses and VSV are consistent with
models of purifying-selection with drift.

## Synergism or Independence of Deleterious Mutations

Although some of the foregoing results argue for Muller's ratchet, they do not
address whether the deleterious mutations that accumulate are independent or synergistic. Furthermore, despite earlier arguments, the ϕ6 studies also do not reject the
possibility that viral populations are immensely large and that synergistic mutations
are responsible for the evolution of segmentation. Thus, it becomes of interest to
know whether deleterious mutations act synergistically or independently in RNA
viruses.

Chao (22) attempted to ascertain the effect of deleterious mutations in RNA viruses by examining the accumulation of mutations in VSV and influenza A viruses
during undiluted passage, which propagates the virus at a high multiplicity of infeciton (MOI, or the number of viruses per host cell). Undiluted passage leads to the
accumulation of mutations because coinfections are common and, as a result, viruses within the same cell (or coinfection group) share gene products and have the
same fitness. With reduced variation in fitness, selection is less effective and mutation pressure becomes the main evolutionary force within the population.

The accumulation of both sequence and temperature-sensitive mutations during
undiluted passage of RNA viruses is well documented (69,70). In fact, a common

procedure for generating live viral vaccines that are attenuated by mutations is to culture the virus under undiluted passage, the celebrated case being Sabin's oral polio vaccine (71). Usually, culturing the virus at low MOI does not lead to attenuation or the accumulation of mutations.

Chao surmised that, as the mutations accumulate during undiluted passage, fitness should drop if the mutations are deleterious. Furthermore, the fashion in which fitness drops should provide information on whether mutations are independent or synergistic. If mutations were independent, the drop should be log-linear, whereas if they were synergistic, it should be log-concave downward (dropping in an accelerating manner).

From published data (28,72), Chao compared the drop in fitness correlates (titer and infectivity of the undiluted cultures) of VSV and influenza A virus and found that both viruses show a log-concave downward decline. The drop, however, is much more rapid in influenza A cultures. Chao interpreted this result to indicate that deleterious mutations interact synergistically in both viruses, but that the synergism is stronger in influenza A virus. Because a stronger synergism increases, in theory, the fitness advantage of a sexual population (see foregoing), Chao also used these results to explain why VSV is asexual (nonsegmented) and influenza A virus is segmented. Vesicular stomatitis virus and influenza A virus have otherwise about the same genome size and mutation rate (22).

Nee (73) objected to the foregoing interpretations on the grounds that Chao underestimated selection on RNAs (discussed earlier). At high MOIs, selection on coinfection groups is relaxed, but that frees both selection on RNAs and mutation pressure. Nee's argument is that the decline in fitness may be due to the increase of DI RNAs and not the accumulation of mutations. The DI RNAs commonly appear and increase during undiluted passage in many viruses, including VSV and influenza A virus (26,28).

The issue of synergism in RNA viruses, therefore, remains unsettled. Whether on RNAs or mutation pressure better accounts for the decline of fitness. A more careful examination of viral clones that appear after undiluted passage may be necessary. If selection on RNAs is more important, most clones should have an intracellular (within coinfection group) advantage in replication or encapsidation. However, in the VSV cultures considered in the foregoing, DI RNAs were never observed, despite attempts by the original investigators to find them (72). Temperature-sensitive mutations did accumulate, but it was not determined whether they had an intracellular advantage. Some, but not all, temperature-sensitive mutations have an intracellular advantage (74).

## ALTERNATIVE MODELS FOR THE EVOLUTION OF SEGMENTATION

There are alternative explanations for the evolution of segmentation in RNA viruses. Lane (75) provided three alternatives, besides the positive-selection models just discussed, and Nee (73,76) presented a fourth. These can be summarized, respectively, as (a) more efficient packaging, (b) more efficient control of translation, (c) increased resistance to damage, and (d) selection on RNAs.

Pressing and Reanney (21) used a comparative approach to argue effectively against points a and b. They note that point a is suspect because it should apply to DNA as well as RNA viruses, and yet segmentation is confined to RNA viruses. Point b results from the suggestion that segmentation may represent an attempt to "accommodate polycistronic RNAs to a biochemical environment [the eukaryotic cell] tailored to process only monocistronic RNAs" (21). Pressing and Reanney counter this view with the observation that the segmented phage $\phi6$ has a prokaryotic host and prokaryotes have polycistronic RNAs.

Points c and d have received more attention and, hence, they will be addressed in more detail.

## RNA Damage

Bernstein and associates (77) recently reemphasized the lethality of RNA damage (e.g., strand breaks) as a model for the evolution of segmentation. In support of the model is the phenomenon of multiplicity of reactivation. For example, when influenza A viruses are inactivated by ultraviolet (UV) radiation (78), live viruses are recovered only if host cells are infected with a high multiplicity of virus. The interpretation is that UV causes a lethal damage to some RNA segments, but segment reassortment recreates an undamaged progeny. At low multiplicities, reactivation is not seen because coinfection and reassortment are rare.

A model based on RNA damage is equivalent to one with lethal mutations, and both are extreme forms of the purifying-selection model with $S = 1$. A damage–lethal mutation model is attractive because it favors segmentation, without requiring population selection (discussed earlier). With lethal mutations or damage, sex is immediately advantageous to an individual virus the genome of which is harmed. Furthermore, the model is also consistent with the earlier observation that larger RNA viruses tend to be segmented.

However appealing a damage–lethal mutation model may be, its importance relative to models of deleterious $(0<S<1)$ mutations is questionable because the spontaneous rate of damage or lethal mutations may be low in RNA viruses. Multiplicity of reactivation is observed after the damage rate is increased with an artificial mutagen. Thus, although damage and lethal mutations may be important for the origin of sex in primitive self-replicating RNA systems (77), where they may be more common, they may not be important for the evolution (origin and maintenance) of sex in RNA viruses, in which deleterious mutations are demonstrably more common.

## Multicomponent Viruses: Sex or Selection on RNAs

The exchange of segments between single-component viruses within a coinfection is easily interpreted as sex. A particle of the virus contains the entire genome, and it is clearly comparable with an individual. Multicomponent viruses, on the other hand, are more perplexing. A particle contains only one segment. Is a particle then the individual? If so, is segmentation sex?

Nee (73,76) has argued that segmentation in multicomponent viruses is not a form of sex. Instead, he suggests that segmentation is the product of intracellular selection for smaller RNAs. Such a form of selection on RNAs favors a segmented genome because segments have a replication or encapsidation advantage over a larger RNA that contains the entire viral genome. By this account, segments are then evolving as selfish RNAs (see foregoing), and a multicomponent virus is best interpreted as a collection of mutualistic viral species. However, a problem with this viewpoint is that selection on RNAs will continuously favor smaller and smaller RNAs. What then prevents a segment from shedding its gene-coding regions and becoming a DI RNA? In such an event, the DI RNAs should drive viral RNAs (segmented and nonsegmented) to extinction and, eventually, themselves to the same fate. Nee (76) suggests that DI RNAs will not evolve because selection on RNAs will be stabilizing and favor the intermediate segment size observed in extant multicomponent viruses and presents as supporting evidence the fact that DI RNAs have never been reported in multicomponent viruses.

Nee's assumption of stabilizing selection can be challenged from several perspectives (79). First, the lack of reported DI RNAs in multicomponent viruses may be simply because no one has looked hard enough under the right conditions. Second, if single-component viruses are in anyway a model, there is no evidence that selection on RNAs is stabilizing. Third, available information on bromoviruses, a three-component virus, supports the possibility that DI RNAs exist in multicomponent viruses. For example, a genetically engineered brome mosaic segment, with a deletion in the coat protein gene has a twofold replicative advantage over the wild-type segment (80). This engineered segment cannot be properly called a DI RNA because its encapsidation properties are not known. Its existence does show, however, that an RNA molecule with DI-like properties can occur at the level of RNA replication in a multicomponent virus.

If selection on RNAs is not stabilizing, what stops the evolution of DI RNAs? As discussed earlier, selection on coinfection groups can occur, and Cole and Baltimore's (30) result that the total yield of a coinfection group is inversely proportional to the frequency of DI RNAs in the group offers the mechanism for the selection. Thus, there are two levels of selection, one on RNAs and one on coinfection groups. Chao (80) has argued that if selection on coinfection groups exists, that, and the demonstrated action of mutation pressure in RNA viruses, show that segmentation in multicomponent virus is a form of sex. The arguments are as follows.

If one follows the definition that *sex* constitutes genetic exchange among individuals (1), segmentation in a multicomponent virus is sex because coinfection groups are equivalent to individuals and the exchange of segments among coinfection groups is genetic. Coinfection groups are like individuals because they contain the viral genome and are units of selection. Mutational pressure explains then, in part, why a coinfection group cannot remain a permanent linkage group. If they did, they would be too large a target for deleterious mutations. Note that selection on coinfection groups, just as individual selection, favors permanent linkage (a nonsegmented virus). This is because a single segment of a multicomponent virus has to be complemented by the other types of segments, and a coinfection group that is not complemented fails to reproduce. A coinfection group containing a nonsegmented virus always reproduces. Thus, the exchange of segments in multicomponent viruses is a form of sex.

## SUMMARY

Because genetic exchange in RNA viruses and eukaryotes may be convergent, segmentation offers an independent model for the evolution of sex. Positive selection can be identified in RNA viruses, but its relevance to the evolution of segmentation remains uncertain. On the other hand, evidence is mounting that purifying-selection may be important. The latter may be, in part, because purifying-selection is easier to identify than positive-selection. Nonetheless, although positive-selection may still have played a role in the evolution of segmentation, it is unlikely that purifying-selection has not been a major player.

## ACKNOWLEDGMENTS

I thank D. Gill, J. Holland, S. Morse, T. Scott, and S. Weaver for helpful comments. This work was supported in part by the NIH and the University of Maryland.

## REFERENCES

1. Michod RE, Levin BR, eds. *The evolution of sex. An examination of current ideas.* Sunderland, Mass: Sinauer Associates; 1988.
2. Meselson M, Weigle JJ. Chromosome breakage accompanying genetic recombination in bacteriophage. *Proc Natl Acad Sci USA* 1961;47:857–868.
3. Ramig RF. Principles of animal virus genetics. In: Fields BN, Knipe DM, eds. *Fundamental virology,* 2nd ed. New York: Raven Press; 1991:95–122.
4. Tomizawa N, Anraku N. Molecular mechanisms of genetic recombination in bacteriophage. I. Effect of KCN on genetic recombination of phage T4. *J Mol Biol* 1964;8:508–515.
5. Horiuchi K. Genetic studies of RNA phages. In: Zinder ND, ed. *RNA phages.* Cold Spring Harbor: Cold Spring Harbor Laboratory; 1975:29–50.
6. Holland J, Spindler K, Horodyski F, Grabau E, Nichol S, VandePol S. Rapid evolution of RNA genomes. *Science* 1982;215:1577–1585.
7. Fiers W, Contreras R, Duerinck F et al. Complete nucleotide sequence of bacteriophage MS2 RNA: primary and secondary structure of the replicase gene. *Nature* 1976;260:500–507.
8. King AMQ, McCahon D, Slade WR, Newman JWI. Recombination in RNA. *Cell* 1982;29:921–928.
9. Makino S, Keck JG, Stohlman SA, Lai MMC. High-frequency RNA recombination of murine coronaviruses. *J Virol* 1986;57:729–737.
10. Romanova LI, Blinov VM, Tolskaya EA, et al. The primary structure of crossover regions of intertypic poliovirus recombinants: a model of recombination between RNA genomes. *Virology* 1986;155:202–213.
11. Jarvis C, Kirkegaard K. *Trends Genet* 1991;7:186–191.
12. Kirkegaard K, Baltimore D. The mechanism of RNA recombination in poliovirus. *Cell* 1986;47:433–443.

13. Matthews REF. Classification and nomenclature of viruses. *Intervirology* 1982;17:1–199.
14. Murphy FA, Kingsbury DW. Virus taxonomy. In: Fields BN, Knipe DM, eds. *Fundamental virology*, 2nd ed. New York: Raven Press; 1991:9–36.
15. Tyler KL, Fields BN. Reoviridae: a brief introduction. In: Fields BN, Knipe DM, eds. *Fundamental virology*, 2nd ed. New York: Raven Press; 1991:583–586.
16. Schiff LA, Fields BN. Reoviruses and their replication. In: Fields BN, Knipe DM, eds. *Fundamental virology*. New York: Raven Press; 1991:587–618.
17. Mindich L, Sinclair JF, Levine D, Cohen J. Genetic studies of temperature-sensitive and nonsense mutants of bacteriophage φ6. *Virology* 1976;75:218–223.
18. Ramig RF, Cross RK, Fields BN. A genetic map of reovirus: assignment of the newly defined mutant groups H, I and J to genome segments. *Virology* 1983;125:299–313.
19. Kilbourne ED. *Influenza*. New York: Plenum Publishing; 1987.
20. Allison RF, Janda M, Ahlquist P. Infectious in vitro transcripts from cowpea chlorotic mottle virus cDNA clones and exchange of individual RNA components with brome mosaic virus. *J Virol* 1988; 62:3581–3588.
21. Pressing J, Reanney DC. Divided genomes and intrinsic noise. *J Mol Evol* 1984;20:135–146.
22. Chao L. Evolution of sex in RNA viruses. *J Theor Biol* 1988;133:99–112.
23. Kilbourne ED. Molecular epidemiology—influenza as archetype. *Harvey Lect* 1979;73:225–258.
24. Fields S, Winter G. Nucleotide sequences of influenza virus segments 1 and 3 reveal mosaic structure of a small viral RNA segment. *Cell* 1982;28:303–313.
25. Bujarski JJ, Kaesberg P. Genetic recombination between RNA components of a multipartite plant virus. *Nature* 1986;321:528–531.
26. Holland J. Defective viral genomes. In: Fields BN, Knipe DM, eds. *Fundamental virology*, 2nd ed. New York: Raven Press; 1991:151–166.
27. Werren JH, Nur U, Wu C-I. Selfish genetic elements. *TREE* 1988;3:297–302.
28. von Magnus P. Incomplete forms of influenza virus. *Adv Virus Res* 1954;2:59–79.
29. Lewontin RC. The units of selection. *Annu Rev Ecol Syst* 1970;1:1–18.
30. Cole CN, Baltimore D. Defective interfering particles of poliovirus. III. Interference and enrichment. *J Mol Biol* 1973;76:345.
31. Felsenstein J. The evolutionary advantage of recombination. *Genetics* 1974;78:737–756.
32. Maynard Smith J. The evolution of recombination. In: Michod RE, Levin BR, eds. *The evolution of sex. An examination of current ideas*. Sunderland, Mass: Sinauer Associates; 1988:106–125.
33. Fisher RA. *The genetical theory of natural selection*. Oxford: Clarendon Press; 1930.
34. Muller HJ. Some genetic aspects of sex. *Am Nat* 1932;66:119–138.
35. Sturtevant AH, Mather K. The interrelations of inversions, heterosis and recombination. *Am Nat* 1938;72:447–452.
36. Charlesworth B. Recombination modification in a fluctuating environment. *Genetics* 1976;23:181–195.
37. Haldane JBS. The effect of variation on fitness. *Am Nat* 1937;71:337–349.
38. Haigh J. The accumulation of deleterious genes in a population—Muller's ratchet. *Theor Popul Biol* 1978;14:251–267.
39. Hopf FA, Michod RE, Sanderson MJ. The effect of reproductive system on mutational load. *Theor Popul Biol* 1988;33:243–265.
40. Muller HJ. The relation of recombination to mutational advance. *Mutat Res* 1964;1:2–9.
41. Maynard Smith J. *The evolution of sex*. Cambridge: Cambridge University Press; 1978.
42. Wagner GP, Gabriel W. Quantitative variation in finite parthenogenetic populations: what stops Muller's ratchet in the absence of recombination? *Evolution* 1990;44:715–731.
43. Eigen M, McCaskill J, Schuster P. Molecular quasi-species. *J Phys Chem* 1988;92:6881–6891.
44. Crow JF. Disease and evolution. *Evolution* 1983;37:863–865.
45. Kondrashov AS. Selection against harmful mutations in large sexual and asexual populations. *Genet Res* 1982;40:325–332.
46. Felsenstein J, Yokoyama S. The evolutionary advantage of recombination II. Individual selection for recombination. *Genetics* 1976;83:845–959.
47. Feldman MW, Christiansen FB, Brooks LD. Evolution of recombination in a constant environment. *Proc Natl Acad Sci USA* 1980;77:4838–4841.
48. Kondrashov AS. Deleterious mutations as an evolutionary factor. 1. The advantage of recombination. *Genet Res* 1984;44:199–217.
49. Felsenstein J. Sex and the evolution of recombination. In: Michod RE, Levin BR, eds. *The evolution of sex. An examination of current ideas*. Sunderland, Mass: Sinauer Associates; 1988:74–86.
50. Williams GC. *Sex and evolution*. Princeton: Princeton University Press; 1975.

51. Laver WG, Webster RG. Ecology of influenza viruses in lower mammals and birds. *Br Med Bull* 1979;35:29–33.
52. Lamb RA, Choppin PW. The gene structure and replication of influenza virus. *Annu Rev Biochem* 1983;52:467–506.
53. Laver WG, ed. *The origin of pandemic influenza viruses.* New York: Elsevier; 1983.
54. Scholtissek C, Rohde W, von Hoyningen V, Rott R. On the origin of the human influenza virus subtypes H2N2 and H3N2. *Virology* 1978;87:13–20.
55. Saitou N, Nei M. Polymorphism and evolution of influenza A virus genes. *Mol Biol Evol* 1986; 3:57–74.
56. Fitch WM, Leiter JME, Li X, Palese P. Positive Darwinian evolution in human influenza A viruses. *Proc Natl Acad Sci USA* 1991;88:4270–4274.
57. Domingo E, Sabo D, Taniguchi T, Weissmann C. Nucleotide sequence heterogeneity of an RNA phage population. *Cell* 1978;13:735–744.
58. Mindich L, Nemhauser I, Gottlieb P, et al. Nucleotide sequence of the large double-stranded RNA segment of bacteriophage φ6: genes specifying the viral replicase and transcriptase. *J Virol* 1988; 62:1180–1185.
59. Gottlieb P, Metzger S, Romantschuk M, et al. Nucleotide sequence of the middle dsRNA segment of bacteriophage φ6: placement of the genes of membrane-associate proteins. *Virology* 1988;163: 183–190.
60. Chao, L. Fitness of RNA virus decreased by Muller's ratchet. *Nature* 1990;348:454–455.
61. Chao L, Tran T, Matthews C. Muller's ratchet and the advantage of sex in the RNA virus φ6. *Evolution* 1992;46:289–299.
62. Crow JF, Kimura M. *An introduction to population genetics theory.* New York: Harper & Row; 1970.
63. Emerson SU. Rhabdoviruses. In: Fields BN, Knipe DM, eds. *Fundamental virology.* New York: Raven Press; 1986:477–490.
64. Nichol ST. Molecular epizootiology and evolution of vesicular stomatitis virus New Jersey. *J Virol* 1987;61:1029–1036.
65. Weaver SC, Rico-Hesse R, Scott TW. Genetic diversity and slow rates of evolution in New World alphaviruses. *Curr Top Microbiol Immunol* 1992;176:99–117.
66. Steinhauer DA, Holland JJ. Direct method for quantification of extreme polymerase error frequencies at selected single base sites in viral RNA. *J Virol* 1986;57:219–228.
67. Ohta T. Extension to the neutral mutation random drift hypothesis. In: Kimura M, ed. *Molecular evolution and polymorphism.* Mishima: National Institute of Genetics; 1977:148–176.
68. Ohta T. Very slightly deleterious mutations and the molecular clock. *J Mol Evol* 1987;26:1–6.
69. Ahmed R, Chakraborty PR, Fields BN. Genetic variation during lytic reovirus infection: high-passage stocks of wild-type reovirus contain temperature-sensitive mutants. *J Virol* 1980;34:285–287.
70. Spindler KR, Horodyski FM, Holland JJ. High multiplicities of infection favor rapid and random evolution of vesicular stomatitis virus. *Virology* 1982;96–108.
71. Almond JW. The attenuation of poliovirus neurovirulence. *Annu Rev Microbiol* 1987;41:153–180.
72. Youngner JS, Jones EV, Kelly M, Frielle DW. Generation and amplification of temperature-sensitive mutants during serial undiluted passages of vesicular stomatitis virus. *Virology* 1981;108:87–97.
73. Nee S. On the evolution of sex in RNA viruses. *J Theor Biol* 1989;138:407–412.
74. Youngner JS, Quagliana DO. Temperature-sensitive mutants of vesicular stomatitis virus are conditionally defective particles that interfere with and are rescued by wild-type virus. *J Virol* 1979; 19:102–107.
75. Lane LC. The RNAs of multipartite and satellite viruses of plants. In: Hall TC, Davies JW, eds. *Nucleic acids in plants,* vol 2. Boca Raton: CRC Press; 1979:65–110.
76. Nee S. The evolution of multicompartmental genomes in viruses. *J Mol Evol* 1987;25:277–281.
77. Bernstein H, Byerly HC, Hopf FA, Michod RE. Origin of sex. *J Theor Biol* 1984;110:323–351.
78. Barry RD. The multiplication of influenza virus. II. Multiplicity reactivaiton of ultraviolet irradiated virus. *Virology* 1961;14:398–405.
79. Chao L. Levels of selection, evolution of sex in RNA viruses, and the origin of life. *J Theor Biol* 1991;153:229–246.
80. French R, Alquist P. Intercistronic as well as terminal sequences are required for efficient amplification of brome mosaic virus RNA3. *J Virol* 1987;61:1457–1465.

# PART IV

## Driving Forces in Evolution II: Natural Selection

*The Evolutionary Biology of Viruses*,
edited by Stephen S. Morse.
Raven Press, Ltd., New York © 1994.

# 12

# Host Determination of Viral Evolution:

## A Variable Tautology

### Edwin D. Kilbourne

*Department of Microbiology, New York Medical College, Valhalla, New York 10595*

Because viruses are the ultimate obligate parasites, the very existence of which depends on host substrates, it appears obvious that host factors must be critical in virus selection and evolution. Therefore, an analysis of the relative importance of host determinants in viral evolution, as essayed here, might seem an exercise both in futility and tautology. Yet, our increasing appreciation of viral diversity, high mutation rates, and intrinsic intraspecies viral heterogeneity seem to fly in the face of the known centuries' old stability of certain virus–host relations and the constancy of disease expression that results from their interactions (1). Thus, particularly with well-established viruses, it probably is useful to assess the mechanisms by which viruses are selected, either by host defense or invitation, and to what extent the limits of viral adaptability constrain the process. Two sides of the same coin, yes, but sometimes the coin stands on edge.

## HOST-DRIVEN VIRUS SELECTION IN CLASSIC VIROLOGY: IMPLICATIONS FOR PRESENT STUDIES

When virology depended on the use of intact animals for isolation and propagation of virus, it was recognized that adaptation of virus to the alien laboratory host required serial passages in that host. It was also observed that adaptation to the new host often resulted in attenuation of viral virulence for the old. Defined in genetic terms, such adaptation implies both the selection in the new host of mutants not formerly dominant or even present in the original virus and, conversely, the loss or declining number of mutants able to replicate efficiently in the host of origin. Such empirically effected loss of virulence led to the development of the yellow fever (2) and live poliovirus vaccines (3).

In the early days of virology, Burnet and Lind (4) demonstrated that adaptation to increased replication and cell damage in the mouse lung by influenza virus required

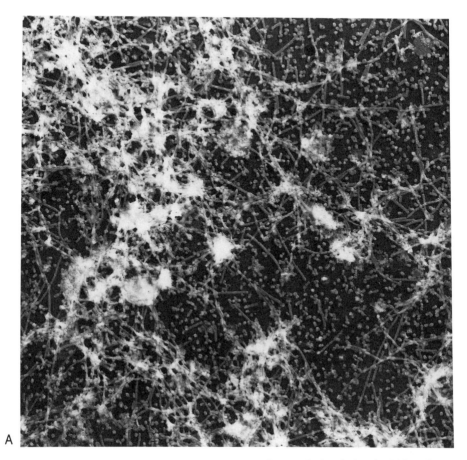

**FIG. 1.** Electron micrographs of (**A**) early passage influenza A virus isolate in chick embryos demonstrating the high proportion of filamentous forms typical of unadapted virus; and (**B**) uniformly spherical virus adapted to replication in the chick embryo ( × 17,500). (From Kilbourne and Murphy, ref. 12, with permission)

several mutational steps in association with environmental selection. Recent studies employing genetic reassortants have shown that mutations in at least four genes were involved in the evolution of mouse lung virulence of A/FM/1/47 (H1N1) virus (5), with each contributing partially to adaptation. Virus selection in the mouse is characterized both by a more rapid growth rate in that host (6) and the capacity to reach higher concentrations of virus in the lung (7,8). Because members of the initial viral population have intrinsically "unequal capacity" for adaptation (9), it is probable that adaptation comprises the selection of both preexisting variants and those arising during serial transfer of virus in the new host. Following mouse lung adaptation of influenza virus, the efficiency of infection in terms of the required

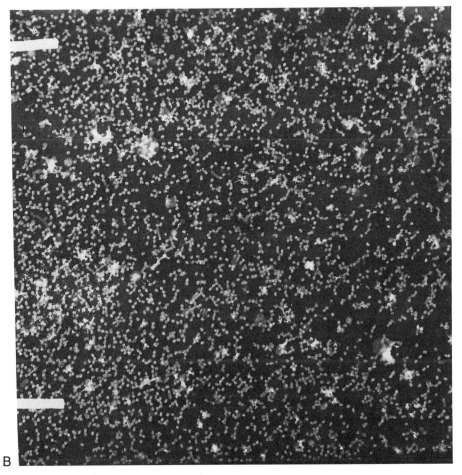

B

FIG. 1. *Continued.*

infecting dose is increased 100-fold (10). This increasing efficiency of viral replication with adaptation to the mouse lung is associated with increased viral virulence and, on occasion, transmissibility in the mouse (Kilbourne ED, unpublished data) (Fig. 1).

It is worth emphasizing that viral adaptation (evolution) under the constraints of laboratory manipulation is not necessarily host animal-specific, but often is organ- and tissue-specific. For example, strains of Coxsackie B virus serially propagated in the mouse brain may lose all capacity to replicate in mouse pancreas or heart—primary replication sites of unselected virus (Kilbourne ED, unpublished).

Despite the polygenic and (usually) gradual nature of virus evolution in adaptation to a new host, a virus may be only a single nucleotide away from the acquisition of lethal virulence (11) or new host specificity.

A dramatic illustration of host selection of adapting virus also is provided by influenza virus, as new isolates from humans are serially propagated in the chick embryo. The marker in this instance is predominantly viral morphologic appearance; the ratio of filamentous to spherical particles changes on passage; with a resulting increase in the spherical form that is preponderant in standard egg-adapted laboratory strains. This morphological trait is correlated with higher total viral yield, is transferable by genetic reassortment (12), is not associated with the envelope proteins, and probably is mediated by M protein gene product(s) of RNA (see Fig. 1).

## INFERENCES FROM THE NATURAL HISTORY OF VIRUSES FOR HOST INFLUENCE ON VIRAL EVOLUTION

### Host Dependent/Directed Evolution-Influence of Host Genotype

Inherent resistance factors in the host influenced the evolution of the insect rhabdovirus, sigma, in *Drosophila melanogaster* populations in the region of Languedoc in southern France. The situation is particularly interesting, because the virus is transmitted transovarially and is not contagious from fly to fly. Therefore, the confounding effects of environment on virus stability are minimized. *Drosophila* populations are polymorphic for two alleles, *O* and *P*, the latter of which interferes with viral replication and transmission. Sigma virus populations comprise two types, I and II, of which type II is better adapted to the *P* resistance allele. Studies of the relative distribution both of the host alleles and the two viral clones in Drosophila populations in contiguous geographic regions demonstrated (a) a highly significant increase in type II virus and (b) an overall increase in the proportion of flies infected, as type II virus appeared to spread to Languedoc from northern France. It is not yet known whether the evolution of the type II virus will result in the elimination of type I, or whether both types will continue to exist after reaching an equilibrium.

Evolution of viral populations in Languedoc appears to have led most recently to a decrease in high type II virus transmission. As noted by the authors of the study (13), ". . . selection of viral clones less efficiently transmitted by males, might be the first and most obvious adjustment to reduce viral invading capacities." This adjustment to select a virus of changed transmissibility is reminiscent of the coevolution of host and virus, of which myxomatosis (discussed later) is the classic example. The implications of this careful field study of viral evolution in nature are many, not the least of which is the potential of a well-established virus to break out of the bonds of host constraints.

The occurrence in *Drosophila* species of host gene migration of the high alcohol dehydrogenase (*Fast*) allele (14) suggests at least the potential for further geographic distribution of the allele for sigma virus resistance and still further host influence on the evolution of the virus.

SELECTION OF ANTIGENIC MUTANTS AS FUNCTION OF POPULATION ANTIBODY

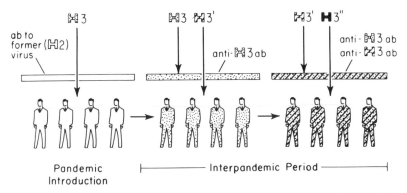

**FIG. 2.** Selection of antigenic mutants as a function of population antibody. New pandemic viral subtype H3 transcends barrier of antibody to unrelated previously prevalent virus H2 and readily infects the population. When a critical percentage of the population has been infected with H3, survival of H3 is impeded, and antigenically changed mutant H3' and later H3'' have survival advantage (minor antigenic variation or antigenic "drift"). (From Kilbourne, ref. 16, with permission)

## Influence of Acquired Immunity

Influenza provides the most striking example of the directing effect of immunoselection in the host on viral evolution. The survival of influenza A and B viruses in humans appears to depend on the continual escape of virus from antibody formed in response to invasion by antecedent strains (15) (Fig. 2). Such escape depends on sequential and accumulating mutations in the genes coding for the two principal external antigens of the virus, the hemagglutinin (HA) and neuraminidase (NA). Of these, the most significant ones affect the HA, the major envelope protein through which neutralization of virus is effected. The accumulating mutations in the HA are not random, but occur principally in regions of the molecule identifiable as antigenic sites: In these regions amino acid changes become progressively "fixed" and, taken together, establish a recognizable evolutionary pathway (17–19). Additional evidence for the role of host antibody in accelerating influenza viral evolution is provided by the apparently discordant evolutionary rates of the HA and NA antigens. The slower antigenic evolution of NA in humans is consistent with the virus-suppressive, but nonneutralizing, effect of antibody to this protein (20).

A unique experiment of nature occurred in 1977 with the return of an H1N1 influenza virus almost identical in its HA sequence to a virus prevalent in 1950. Comparative studies of the direction and rate of evolution of the 1950 and 1977 viruses revealed (a) a faster rate of evolution in the earlier period in a population with greater background immunity to the virus (i.e., imposing HA selection pressure) and (b) different evolutionary pathways, as shown by fixation of different mutations in key antigenic sites of the HA molecule. Another important observation was that diverging strains might be antigenically related (through a single epitope mutation), but evolutionarily distinct (21).

It is a paradoxical, but, I think, reasonable conclusion that influenza viruses, although episodically pervasive are, in fact, ephemeral entities, with a precarious existence, relying on the ancient strategy of Proteus for their survival. As with no other viruses, evolution of viral genotype is driven by changing host immunophenotype. Were access to the extended gene pool of animal influenza A viruses interdicted, this currently ubiquitous virus might, indeed, be headed for extinction. Continuing reassortment to provide the gene combinations requisite for virulence, transmissibility, and continued propagation of the virus is by no means a certainty.

Antigenic variation in field isolates of Rous sarcoma virus (RSV; 22), poliovirus (23), human immunodeficiency virus (HIV; 24), mumps (25), rabies (26), and dengue viruses (27), among others, has been documented, but epidemiologically significant viruses do not seem to evolve by sequential drift, as does influenza. Surprisingly, one exception may be canine parvovirus (CPV), a DNA virus that has shown a progressive evolution and antigenic drift in recent years, with replacement in the United States of type 2a with type 2b on the same phylogenetic pathway. Previously, in 1979–1981, type 2a had supplanted the initially predominant CPV-2. One amino acid change in a VP-2 capsid protein antigenic site is the apparent determinant of the most recent evolutionary step. The variation rate of the responsible *VP1/VP2* gene over a 12-year span is 10- to 100-fold lower than that of the influenza virus HA gene (28). Because parvovirus DNA is replicated by host DNA polymerases with low error rates, these studies imply that the high polymerase error rate characteristic of RNA viruses is not a definitive requirement for epidemiologically significant antigenic change and viral emergence. Gibbs (29) has pointed out the limited data available on sequential isolates of DNA viruses with reference to their rate of evolution.

Some directionality in the antigenic drift of HIV-1 isolates among familial contacts is suggested by a continuous sequence of nucleotide diversity among isolates (30). Other studies indicate that dominant epitopes can change during the course of HIV infection, an indication that immunoselection and the emergence of neutralization escape-mutants may occur (31). On the other hand, immunodirection to a predominant major variant seems not to occur with SIV infection of African green monkeys in the presence of a very high rate of coding mutations (32).

Neutralization-resistant variants seem to emerge during the entire course of HIV-1 infection in humans (24). However, the biological phenotype of HIV-1 clones may change at different stages of infection as monocytotropic variants are replaced by T-cell–tropic populations (33). In a clustered outbreak of HIV infection related to blood donation to two premature infants, a low level of envelope protein amino acid divergence was observed in sequential isolations. These apparently atypical findings may reflect the reduced immunoselection response in the neonates, but leave unexplained the similar limited variation in the adult donor (of both blood and virus). The donor, however, was not studied for the full course of the disease

(34). Furthermore, direct isolation of HIV DNA from the mononuclear cells of individual patients revealed a different distribution of sequences from each patient, with subsequent multidirectional antigenic variation. The evolution of the viral V3 region sequences was thought to be independent of the level of antigen expression and to be host dependent (35).

### Host-Mediated Nonimmunological Viral Selection

The older literature is replete with examples of viral change when transferred to laboratory hosts. Recently, changes at certain sites in the HA of influenza viruses have been associated with adaptation to the chick embryo, suggesting an identifiable molecular basis for host adaptation (36). However, replication in the egg can occur in the absence of "egg-adaptive" mutations so that egg-grown viruses can be indistinguishable from those found in clinical isolates by polymerase chain reaction (PCR) technology (37). Adaptation of egg-grown viruses to Madin–Darby canine kidney (MDCK) cells apparently directs the emergence of influenza virus HA mutants that are altered in their susceptibility to host cell-mediated proteolytic cleavage (38). However, similar mutations at certain of these sites are found in immunoselected field strains as well as in virus grown in other laboratory hosts (36,37).

The term *genetic dimorphism* (39), has been coined to describe viral populations in which two variants appear to be in equilibrium, almost as alleles, with one or the other favored in one or another host (or environmental situation). These are analogous to the alternative adaptations of evolution, in general, as discussed by West-Eberhard (40), or they may represent examples of the occurrence of mutator mutations, as recently described by Suarez et al. (41). Viral dimorphism has been most clearly demonstrated with two *L* and *H* HA mutants of swine influenza virus in which the *H* form predominates in the chick embryo and the *L* form in the natural host (42). So compelling is host selection in this instance that infection of swine with cloned *H* virus results within 3 to 4 days in the mutation, emergence, and eventual preponderance of the *L* mutant in animals so infected (43). The *L* and *H* mutants are distinguished by a single amino acid (glycine to glutamic acid) change at position 155, which borders the receptor pocket of the HA. This change is identifiable with monoclonal antibodies (i.e., is antigenic) and pleiotropically affects viral replication (44). This and evidence with other viruses emphasize that fortuitous or coincidental antigenic change can occur in the absence of immunoselection, a fact to be reckoned with in assessing change in antigen phenotype during viral evolution. With influenza B viruses, evidence for sequential immunoselection is lacking and, unlike their influenza A counterparts, their antigenic variation is sporadic, nonprogressive, and attended by a lower rate of amino acid-coding mutations (45). In this situation it is likely that the appearance of antigenic variants reflects a combination of some immunoselection coupled with fortuitous pleiotropic changes resulting from amino acid substitutions favorable for viral survival. In the laboratory, immunoselection pleiotropically may result in the emergence of virus with higher growth capacity in the chick embryo, as competing low-yielding mutants are suppressed (42).

A probable example of the influence of host on viral evolution is that provided by the demonstration of species-specific divergence of SIV virus in West and East African green monkeys, respectively. By polymerase chain reaction (PCR) techniques, which allow direct assessment of viral gene sequences without intervening cultivation in laboratory hosts, it was shown that the phylogenetic branching order of the monkey species parallels that of their respective viruses (46). A similar coincident divergence of virus and host had been described earlier for three papovaviruses (47).

### Viral Evolution Less Dependent on Host Influence

As cautioned in the introduction to this chapter, it is unlikely that evolution of viruses is ever truly host-independent. However, the occurrence of nonparallel divergence of virus lines from their host species suggests that, with certain viral–host pairs, host selection of virus may be minimal. This seems to be true of hepadnaviruses for which divergence, measured by synonymous substitution rates $(4.57 \times 10^{-5}$ per site per year), appears to have taken place much more recently than the divergence of their host species (birds, lower mammals, and humans) (48). Rhabdoviruses, on the other hand, serve as examples of viruses that have preserved their gene order and some regulatory sequences, but otherwise have diverged considerably in plant and animal hosts (49).

### Phylogenetic Implications of Viral Evolution

Thus far we have discussed the evolution of viruses with the implication that each evolves as a unit. This is, of course, a simplistic view, because the component genes and proteins of viruses can and do evolve at different rates. Relative to the present discussion of host influences, these rates may be negligible for such ancient and well-conserved enzymes as the reverse transcriptases of retroviruses, in contrast to their *env* and *gag* gene products (subject to host immunosurveillance) that are changing so rapidly that it may prove difficult, if not impossible, to trace their origins (50). It is also probable that the evolutionary rates of vertically transmitted viruses tied to host DNA replication are much less than those of horizontally transmitted infectious viruses, for which greater opportunity for selection is present. The overall evaluation of retroviruses derived from genome sequences implies that exogenous retroviruses "exist as short lived bursts upon a backdrop of germline-encoded endogenous viruses" (50). But even the exogenous virus human T-cell lymphotropic virus (HTLV)-1, in which both intra- and interstrain variability are demonstrable, has proved sufficiently stable over time to serve as a potential population marker in anthropological studies (51).

Even RNA viruses with very dissimilar genomic organization may exhibit striking similarities in amino acid sequence among their nonstructural proteins (52).

## MECHANISMS OF HOST INFLUENCE ON VIRAL EVOLUTION

### Viral Genetic Polymorphism

Elsewhere in this book the intrinsic polymorphism within viral populations is discussed extensively. This concept is a premise essential for the present discussion of host influence on viral evolution. It is also assumed that at least some minority mutants within such mixed populations have the potential to evolve to preponderance if removed from competition with the majority phenotype (43,53), and that intrapopulation variants can differ in individual mutation rates (41).

### Viral Selection as a Reciprocal of Host Immune Response (Change in Immunophenotype)

In short-lived infections (e.g., influenza), there is little opportunity for the emergence during a single infection of virus antigenically dissimilar to the original infecting virus. In fact, the disappearance of virus coincides with the first appearance of circulating antibody about 7 days after infection, although probably not as a consequence (54). Therefore, immunoselection is presumed to occur with serial transmission of virus within members of a partially immune population, leading to the emergence of antigenically changed mutants better able to evade neutralization by host antibody. In the case of influenza, the evolution of an epidemiologically significant strain, equipped to challenge a previously infected population, appears to require more than one change at critical sites on the viral HA (17,18), leading to antigenic difference of more than 50% (20). Furthermore, in contrast with what occurs in other viral infections, the influenza A strain previously dominant tends to disappear within a year or so, being essentially replaced by the evolving antigenic variant.

In chronic, persistent, relapsing infections, antigenic variants can arise during infection in the individual patient (see following section on nonimmunological selection). Immunoselection is mediated not only by antibody, but in vitro has followed lymphocytic choriomeningitis (LCM) virus point mutations within cytotoxic T-cell reactive epitopes (55). In vivo, cytotoxic T lymphocytes (CTLs) can recognize viral sequences that antibodies cannot detect (56). With influenza viruses, there may be overlap of CTL and B-cell antigenic sites (57).

It is important to note that immunoselection influences not only the evolution of those viral antigens directly subject to immunological pressure, but that other "passenger" genes of the virus may concomitantly be selected as they are carried within the escape mutant (29,58). This may be true for the changing *NS* gene of influenza A virus, against which no serological pressure exists (59).

Yet another aspect of immunoselection is its possible pleiotropic effect. This has already been noted with a swine influenza virus high-yield *HA* mutant that emerges with antibody suppression of a coexisting low-yield phenotype virus in a mixture of dimorphic variants (42).

### Nonimmunological Host Selection of Virus

#### *Viral Mutations in the Immunologically Suppressed Host*

In those experiments of nature in which the immune response is effectively re-moved as a selective host force, viral evolution does not cease, but rather, appears to falter or follow different pathways. Infections with HIV already have been dis-cussed as illustrating increased viral heterogeneity with the persistence of chronic infection. Furthermore, the fate of viral populations in ordinarily acute infections occurring in immunocompromised hosts is one of a seemingly undirected increase in diversity. This has been observed most frequently with chronic human rotavirus infection, in which extensive genome rearrangements have been found. These rear-rangements reflect changes in polyacrylamide gel electrophoresis (PAGE) migration patterns of the segmented double-stranded (ds)RNA of the virus genome. In these viruses, normal RNA segments may be missing from the gel profile, and novel bands of dsRNA are found that probably are concatameric forms of segment-spe-cific sequences. The protein products of these atypical viruses isolated during 1 year of chronic infection in an immunosuppressed child were largely unaltered. Different genotypes appeared at different frequencies in different isolates, with progressive replacement; thus, they seemed to be in dynamic equilibrium (60).

Of special interest are studies of antigenic and genetic variation of influenza A (H1N1) virus in a persistently infected child with severe combined immunodefi-ciency syndrome. During the course of 10 months of active infection, nonprogres-sive changes in virus phenotype and genotype were demonstrated. The genetic di-versity of the isolates was attributed to shifts in population equilibrium of replicating variant genomes in the virtual absence of immunoselection pressure (61). This case report adds further weight to prevalent assumptions that the sequen-tial mutations of influenza viruses are, in fact, driven by positive-Darwinian selec-tion (19). In this and other cases in which the infection-restricting effects of host immune response are removed, the ensuing protracted infection appears to present abundant opportunity in the relatively passive host for the emergence and testing of mutants previously absent or suppressed. Some of these may, indeed, show anti-genic changes, but such changes are linked or coincident with altered replication properties of the virus and not immunoselected. However, we cannot assume that the host is a passive substrate in these situations simply because it is not always possible to associate shifts in viral population equilibrium with demonstrable envi-ronmental changes within the host.

In chronic LCM virus infection of the mouse, the confinement of sustained viral replication to lymphoid tissue is associated with minimal genetic drift and emphasizes the role of host tissue in viral selection and evolution. As noted by the authors of the study, this example is contrary to the dogma that chronic infection with the RNA viruses necessarily leads to extensive genetic variation (62). However, with chronic natural infection with foot-and-mouth disease virus (FMDV) in immuno-competent hosts, the dogma appears to be sustained (63), a finding reinforced by studies in cell culture (64).

Recent evidence argues that the in vivo selection of variants from the notoriously unstable equine infectious anemia virus (EIAV) is the result of nonimmunological selection processes. Clinical relapses in association with recurrent viremia occurred before the appearance of neutralizing antibody (65).

There is interesting in vitro evidence that the nature of the host cell can influence the nature and frequency of measles viral mutations by fostering biased hypermutation (U→C transitions), although the relevance of this phenomenon to natural selection has not been demonstrated. This effect appears to be mediated by increased RNA-modifying and -unwinding activity in neural cells (66).

Finally, in considering the implications of selection and prevalence of the most fit members of a viral population, the chance or fortuitous nature of such ascendancy should be considered. This is emphasized by the studies of Domingo (67), who demonstrated that minority members of a viral quasispecies have the potential to become dominant as the majority phenotype and, given the opportunity, may do so.

### Mutation-Driven Evolution

There is evidence that the intrinsic nature of the virus (e.g., with reference to its mutation rate) may be of overriding importance in evolution. Temin (68) views the evolution of retroviral cancer genes as a mutation-driven, rather than a selection-driven, process. He points out that a variant (precancer) cell must accumulate multiple mutations before a change in fitness is achieved. Selection, then, is operative on cells with more than a single mutation.

### Host and Organ-Specific Selection

It has long been recognized that viral variants are selected on transfer of virus from one host to another in a laboratory setting. The NWS strain of influenza A virus, markedly neurotropic in mice, lost virulence when propagated in various cell cultures (69). Cultivation of influenza B virus in eggs selects viral subpopulations antigenically distinct from virus from the same source grown in mammalian cell cultures (70). Similar host-related selection has been shown with influenza A viruses and is associated with specific nucleotide substitutions (36). Changes in the receptor-binding properties of the influenza A virus H1 HA can result from either the direct alteration of the HA by host cell glycosylation or from mutations in the *HA* gene. Such adaptive changes may enhance viral heterogeneity and so contribute to survival of the virus in nature (71). Indeed, the respiratory tract of a patient with

influenza may harbor more than one antigenic variant (72,73). In the case of papillomaviruses (which are specific for particular epithelial cells), viral evolution appears to be linked not only to host species, but to the nature of the target cell (74).

## Host Population Effects

It seems obvious that the survival (hence, continuing selection and evolution) of
species-specific contagious viruses will be influenced by the density and variability
of host populations. Yet this aspect of viral evolution has been largely ignored,
except as it relates to coevolution of the host and virus, as discussed later. However,
the general dogma that evolution necessarily tends to avirulence has been challenged. The complexities of host–parasite interactions have been searchingly analyzed by May and Anderson (75). Theirs and other studies remind us that the host,
like the virus, is not a monolith, but comprises a population the members of which
vary constantly in number and susceptibility. Such host population variation is obviously a factor in viral evolution.

## COEVOLUTION OF VIRUS AND HOST

Not only does the host influence the direction of viral evolution, but, recip-
rocally, the host must adapt to virus as well. Yet, given the enormous difference in
generation times of viruses and mammalian hosts, it should not be surprising that
little but anecdotal evidence exists to directly support the notion of host adaptation
to virus. The severe impact of smallpox and measles on ostensibly virgin human
populations in the New World is often cited as evidence that the invading European
population was, by comparison, relatively resistant. Yet, the pervasive and severe
effects of measles in a population long isolated from contact with the virus was well
documented in the notorious Faroe Islands epidemic of 1846 (76), which involved
people of European ancestry. No discussion of coevolution fails to evoke the singular (and perhaps single), well-studied instance of viral–host coadaptation provided
by human-induced myxomatosis in Australia (see Chapter 14). In an ever-mounting
series of assaults on nature, early colonists introduced first sheep for domestic use,
then European rabbits to provide sport, then finally myxomavirus to conquer the
rabbits, which by then were threatening the sheep and wool industry by overgraz-
ing. The subsequent events have been cogently described in a classic monograph by
Fenner and Ratcliffe (77). Suffice it to point out here that after several introductions
of this highly lethal virus, coevolution led to the selection of a more resistant rabbit
and a less virulent virus—less virulent, but not avirulent, because a critical level of
viremia was necessary to ensure viral transmission by an insect vector.

It might be expected that sustained contact of virus and host would most favor
their coevolution. Indeed, recent studies of molecular variants of papillomavirus
type 16 from four continents suggest an ancient pandemic spread of the virus and its
coevolution with disparate human populations (78). In this study, African and Eura-

sian branches of the virus were distinguished, and representatives of both branches were found in Brazil, subject 200 years ago both to European colonization and importation of African slaves.

With in vitro cell culture systems, coevolution of virus and cells has been documented repeatedly, and the mechanisms involved have differed. Following persistent infection of L cells with a lymphotropic mouse parvovirus, cells restrictive to virus replication emerged as virus mutants able to replicate in the restrictive cells also evolved. In this instance, the evolved virus, rather than being avirulent, proved to be more virulent in the original parental cells (79). In this and an earlier study in L cells, it was possible to reconstruct and reinitiate the carrier state by combination of adapted viral and cell mutants (80). Coevolution of cells and virus in a persistent FMDV infection of cloned baby hamster-kidney (BHK) 21 cells resulted in evolution of a mutant cell in which FMDV RNA replication was restricted and, concomitantly, a mutant virus which, like Ron and Tal's parvovirus, was more cytolytic in the parental cells (81).

Such experiments are valuable in addressing a neglected dimension of cell–virus interaction and in defining mechanisms of cellular resistance and viral virulence, but are perhaps less relevant to coevolution in nature than are the papillomavirus studies.

## HUMAN INTERVENTION AS A FACTOR IN VIRUS EVOLUTION

The influence of human intervention on viral evolution goes beyond man's unwitting role as infected host. It includes also the effects of vaccines and antiviral agents on the selection of viral mutants as a cognitive response to viral threats, the selection in the laboratory of mutations, and the deliberate construction of new viruses.

### Vaccines

It is unlikely that the use of viral vaccines has yet been sufficiently widespread to significantly influence the evolution of wild-type viruses. In the case of rapidly evolving viruses, such as influenza A, it has been feared that worldwide mass immunization might serve to accelerate the selection of an antigenically novel, potentially pandemic virus. Unfortunately, (or fortunately) that efficiency of immunization seems most unlikely and, in any event, with present vaccines would closely simulate the immunological pressure that follows natural infection. However, new vaccine strategies, including the use of single proteins or oligopeptides could drive selection of mutants that are not in the mainstream of cumulative evolution. With live virus vaccines, competition and interference between vaccine and wild-type viruses might modulate evolution of wild-type virus in unexpected ways. Interfering, cold-adapted "dominant-negative" mutants, in fact, have already been described (82).

## Antiviral Agents

More likely to exert direct effects on viral evolution are the man-made antiviral drugs. Just as the host immunophenotype can be changed or created by artificial immunization, so can the internal selective environment be altered by antiviral agents. Clinical resistance to acyclovir was reported soon after the introduction of this drug for herpes simplex therapy (83). A strain of fowl plague virus resistant to ribavirin (Virazole) and another resistant to rimantadine have also been reported (84). Although most human influenza A viruses in nature remain sensitive to aman-tadine (85), a recent disturbing report describes both the emergence of amantadine-resistance in isolates of patients receiving prophylaxis and the possible transmission of these resistant strains to other patients (86). Here, we have a tandem host effect on viral evolution: (a) cognitive action of one host in prescribing the drug and (b) the unwitting selection of virus by the treated patient.

## Evolutionary Implications of Viral Engineering

It is important that in trying to explore the mechanisms of viral virulence and variation (as a kind of cognitive host response) (87), we do not create agents that without prolonged adaptation can instantly transcend normal host defenses. The adaptation of a mammalian (seal) influenza A virus to chicken embryo cells yielded a variant better disposed to host proteolytic cleavage of the HA and, therefore, lethal for chickens (88). Similarly, induction of a high sensitivity to proteolytic activation of a human influenza A virus was accomplished by site-specific muta-genesis (89). It has been shown by the creation of antigenic chimeras by random insertion mutagenesis that heterologous viral proteins can be incorporated into the structural proteins of Sindbis virus (90), an enveloped animal virus. The partial protective effect engendered by vaccination or infection with such a hybrid might predispose to differential, nonconcordant evolution of viral antigens, as occurs in nature with HA and neuraminidase antigens of influenza virus (20).

Finally, expansion of the host range of recombinant viruses offers a further mech-anism of, and dimension to, viral evolution (91).

## HOST FACILITATION AND CONSTRAINT OF VIRAL EVOLUTION: A SUMMARY

Host and virus are bound together by the tyranny of specificity that acts to con-strain both the evolution of virus and the resistance of the host. Viruses that escape this tyranny by adopting a second host will face the new constraints imposed by the demands of dual specificity.

Such an evolutionary constraint in the absence of immuno-selection pressure, as with Ross River virus in a nonimmune population, was associated with a surprisingly low evolutionary rate of the virus (92). Similarly, the regional genetic stability of the unstable negative-stranded RNA virus, VSV, has been explained by the hypothetical necessity for the virus to replicate with high efficiency in both insect and mammalian hosts. Furthermore, sequestration of the virus from mammalian immune responses during its residence in the insect host could further reduce selective evolutionary pressures (93). Also acting to constrain viral variability are properties intrinsic to the virus itself, such as functional constraints reflecting the probability of fixation of favorable mutations (94). Thus, mutants initially dominant in wild-type FMDV after suppression by monoclonal antibody, reasserted their dominance after serial passage in cell cultures (64). In nature, avian influenza viruses show far less variation and less continuity of evolution than comparable mammalian influenza virus lineages. It is suggested that the survival of the avian viruses favors those that have maintained the original phenotype (95).

Clearly, the most powerful potential force directing viral evolution is the host's immunological response. Yet the antigenic stability in nature of most viruses appears to have persisted intact at least during the decades that constitute the era of modern virology. Yet in the laboratory, neutralization escape mutants can readily be found for all viruses. Therefore, the paradox of high viral mutation rates and (usually) high antigenic stability in vivo requires explanation. Part of this explanation may lie in variation in the kinetics of infection and in the adequacy and persistence of the host's immunological response.

When examining influenza, the most notable example of viral antigenic instability, we see an infection probably initiated by only a few infectious units (96), short-lived, confined to respiratory epithelium, and characterized by subsequent imperfect immunity. Therefore, in primary infection, little opportunity for immunoselection from a large initial viral population exists. Rather, such selection most effectively occurs with reinfection of and passage in partially immune hosts. These events lead to a self-propagating sequence of pervasive epidemics, leading to widespread partial immunity, followed by emergence and preponderance of viral populations best equipped to survive. It is notable that the influenza virus H1 antigen, rapidly evolving in humans, is relatively stable in swine, perhaps because its propagation depends on infection principally of young (nonimmune) pigs in which immunoselection pressures are less (97). In those infections, respiratory or otherwise, in which such viral immunoselection is not usually evident in nature (e.g. measles), the opposite conditions obtain; that is, infection is relatively protracted, involves multiple viscera, and engenders more solid immunity, so that the opportunity for virus to flourish in the presence of antibody during subsequent challenge is reduced or absent as reinfection is diminished.

By inference then, humans and influenza virus live in precarious balance. Were it not for the extended viral gene pool provided by animal influenza A viruses, sustained universal artificial immunization to provide really effective and pervasive immunity, might actually lead to eradication of human influenza viruses. Thus the human host has the potential to direct the evolution of his viruses to the point of extinction.

# REFERENCES

1. Kilbourne ED. New viruses and new diseases: mutation, evolution and ecology. *Curr Opin Immunol* 1991;3:518–524.
2. Theiler M, Smith HH. The use of yellow fever virus modified by in vitro cultivation for human immunization. *J Exp Med* 1937;65:782–800.
3. Sabin A, Hennessen WA, Winsser J. Studies of variants of poliomyelitis virus. I. Experimental segregation and properties of avirulent variants of three immunologic types. *J Exp Med* 1954;99: 551–576.
4. Burnet FM, Lind PE. An analysis of the adaptation of an influenza virus to produce lesions in the mouse lung. *J Exp Biol Med Sci* 1954;32:711–720.
5. Brown EG. Increased virulence of a mouse-adapted variant of influenza A/FM/1/47 virus is controlled by mutations in genome segments 4, 5, 7, and 8. *J Virol* 1990;64:4523–4533.
6. Ledinko N. Influenza B virus. An analysis of the process of adaptation of influenza virus B of recent human origin to the mouse lung. *J Gen Microbiol* 1956;15L:47–60.
7. Davenport FM, Francis T. A comparison of the growth curves of adapted and unadapted lines of influenza virus. *J Exp Med* 1951;93:129–137.
8. Schulman J. Effects of immunity on transmission of influenza: experimental studies. *Prog Med Virol* 1970;12:128–160.
9. Davenport FM. The inequality of potential in influenza virus for adaptation to mice. *J Immunol* 1954;72:485–488.
10. Ginsberg HS. Comparison of quantity of egg and mouse-adapted influenza viruses required to infect each host. *Proc Soc Exp Biol Med* 1953;84:249–252.
11. Kawaoka Y, Webster RG. Molecular mechanism of acquisition of virulence in influenza virus in nature. *Microb Pathogen* 1988;5:311–318.
12. Kilbourne ED, Murphy JS. Genetic studies of influenza viruses. I. Viral morphology and growth capacity as exchangeable genetic traits. Rapid in ovo adaptation of early passage Asian strain isolates by combination with PR8. *J Exp Med* 1960;111:337–406.
13. Fleuriet A, Periquet G, Anxolabehere D. Evolution of natural populations in the *Drosophila melanogaster* sigma virus system I. Languedoc (southern France). *Genetica* 1990;81:21–31.
14. Simmons GM, Kreitman ME, Quattlebaum WF, Miyashita N. Molecular analysis of the alleles of alcohol dehydrogenase along a cline in *Drosophila melanogaster*. I. Maine, North Carolina, and Florida. *Evolution* 1989;43:393–409.
15. Kilbourne ED. *Influenza*. New York: Plenum Publishing; 1987:137–138.
16. Kilbourne ED, ed. *The influenza viruses and influenza*. New York: Academic Press; 1975:255.
17. Air GM, Laver WG, Webster RG. Antigenic variation in influenza viruses. *Contr Microbiol Immunol* 1987;8:20–59.
18. Wiley DC, Wilson IA, Skehel JJ. Structural identification of the antibody-binding sites of Hong Kong influenza haemagglutinin and their involvement in antigenic variation. *Nature* 1981;289:373–378.
19. Fitch WM, Leiter JME, Li X, Palese P. Positive Darwinian evolution in human influenza A viruses. *Proc Natl Acad Sci USA* 1991;88:4270–4274.
20. Kilbourne ED, Johansson BE, Grajower B. Independent and disparate evolution in nature in influenza A virus hemagglutinin and neuraminidase. *Proc Natl Acad Sci USA* 1990;87:786–790.
21. Raymond FL, Caton AJ, Cox NJ, et al. The antigenicity and evolution of influenza H1 haemagglutinin, from 1950–1957 and 1977–1983: two pathways from one gene. *Virology* 1986;148:275–287.
22. Sullender WM, Mufson MA, Anderson LJ, Wertz GW. Genetic diversity of the attachment protein of subgroup B respiratory syncytial viruses. *J Virol* 1991;65:5425–5434.
23. Kinnunen L, Huovilainen A, Poeyry T, Hovi T. Rapid molecular evolution of wild type 2 poliovirus during infection in individual hosts. *J Gen Virol* 1990;71:317–324.
24. Von Gegerfelt A, Albert J, Morfeldt-Manson L, Broliden K, Fenyo EM. Isolate-specific neutralizing antibodies in patients with progressive HIV-1-related disease. *Virology* 1991;185:162–168.
25. Yamada A, Takeuchi D, Tanabayashi K, Hishiyama M, Sugiura A. Sequence variation of the P gene among mumps virus strains. *Virology* 1989;172:374–376.
26. Botros BAM, Salib AW, Mellick PW, et al. Antigenic variation of wild and vaccine rabies strains of Egypt. *J Med Virol* 1988;24:153–159.

27. Rico-Hesse R. Molecular evolution and distribution of Dengue viruses type 1 and 2 in nature. *Virology* 1990;174:479–493.
28. Parrish CR, Aquadro CF, Strassheim ML, et al. Rapid antigenic-type replacement and DNA sequence evolution of canine parvovirus. *J Virol* 1991;65:6544–6552.
29. Gibbs A. Molecular evolution of viruses; "trees," "clocks" and "modules." *J Cell Sci* 1987;7 (suppl):319–337.
30. Burger H, Weiser B, Flaherty K, et al. Evolution of human immunodeficiency virus type 1 nucleotide sequence diversity among close contacts. *Proc Natl Acad Sci USA* 1991;88:11236–11240.
31. Nara PL, Smit L, Dunlop N, et al. Emergence of viruses resistant to neutralization by V3-specific antibodies in experimental human immunodeficiency virus type 1 IIIB infection of chimpanzees. *J Virol* 1990;64:3779–3791.
32. Baier M, Dittmar MT, Cichutek K, Kurth R. Development in vivo of genetic variability of simian immunodeficiency virus. *Proc Natl Acad Sci USA* 1991;88:8126–8130.
33. Schuitemaker H, Koot M, Kootstra NA, et al. Biological phenotype of human immunodeficiency virus type 1 clones at different stages of infection: progression of disease is associated with a shift from monocytotropic to T-cell–tropic virus populations. *J Virol* 1992;66:1345–1360.
34. McNearney T, Westervelt P, Thielan BJ, et al. Limited sequence heterogeneity among biologically distinct human immunodeficiency virus type 1 isolates from individuals involved in a clustered infectious outbreak. *Proc Natl Acad Sci USA* 1990;87:1917–1921.
35. Wolfs TFW, Zwart G, Bakker M, et al. Naturally occurring mutations within HIV-1 V3 genomic RNA lead to antigenic variation dependent on a single amino acid substitution. *Virology* 1991;185: 195–205.
36. Robertson JS, Nicolson C, Bootman JS, et al. Sequence analysis of the haemagglutinin (HA) of influenza A (H1N1) viruses present in clinical material and comparison with the HA of laboratory-derived virus. *J Gen Virol* 1991;72:2671–2677.
37. Rajakumar A, Swierkosz EM, Schulze IT. Sequence of an influenza virus hemagglutinin determined directly from a clinical sample. *Proc Natl Acad Sci USA* 1990;87:4154–4158.
38. Rott R, Orlich M, Klenk H-D, et al. Studies on the adaptation of influenza viruses to MDCK cells. *EMBO J* 1984;3:3329–3332.
39. Mowshowitz S, Kilbourne ED. Genetic dimorphism of the neuraminidase in recombinants of H3N2 influenza virus. In: Barry RD, Mahy BWJ, eds. *Negative strand viruses*, vol 2. London: Academic Press: 1975:765–775.
40. West-Eberhard MJ. Alternative adaptations, speciation, and phylogeny (a review). *Proc Natl Acad Sci USA* 1986;83:1388–1392.
41. Suarez P, Valcarcel J, Ortin J. Heterogeneity of the mutation rates of influenza A viruses: isolation of mutator mutants. *J Virol* 1992;66:2491–2494.
42. Kilbourne ED. Genetic dimorphism in influenza viruses: characterization of stably associated hemagglutinin mutants differing in antigenicity and biological properties. *Proc Natl Acad Sci USA* 1978;75:6258–6262.
43. Kilbourne ED, Easterday BC, McGregor S. Evolution to predominance of swine influenza virus hemagglutinin mutants or predictable phenotype during single infections of the natural host. *Proc Natl Acad Sci USA* 1988;85:8098–8101.
44. Both GW, Shi CH, Kilbourne Ed. Hemagglutinin of swine influenza virus: a single amino acid change pleiotropically affects viral antigenicity and replication. *Proc Natl Acad Sci USA* 1983;80: 6996–7000.
45. Air GM, Gibbs AJ, Laver WG, Webster RG. Evolutionary changes in influenza B are not primarily governed by antibody selection. *Proc Natl Acad Sci USA* 1990;87:3884–3888.
46. Allan JS, Short M, Taylor ME, et al. Species-specific diversity among simian immunodeficiency viruses from African green monkeys. *J Virol* 1991;65:2816–2828.
47. Soeda E, Maruyama T, Arrand JR, Griffin BE. Host-dependent evolution of three papova viruses. *Nature* 1980;285:165–167.
48. Orito E, Mizokami M, Ina Y, et al. Host-independent evolution and a genetic classification of the hepadnavirus family based on nucleotide sequences. *Proc Natl Acad Sci USA* 1989;86:7059–7062.
49. Peters D. Divergent evolution of Rhabdoviridae and Bunyaviridae in plants and animals. In: *Seminars in Virology*, vol 2. The Netherlands: WB Saunders; 1991:27–37.
50. Doolittle RF, Feng D-F, Johnson MS, McClure MA. Origins and evolutionary relationships of retroviruses. *Q Rev Biol* 1989;64:1–30.
51. Gessain A, Gallo RC, Franchini G. Low degree of human T-cell leukemia/lymphoma virus type I

genetic drift in vivo as a means of monitoring viral transmission and movement of ancient human populations. *J Virol* 1992;66:2288–2295.

52. Haseloff J, Goelet P, Zimmern D, et al. Striking similarities in amino acid sequence among non-structural proteins encoded by RNA viruses that have dissimilar genomic organization. *Proc Natl Acad Sci USA* 1984;81:4358–4362.

53. Martinez MA, Carrillo C, Gonzalez-Candelas F, et al. Fitness alteration of foot-and-mouth disease virus mutants: measurement of adaptability of viral quasispecies. *J Virol* 1991;65:3954–3957.

54. Imasaki T, Nozima T. Defense mechanisms against primary influenza virus infection in mice. I. The role of interferon and neutralizing antibodies and thymus dependence of interferon and antibody production. *J Immunol* 1977;118:256–263.

55. Aebischer T, Moskophidis D, Hoffman Rohrer U. In vitro selection of lymphocytic choriomeningitis virus escape mutants by cytotoxic T lymphocytes. *Proc Natl Acad Sci USA* 1991;88:11047–11051.

56. Oldstone MBA. Molecular anatomy of viral persistence. *J Virol* 1991;65:6381–6386.

57. Hioe CE, Dybdahl-Sissoko N, Philpott M, Hinshaw VS. Overlapping cytotoxic T-lymphocyte and B-cell antigenic sites on the influenza virus H5 hemagglutinin. *J Virol* 1990;64:6246–6251.

58. Birky CW, Walsh JB. Effects of linkage on rates of molecular evolution. *Proc Natl Acad Sci USA* 1988;85:6414–6418.

59. Buonagurio DA, Nakada S, Parvin JD, et al. Evolution of human influenza A viruses over 50 years: rapid, uniform rate of change in *NS* gene. *Science* 1986;232:980–982.

60. Hundley FM, McIntyre BC, Beards G, et al. Heterogeneity of genome rearrangements in rotaviruses isolated from a chronically infected immunodeficient child. *J Virol* 1987;61:3365–3372.

61. Rocha E, Cox NJ, Black RA, et al. Antigenic and genetic variation in influenza A (H1N1) virus isolates recovered from a persistently infected immunodeficient child. *J Virol* 1991;65:2340–2350.

62. Ahmed R, Hahn CS, Somasundaram T, et al. Molecular basis of organ-specific selection of viral variants during chronic infection. *J Virol* 1991;65:4242–4247.

63. Gebauer F, De La Torre JC, Gomes I, et al. Rapid selection of genetic and antigenic variants of foot-and-mouth disease virus during persistence in cattle. *J Virol* 1988;62:2041–2049.

64. Gonzalez MJ, Saiz JC, Laor O, Moore DM. Antigenic stability of foot-and-mouth disease virus variants on serial passage in cell culture. *J Virol* 1991;65:3949–3953.

65. Carpenter S, Evans LH, Sevoian M, Chesebro B. Role of the host immune response in selection of equine infectious anemia virus variants. *J Virol* 1987;61:3783–3789.

66. Rataul SM, Hirano A, Wong TC. Irreversible modification of measles virus RNA in vitro by nuclear RNA-unwinding activity in human neuroblastoma cells. *J Virol* 1992;66:1769–1773.

67. Domingo E. RNA virus evolution and the control of viral disease. *Prog Drug Res* 1992;93–133.

68. Temin HM. Evolution of cancer genes as a mutation-driven process. *Cancer Res* 1988;48:1697–1701.

69. Janda Z, Vonka V. Variation in neurovirulence of NWS influenza virus after repeated passages in different tissue culture systems. *Arch Gesamte Virusforschung* 1968;24:197–202.

70. Schild GC, Oxford JS, de Jong JC, Webster RG. Evidence for host–cell selection of influenza virus antigenic variants. *Nature* 1983;303:706–709.

71. Aytay S, Schulze IT. Single amino acid substitutions in the hemagglutinin can alter the host range and receptor binding properties of H1 strains of influenza A virus. *J Virol* 1991;65:3022–3028.

72. De Jong JC, De Ronde-Verloop FM, Veenedaal-Van Herk M, et al. Antigenic heterogeneity within influenza A (H3N2) virus strains. *Bull WHO* 1988;66:47–55.

73. Katz JM, Webster RG. Antigenic and structural characterization of multiple subpopulations of H3N2 influenza virus from an individual. *Virology* 1988;165:446–456.

74. Cole ST, Danos O. Nucleotide sequence and comparative analysis of the human papillomavirus type 18 genome. *J Mol Biol* 1987;193:599–608.

75. May RM, Anderson RM. Population biology of infectious diseases: part II. *Nature* 1979;280:455–461.

76. Panum PL. *Observations made during the epidemic of measles on the Faroe Islands in the year 1846*. New York: American Publishing Association; 1940.

77. Fenner F, Ratcliffe FN. *Myxomatosis*. London: Cambridge University Press; 1965.

78. Chan S-Y, Ho L, Ong C-K, et al. Molecular variants of human papillomavirus type 16 from four continents suggest ancient pandemic spread of the virus and its coevolution with humankind. *J Virol* 1992;66:2057–2066.

79. Ron D, Tal J. Coevolution of cells and virus as a mechanism for the persistence of lymphotropic minute virus of mice in L-cells. *J Virol* 1985;55:424–430.

80. Ahmed R, Canning WM, Kauffman RS, et al. Role of the host cell in persistent viral infection: coevolution of L cells and reovirus during persistent infection. *Cell* 1981;25:325–332.
81. De La Torre JC, Martinez-Salas E, Diez J, et al. Coevolution of cells and viruses in a persistent infection of foot-and-mouth disease virus in cell culture. *J Virol* 1988;62:2050–2058.
82. Whitaker-Dowling P, Maassab HF, Youngner JS. Dominant-negative mutants as antiviral agents: simultaneous infection with the cold-adapted live-virus vaccine for influenza A protects ferrets from disease produced by wild-type influenza A. *J Infect Dis* 1991;164:1200–1208.
83. Field HJ, Wildy P. Clinical resistance of herpes simplex virus to acyclovir. *Lancet* 1982;1:1125.
84. Indulen MK, Feldblum RL. Obtaining of a Virazole-resistant fowl plague virus mutant. *Acta Virol* 1982:109.
85. Belshe RB, Burk B, Newman F, et al. Resistance of influenza A virus to amantadine and rimantadine: results of one decade of surveillance. *J Infect Dis* 1989;159:430–435.
86. Mast EE, Harmon MW, Gravenstein S, et al. Emergence and possible transmission of amantadine-resistant viruses during nursing home outbreaks of influenza A (H3N2). *Am J Epidemiol* 1991;134: 988–997.
87. Kilbourne ED. Genetic variation among influenza viruses. In: Nayak DP, ed. *Human adaptation to influenza viral evolution*. New York: Academic Press; 1981:481–488.
88. Li S, Orlich M, Rott R. Generation of seal influenza virus variants pathogenic for chickens, because of hemagglutinin cleavage site changes. *J Virol* 1990;64:3297–3303.
89. Ohuchi R, Ohuchi M, Garten W, Klenk H-D. Human influenza virus hemagglutinin with high sensitivity to proteolytic activation. *J Virol* 1991;65:3530–3537.
90. London SD, Schmaljohn AL, Dalrymple JM, Rice CM. Infectious enveloped RNA virus antigenic chimeras. *Proc Natl Acad Sci USA* 1992;89:207–211.
91. Kondo A, Maeda S. Host range expansion by recombination of the baculoviruses *Bombyx mori* nuclear polyhedrosis virus and *Autographa californica* nuclear polyhedrosis virus. *J Virol* 1991; 65:3625–3632.
92. Burness ATH, Pardoe I, Faragher SG, et al. Genetic stability of Ross River virus during epidemic spread in nonimmune humans. *Virology* 1988;167:639–643.
93. Nichol ST. Genetic diversity of enzootic isolates of vesicular stomatitis virus New Jersey. *J Virol* 1988;62:572–579.
94. Wilson AC, Carlson SS, White TJ. Biochemical evolution. *Annu Rev Biochem* 1977;46:573–639.
95. Bean WJ, Schell M, Katz J, et al. Evolution of the H3 influenza virus hemagglutinin from human and nonhuman hosts. *J Virol* 1992;66:1129–1138.
96. Alford RH, Kasel JA, Gerone PJ, Knight V. Human influenza resulting from aerosol inhalation. *Soc Exp Biol Med* 1966;122:800–804.
97. Luoh S-M, Mcgregor MW, Hinshaw V. Hemagglutinin mutations related to antigenic variation in H1 swine influenza viruses. *J Virol* 1992;66:1066–1073.

The Evolutionary Biology of Viruses,
edited by Stephen S. Morse.
Raven Press, Ltd., New York © 1994.

# 13

# Evolution of the Poxviruses, Including the Coevolution of Virus and Host in Myxomatosis

Frank Fenner and *Peter J. Kerr

*John Curtin School of Medical Research, Australian National University, Canberra 2600, Australia and *CSIRO Division of Wildlife and Ecology, Canberra 2912, Australia*

The family Poxviridae consists of two subfamilies: Chordopoxvirinae, the poxviruses of vertebrates, and Entomopoxvirinae, the poxviruses of insects. The member viruses are the largest and the most complex of all viruses and have a large ovoid or brick-shaped virion, and a genome consisting of a single molecule of double-stranded (ds)DNA, varying in size in different genera from 130 to 375 kilobase parrs (kbp). Unlike all other DNA viruses, except African swine fever virus, they replicate in the cytoplasm and, to accomplish this, they encode numerous enzymes. In this paper we discuss three aspects of the evolution of viruses of the subfamily Chordopoxvirinae: (a) biogeography and sequence comparisons as clues to their evolution; (b) a more detailed consideration of the biogeography and host specificity of viruses of the genus *Leporipoxvirus*; and (c) a natural experiment of the coevolution of myxoma virus (a leporivirus) and the new host in which it has spread in Australia and Europe since the 1950s, the European rabbit, *Oryctolagus cuniculus*. Detailed references to work on myxomatosis are provided in Fenner and Ratcliffe (1) and Fenner and Ross (2).

## EVOLUTION OF THE CHORDOPOXVIRINAE

In the absence of sequence data, such as those available for the papillomaviruses (see Chapter 8), the present distribution and geologic history of the reservoir hosts of the poxviruses, especially those with a narrow host range, provide a clue to their antiquity.

**TABLE 1.** *Host range and geographic distribution of genera and unclassified members of the subfamily Chordopoxvirinae*

| Genus and species | Reservoir host | Geographic distribution |
|---|---|---|
| *Orthopoxvirus* | | |
| Camelpox virus | Camels | Africa, Asia |
| Cowpox virus | Rodents | Europe, western Asia |
| Ectromelia virus | Rodents | Europe |
| Monkeypox virus | Squirrels | Western and central Africa |
| Raccoonpox virus | Raccoons | Eastern United States |
| Skunk poxvirus | Skunks | United States |
| Tatera poxvirus | Gerbils | Western Africa |
| Uasin Gishu virus | Unknown | Eastern Africa |
| Vaccinia virus | Unknown | Worldwide |
| Variola virus | Humans | Worldwide (now extinct) |
| Volepox virus | Voles | Western United States |
| *Parapoxvirus* | | |
| Ausdyck virus | Camels | Africa, Asia |
| Orf virus | Sheep | Worldwide |
| Pseudocowpox virus | Cattle | Worldwide |
| Red deer poxvirus | Red deer | New Zealand |
| Sealpox virus | Seals | Worldwide |
| *Capripoxvirus* | | |
| Sheep-pox virus | Sheep | Asia, Africa |
| Goatpox virus | Goats | Asia, Africa |
| Lumpyskin disease virus | Buffalo | Africa |
| *Suipoxvirus* | | |
| Swinepox virus | Swine | Worldwide |
| *Leporipoxvirus* (see Table 2) | | |
| *Avipoxvirus* | | |
| Many species | Birds | Worldwide |
| *Yatapoxvirus* | | |
| Tanapox virus | ?Rodents | Eastern and central Africa |
| Yabapoxvirus | ?Monkeys | Western Africa |
| *Molluscipoxvirus* | Humans | Worldwide |
| Molluscum contagiosum virus | Humans | Worldwide |
| Unclassified | | |
| Macropod poxvirus | Kangaroos Quokkas | Australia |
| Crocodilian poxvirus | Crocodiles | Australia Southern Africa |
| | Caimans | Florida |

## Biogeography of Viruses of the Subfamily Chordopoxvirinae

The subfamily Chordopoxvirinae of the family Poxviridae contains eight genera and some unclassified viruses (Table 1). Except for the genus *Parapoxvirus*, the virions of all member viruses are similar morphologically, but show very limited antigenic cross-reactivity between genera, and no cross-protection, except between the different species of each genus. Many poxviruses of vertebrates are now known worldwide, because they cause diseases of humans, domestic animals, or birds. Very little is known of the distribution or even the existence of viruses of wildlife that do not infect humans or their domestic animals (except for the arboviruses, many of which have been recovered from collections of insects). Some inkling of the geologic age of the subfamily is provided by the geographic distribution of different genera and species, excluding those in which the host animals have been spread around the world by domestication.

Among the orthopoxviruses, cowpox and ectromelia viruses, both of which have rodent reservoir hosts, are restricted to Europe, with cowpox extending east as far as Turkmenia. Monkeypox, taterapox, and Uasin Gishu viruses have been found only in tropical Africa, and raccoonpox, skunkpox, and volepox viruses are restricted to North America. The capripoxviruses have a narrow host range and are now restricted to sheep and goats in Africa and Asia; they probably evolved with these animals. The avipoxviruses, parapoxviruses, and swinepox virus now have a worldwide distribution, but the two yatapoxviruses have been found only in tropical Africa (their reservoir hosts are unknown), and the macropod poxvirus appears to be an indigenous virus of Australian marsupials. Most leporipoxviruses (see later discussion) occur in leporids (order Lagomorpha) or squirrels (order Rodentia) in the Americas, but one species is found in hares in Europe and what is possibly another species in African hares. Finally, poxviruses have been visualized with the electron microscope in lesions of African, Australian, and American crocodilians, but these viruses have not yet been further studied. Avipoxviruses appear to be specific for birds, but occur worldwide, in avian species belonging to many different orders. Thus, there appear to be poxviruses specific for crocodilians in three continents, and some species of poxvirus appear to be specific to animals of Africa, Australia, North America, or the Eurasian landmass. This suggests that the chordopoxviruses as a subfamily must have an evolutionary origin at least as old as the reptiles. Viruses of different genera may have evolved with the birds, marsupials, and the placental mammals in which they are now found.

## Molecular Comparisons of Poxvirus Genomes

The molecular evolution of poxviruses has recently been reviewed (3); a brief overview will be given here to illustrate some of the mechanisms involved. The poxvirus genome is a linear molecule of double-stranded DNA, with inverted terminal repetitions (ITR) and hairpin-loop structures at the ends. Replication occurs in the cytoplasm of infected cells and involves the formation of concatameric intermediates (reviewed in 4). Blocks of tandem-repeat sequences within the ITRs are a feature of the orthopoxviruses (5) and parapoxviruses (6,7), and imperfect, palindromic-repeat sequences have been described in the termini of Shope fibroma virus (8). Alterations in copy numbers of these repeat sequence elements can produce microheterogeneity at the ends of the genome, without alterations in phenotype (5).

The genomes of all the orthopoxviruses, except Uasin Gishu disease virus and skunkpox virus, have been examined by restriction mapping (9,10). The results of these studies support the species assignment based on biological properties. Restriction-mapping studies have also been reported for parapoxviruses (6,11), capripox-viruses (12), leporipoxviruses (13,14), avipoxviruses (15), yatapoxviruses (16), and molluscipoxviruses (17).

Comparison of restriction maps within genera gives an index of the degree and location of sequence divergence between viruses in that genus. For example, differences in the orthopoxviruses arise mostly in the termini of the genome; the central region of the genome tends to be highly conserved. This is reflected in the gene organization of the poxviruses, with essential genes located in the central regions of the genome and genes involved in functions, such as host range and virulence, toward the termini.

Recombination between different strains of vaccinia virus was first conclusively demonstrated in 1959 (18) and has been extensively studied since then. Restriction mapping has demonstrated unsuspected recombination between poxviruses. Three examples are known: in the leporipoxviruses, a segment of Shope fibroma DNA has recombined into the myxoma genome, giving rise to a virus, termed malignant rabbit fibroma virus, with properties of both parental types (19,20). Similarly, the capripoxvirus Yemen goat 1 appears to have arisen from a recombination event between a goat-type and a cattle-type genome (12). Finally, a strain of "fibroma virus" that shared antigenic determinants with Shope's fibroma virus and vaccinia virus was shown to be a recombinant of which the genome was mainly derived from vaccinia virus, but contained a few fibroma virus sequences (21).

The Copenhagen strain of vaccinia virus is the only poxvirus for which the genome has been completely sequenced (22), although the complete sequence of the variola genome should soon be available. However, there is a large body of sequence data derived from other vaccinia strains and other orthopoxviruses. Comparisons of the vaccinia sequences indicate that, apart from the variability in the terminal regions of the genome, the most common mutations are silent nucleotide substitutions or single amino acid changes. Terminal regions of orthopoxviruses are quite variable, with extensive deletions, duplications, and transpositions. Joklik et al. (23) have proposed that these could arise by recombination between opposite ends of two genomes. As the terminal regions contain genes associated with host range and virulence, mutations that delete genes, duplicate genes, create new open-reading frames (ORFs), or bring genes under the control of different promoters have the potential to create new poxvirus strains or strains with altered host ranges (23). An example of evolution at the gene level is the remnants within the vaccinia sequence of several genes that have functional equivalents in cowpox. These include a truncated gene for the A-type inclusion protein and an interrupted gene for a serine protease inhibitor (22). Also of interest is the presence of only a single copy of the vaccinia growth factor gene in the Copenhagen strain, whereas this gene is within the inverted terminal repeats of the WR strain of vaccinia and, therefore, duplicated (22).

A fascinating aspect of the poxvirus genome is the number of genes that have cellular homologues. Examples include genes with homologies to thymidine kinase, epidermal growth factor, complement regulatory proteins, and serine protease in-

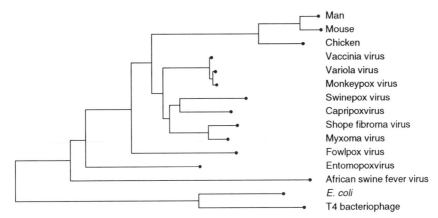

**FIG. 1.** Dendrogram illustrating the similarities and differences between the sequences of viral and cellular thymidine kinase genes. (From Blasco, ref. 3, with permission.)

hibitors (3,22). The origin of such genes is unknown. Since poxviruses replicate in the cytoplasm, they could be derived from the cellular genome only by reverse transcription of mRNA and subsequent recombination. Alternatively, they may have been derived from whatever constituted the ancestor of the poxviruses. Sequence comparisons of the thymidine kinase genes for viruses belonging to five genera (Fig. 1) indicate that the ancestral poxvirus probably acquired the gene before the divergence of mammals and birds (3). The thymidine kinase gene of fowlpox virus has diverged from those of the other chordopoxviruses and its location in the genome is different. From this, it has been suggested that the acquisition of the fowlpox thymidine kinase gene was independent of its acquisition in other chordopoxviruses (3), or it may have moved within the fowlpox virus genome by recombination during viral replication. It would be interesting, in this light, to obtain sequence data for this gene from the recently reported poxviruses of crocodilians (24).

## EVOLUTION OF THE LEPORIPOXVIRUSES

The natural hosts of leporipoxviruses are either leporids (*Sylvilagus* sp., *Lepus* sp.). or squirrels (*Sciurius* sp.) (Table 2). The Brazilian and Californian strains of myxoma virus show substantial differences in their host ranges in American leporids, most species of which are not susceptible to infection with fibroma virus. *Oryctolagus cuniculus*, which is not a natural host to any leporipoxvirus, appears to be the only animal that is susceptible to infection with all leporipoxvirus species so far tested.

All members of the genus, except hare fibroma virus, a virus of *Lepus europaeus*, and what may be a leporipoxvirus of *L. capensis* in Kenya (25), are indigenous to

**TABLE 2.** *Host and geographic ranges of viruses of the genus* Leporipoxvirus[a]

| Virus species | Reservoir host | Other leporids that develop infectious lesions[b] | Leporids that develop noninfectious lesions[b] | Insusceptible leporids | Geographic range | Ref. |
|---|---|---|---|---|---|---|
| Myxoma virus: Brazil | *S. brasiliensis* | *S. audubonii* *S. floridanus* *S. nuttallii* *O. cuniculus* *L. europaeus*[c] | *S. bachmani* | *L. californicus* | South and central America | 1,27 |
| Myxoma virus: California | *S. bachmani* | *O. cuniculus* | *S. audubonii* *S. floridanus* *S. nuttallii* *S. idahoensis* *L. europaeus* | *S. brasiliensis* *L. californicus* | Western United States | 28 |
| Rabbit fibroma virus | *S. floridanus* | | *O. cuniculus*[d] | *S. bachmani* *S. audubonii* *S. nuttallii* *L. europaeus* | Eastern United States | 1 |
| European hare fibroma virus | *L. europaeus* | *L. californicus* *O. cuniculus*[e] | | | Europe | 1 |
| African hare fibroma virus[f] | *L. capensis* | ? | ? | ? | Kenya | 25 |
| Squirrel fibroma virus | *Sciurls carolinensis* | *Marmota monax*[e,g] *O. cuniculus*[e,h] | | | Eastern United States | 1 |
| Western squirrel fibroma virus | *Sciurus griseus griseus* | ? | ? | ? | California | 29 |

[a]Abbrev: *L., Lepus*; *O., Oryctolagus*; *S., Sylvilagus*.
[b]Infectious lesions indicate transmission after mosquito probing.
[c]Lesions in only a very rare animal.
[d]Very occasional positive transmission by mosquito probing.
[e]Not tested for transmissibility.
[f]Not yet shown to be a leporipoxvirus.
[g]Woodchuck.
[h]After passage through woodchuck.

the Americas, and their evolution is probably linked with that of the Leporidae, subfamily Leporinae, which reached its greatest diversity in North America (26).

The natural hosts of myxoma virus are *Sylvilagus* rabbits, of which there are 12 species in the Americas. Since 1896, the disease myxomatosis has been known as a lethal infection of European laboratory and hutch rabbits, "spontaneous" outbreaks having been reported in Argentina, Brazil, Colombia, Panama, Uruguay, Venezuela and the western United States (Fig. 2). In 1943, Aragão (30) showed that the wildlife reservoir of the virus in South America was the tropical forest rabbit *Sylvilagus brasiliensis*, in which the virus caused only a trivial localized fibroma. He also showed that the virus was transmitted mechanically by insects that probed through the skin lesions. Twenty years later Marshall et al. (32) showed that the natural host of the strain of myxoma virus that occurred in California was the brush

**FIG. 2.** Map of the Americas showing the geographic distribution of feral European rabbits and the reservoir hosts of myxoma and rabbit fibroma viruses, and the places where spontaneous outbreaks of myxomatosis have occurred in hutch or laboratory European rabbits. Myxomatosis did not occur spontaneously in Chile, but regular inoculation campaigns with myxoma virus were carried out there from the 1960s. (From Marshall, ref. 31.)

**FIG. 3.** Small fibroma produced in the Californian brush rabbit (*Sylvilagus bachmani*) by the Californian strain of myxoma virus. The Brazilian strain of myxoma virus produces a similar lesion in its natural host, *S. brasiliensis*. (Courtesy Dr. D. C. Regnery.)

rabbit, *S. bachmani* (Fig. 3). From records of the occurrence of outbreaks of myxomatosis in domesticated rabbits (see Fig. 2), it seems likely that the leporids *S. bachmani* and *S. brasiliensis* carry the virus throughout their respective geographic ranges.

Restriction mapping has still to be done on the genomes of myxoma viruses recovered from different parts of the Americas, but studies many years ago showed that the Californian and most South American strains differed considerably in their biological properties and in the gel-diffusion patterns exhibited by their soluble antigens (33). Strains from Panama and some from Colombia had gel-diffusion patterns similar to those of Californian strains, although the clinical picture of infection of *Oryctolagus* by all South and Central American strains was similar and differed from that produced by Californian strains. These differences, and the high level of adaptation of the Californian strain to *S. bachmani*, which was the only one of six species of western North American leporid in which it produced transmissible lesions (28), suggest that strains of myxoma virus occurring in different parts of America have undergone long periods of evolution in their respective leporid hosts.

The disease produced in *O. cuniculus* by both South American and Californian strains of myxoma virus is very severe, with very few laboratory rabbits surviving longer than 13 days after infection. This feature, and their high host specificity, led in 1950 to the use of a strain of myxoma virus deriving from Brazil to control introduced European rabbits in Australia, and of another Brazilian strain for rabbit control on an isolated estate in France in 1952, whence it spread all over Europe.

### European Rabbits in New Ecosystems: America and Australia

As a background to the latter part of this chapter, we need to consider briefly the history of the European rabbit as an emerging pest when it invaded new ecosystems. This happened to a limited extent in the Americas, on a few offshore islands, and in Chile and Patagonia (see Fig. 2). In part, the abundant predators on the American mainland played a role in preventing an explosion of rabbit numbers. However, the presence of myxoma virus in the leporids in otherwise favorable habitats (but not in Chile or the offshore islands) may have been even more important, preventing wild *Oryctolagus* rabbits from colonizing the American mainland, in much the same way that trypanosomiasis kept cattle out of large regions in Africa. In Australia, however, wild *Oryctolagus* rabbits met with no competition from native predators or local viruses, and soon after their introduction, in 1859, they rapidly spread over the southern half of the continent and became Australia's principal agricultural pest. Their effect on pastures was dramatic, and they were even more devastating in causing severe erosion problems in the semidesert, which constitutes the greater part of inland Australia.

Rabbits are also a serious pest in many parts of Europe, although in some countries, notably France, they are valued as game animals, and the breeding of rabbits as a food source is widespread.

### Myxoma Virus for the Biological Control of *Oryctolagus cuniculus* in Australia

Although attempts at biological control of the rabbit pest were attempted over a hundred years ago by none less than Louis Pasteur (34), and although myxomatosis was suggested as a method of rabbit control by the Brazilian microbiologist Aragão as early as 1918, it was not until 1950 that serious attempts were made to infect Australian wild rabbits with the virus. Field trials begun in the Murray Valley in May 1950 continued through until November, that is, from autumn to late spring. The results were disappointing. Myxomatosis spread within rabbit warrens, but the numbers of rabbits in these warrens continued to increase, and there was no spread of the virus between warrens. However, in December 1950, hundreds of rabbits were found affected with myxomatosis in places many miles from the trial sites. Over the summer of 1950–1951 the disease spread dramatically over the whole of southeastern Australia, principally along the major rivers, where mosquitoes were most numerous. Few cases were seen during the winter months, but the following summer was wet and the disease spread both along and between the rivers. Investigations at a trial site, Lake Urana, showed that the case–fatality rate was fantastically high, over 99% (Fig. 4). A year later it was found that a less virulent variant had replaced the original highly virulent strain. This change in virulence set the stage for investigations into the evolutionary changes in myxomatosis, carried out in collaboration by workers in the Commonwealth Scientific and Industrial Research Organization (CSIRO) and the Australian National University (ANU) over the next 15 years, and continued after 1965 by scientists at the CSIRO Division of Wildlife and Ecology and the Keith Turnbull Research Station in Victoria. The changes in the virulence of the virus, the resistance of the rabbit, and the disease that was the result of this evolving interaction are summarized in the following sections.

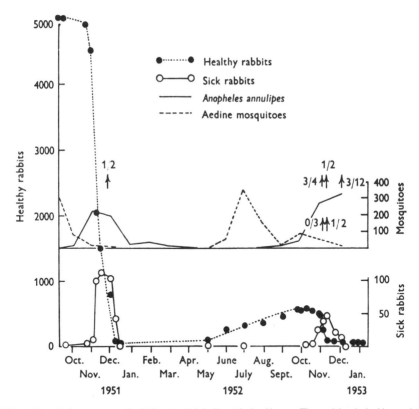

**FIG. 4.** Population counts of rabbits at a trial site at Lake Urana. The epidemic in November–December 1951 was initiated by inoculating several hundred rabbits with myxoma virus of grade I virulence (Moses strain) in September and October and releasing them. The *arrows* indicate the occasions on which virus was isolated from batches of mosquitoes. *Anopheles annulipes,* but not aedine mosquitoes, were efficient vectors. The first epidemic was caused by the grade I virus, and the calculated case–fatality rate was 99.8%. The second epidemic was caused by a naturally occurring mutant virus of grade III virulence; the case–fatality rate in genetically unselected laboratory rabbits was 90%. (From Myers et al., ref. 35, with permission.)

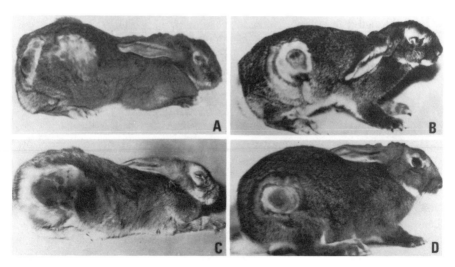

**FIG. 5.** The appearance of normal (genetically unselected) laboratory rabbits infected with various strains of myxoma virus. **A:** Lausanne strain, grade I virulence, 10 days after inoculation. **B:** Standard laboratory (Moses) strain, grade I virulence, 10 days after inoculation. **C:** Field strain of grade III virulence, 21 days after inoculation. **D:** Field strain of grade V virulence, 16 days after inoculation.

## Changes in the Virulence of Myxoma Virus

The first problem facing the ANU team was to devise a way by which it would be possible to determine the virulence of field strains using only a small number of rabbits. Tests with a few strains of widely differing virulence showed that the mean survival times in groups of six rabbits were correlated with death rates tested in hundreds of rabbits. With use of this method, five "virulence grades" were established. Typical clinical pictures of three of the prototype strains and the highly virulent strain "Lausanne," which was used to introduce the disease into Europe, are shown in Fig. 5. The virus first used in Australia, which was called the "standard laboratory strain" (see Fig. 5B), was provided by Richard Shope of the Rockefeller Institute of Medical Research, and had been derived from a strain originally recovered by Moses at the Oswaldo Cruz Institute in 1911. In laboratory rabbits, it was always lethal, with a mean survival time of 11 to 13 days. Such strains were designated as being of grade I virulence. What became the most common kind of virus to be recovered from field cases since the first release (see Fig. 5C) was virus of grade III virulence, with a case–fatality rate of 70% to 95% and a mean survival time of 17 to 28 days. A few strains killed fewer than 50% of infected rabbits and produced a much milder disease; these were designated grade V strains (see Fig. 5D). There was also available for experimental study a much less virulent strain that had been developed by serial passage of the Moses strain by the intracerebral inoculation of laboratory rabbits—the "neuromyxoma" strain of Hurst (36).

After the outbreak at Lake Urana in the summer of 1951–1952, an occasional infected rabbit was seen there every month throughout the winter, and when mosquitoes became numerous again in the spring another epidemic occurred (see Fig. 4), but this time the strains recovered were all of grade III virulence.

### *Natural Selection for Transmissibility*

The disappearance of the original very virulent strain and its replacement by a grade III strain was a consequence of mechanical transmission by mosquitoes, which acquire virus by probing through the skin lesions. Grade I, grade III, and grade IV strains produced large amounts of virus in the superficial parts of these tumors and all were well transmitted by mosquitoes that were allowed to probe and then feed on a normal rabbit (Fig. 6). Very highly attenuated strains, such as the laboratory variant called neuromyxoma virus, did not multiply to high titer in the skin and had a low rate of mosquito transmission. The key to selection of the grade III strains was that they caused a disease that was highly infectious for mosquitoes for up to 2 weeks and occasionally longer, whereas the rabbits infected with grade I

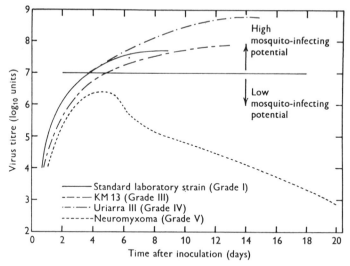

**FIG. 6.** Correlation between the titer of myxoma virus in skin lesions in *Oryctolagus cuniculus* and the suitability of the lesion as a source of virus for mechanical transmission by mosquitoes. Except for a few grade V strains, from which some infected rabbits rapidly recovered, the grade I virus used for inoculation campaigns and all field strains produced lesions that were highly infectious for rabbits. However, the rapid death caused by the highly virulent strains removed the infectious source within a few days, whereas strains of grade III virulence produced lesions that were infectious for 2 weeks and occasionally much longer. With more-attenuated strains, the lesions healed within 10 days. The laboratory variant *neuromyxoma* was so attenuated and replicated so poorly in the skin that moxquito transmission was rarely achieved. (From Fenner and Ratcliffe ref. 1, with permission.)

strains were dead within about 4 days of becoming infectious. The lesions in rabbits infected with grade V strains, besides containing fewer viruses, healed rapidly and were thus not available for mosquito transmission for more than a few days.

The validity of this interpretation was revealed, unwittingly, in a field experiment designed to study the effectiveness of the highly virulent strain of virus that had been recently recovered from a naturally infected laboratory rabbit in Brazil, and had been used to introduce myxomatosis into Europe in 1952 (the Lausanne strain; see Fig. 5A). Unexpectedly, soon after the inoculation of wild rabbits with this virus at the field site, an enzootic grade III strain was recovered from a captured rabbit. The lesions that the Lausanne strain produced after intradermal inoculation were much more protuberant than those produced by the Moses strain that had previously been used in Australia (and the less virulent strains derived from it), so it was possible to use a simple intradermal inoculation test to distinguish between the Lausanne virus and the virus that happened to be spreading naturally in the population—a grade III strain. The results of tests carried out through the transmission season showed that, although the highly virulent Lausanne virus, which had been introduced in a small area on a massive scale, caused many early deaths, by the end of the season it was completely replaced by the endemic grade III strain (Fig. 7), and this grade III strain was the only kind of virus recovered in the outbreak that occurred in the following summer. Similar results were reported by Shepherd and Edmonds (38) in the presence of rabbit fleas (*Spilopsyllus cuniculus*), which had been deliberately introduced into Australia in 1968 and extensively distributed in

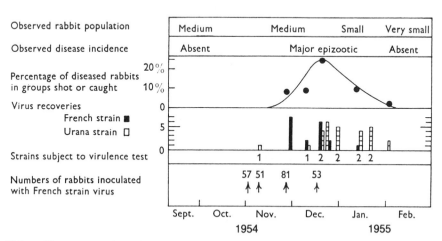

**FIG. 7.** The epizootic that followed the introduction of the Lausanne ("French") strain (see Fig. 5A) into the wild rabbit population at a trial site in 1954–1955. A single rabbit suffering from infection caused by a naturally occurring strain of grade III virulence (Urana) was captured early in November, when the second group of rabbits was being inoculated with the French strain of virus. Early deaths were due to the French strain, but after the end of December, almost all strains of virus obtained were similar to the naturally occurring strain of grade III virulence. (From Fenner et al., ref. 37, with permission.)

the 1970s. The Lausanne strain, introduced on fleas, persisted for only 10 weeks in the presence of field strains of the virus.

### Restriction Patterns of Field Strains

Despite the widespread introduction of the Lausanne strain on fleas since 1970, examination of the restriction patterns of recent field isolates of myxoma virus has demonstrated that they are derived from the standard laboratory strain rather than Lausanne (Kerr PJ and Saint K, unpublished). The overall restriction patterns of field isolates are highly conserved, but genetic differences can be demonstrated between isolates by using a wide range of restriction enzymes. Although only a limited number of isolates have been examined, some patterns emerge; conserved genetic markers that distinguish some field isolates from the standard laboratory strain occur in geographically widely separated isolates, which appear to have undergone further local divergence. There is evidence for the local maintenance of particular, unique genetic markers over long periods. Similarly, viruses with distinct restriction profiles have been isolated during epidemics separated by only a few kilometers.

### Selection for Resistance in Rabbits

The existence of virus strains that allowed 10% or more of infected rabbits to survive and breed provided an opportunity for Darwinian selection for increased resistance to occur in the rabbit population. To investigate genetic changes in the virus, large numbers of strains had been obtained from field cases each year and tested in laboratory rabbits from a stock that had never been exposed to myxomatosis. To test for changes in innate resistance, the process was reversed. Young rabbits were captured each spring, before myxomatosis had broken out, and raised to young adulthood in the laboratory, by which time they had lost maternal antibody. They were then tested for their innate resistance to myxomatosis by inoculating them with a particular sample of grade III virus, many ampoules of which had been stored frozen. Over the next 7 years the results were dramatic—the case–fatality rate fell from 90% to about 30% (Table 3). With continuing selection for genetic resistance, it was necessary to change to grade I virus for tests of increasing resistance, which was shown to increase most rapidly in a region (Mallee) where weather conditions favored frequent epidemics (Table 4). Interestingly, tests in these resistant rabbits showed that the strain of virus used in Europe (see Fig. 5A) was substantially more virulent than the Australian grade I strain (see Fig. 5B), although the virulence of these two strains had appeared to be the same when they were tested in laboratory rabbits.

### Coevolution of Virus and Host

What of the virulence of field strains of virus over the next 20 years? Although, after 1959, the tests were conducted in other laboratories and on a smaller scale,

**TABLE 3.** *Responses to inoculation, with Grade III virulent myxoma virus, of nonimmune wild rabbits captured in succeeding years, after exposure of the population to various numbers of epidemics*

| Number of epidemics | Case–fatality rate (%) | Clinical signs | | |
|---|---|---|---|---|
| | | Severe | Moderate | Mild |
| 0 | 90 | 93 | 5 | 2 |
| 2 | 88 | 95 | 5 | 0 |
| 3 | 80 | 93 | 5 | 2 |
| 4 | 50 | 61 | 26 | 12 |
| 5 | 53 | 75 | 14 | 11 |
| 7 | 30 | 54 | 16 | 30 |

(From Fenner and Ross, ref. 2.)

assays of field strains demonstrated the continuing dominance of grade III and the somewhat less virulent grade IV strains (Table 5).

What might be expected from the dynamic interaction between the changing virulence and the developing genetic resistance? In 1965, it had been suggested that if selection of the virus strain was dependent on transmissibility, then, as rabbits became more resistant, somewhat more virulent viruses than grade III might be better transmitted. Subsequently, analysis of the virulence of strains of virus recovered from wild rabbits from two parts of the Victoria where the rabbits had different levels of genetic resistance (see Table 5) appeared to support this hypothesis (Table 6). Although grade III strains remained dominant through the 20-odd years over which tests were carried out, a higher proportion of more virulent strains were recovered from the Mallee region, where the rabbits had the highest innate resistance.

**TABLE 4.** *Responses to inoculation, with Grade I virulent myxoma virus, of nonimmune wild rabbits captured in various parts of Victoria where the exposure to myxomatosis differed*

| Place and time of capture | Case–fatality rate (%) | |
|---|---|---|
| | Standard laboratory strain (see Fig. 5B) | Lausanne strain (see Fig. 5A) |
| Gippsland[a] | | |
| 1961–66 | 94 | |
| 1967–71 | 90 | |
| 1972–75 | 85 | ~100 |
| 1975–81 | 79 | ~100 |
| Mallee[b] | | |
| 1961–66 | 68 | |
| 1967–71 | 66 | |
| 1972–75 | 67 | ~100 |
| 1975–81 | 60 | 98 |

(Data from Fenner and Ross ref. 2.)
[a]Gippsland is a part of Victoria where epidemics of myxomatosis are infrequent.
[b]Mallee is a part of Victoria where there have been annual outbreaks of myxomatosis since 1951.

**TABLE 5.** *The virulence of strains of myxoma virus recovered from wild rabbits in Australia between 1951 and 1981, calculated on the basis of survival times (expressed as percentages)*

| Virulence grade<br>Case–fatality rate (%)<br>Mean survival times (d) | I<br>>99<br><13 | II<br>95–99<br>14–16 | III<br>70–95<br>17–28 | IV<br>50–70<br>29–50 | V<br><50 | Number of<br>samples |
|---|---|---|---|---|---|---|
| 1950–51 | 100[a] | | | | | 1 |
| 1952–55 | 13.3 | 20.0 | 53.3 | 13.3 | 0 | 60 |
| 1955–58 | 0.7 | 5.3 | 54.6 | 24.1 | 15.5 | 432 |
| 1959–63 | 1.7 | 11.1 | 60.6 | 21.8 | 4.7 | 449 |
| 1964–66 | 0.7 | 0.3 | 63.7 | 34.0 | 1.3 | 306 |
| 1967–69 | 0 | 0 | 62.4 | 35.8 | 1.7 | 229 |
| 1970–74 | 0.6 | 4.6 | 74.1 | 20.7 | 0 | 174 |
| 1975–81 | 1.9 | 3.3 | 67.0 | 27.8 | 0 | 212 |

(From Fenner and Ross, ref. 2.)
[a]Although only one field strain was tested, this extrapolation is justified by the very high mortality rates in the initial outbreaks.

## Modeling of Coevolution of Rabbits and Myxoma Virus

The myxoma–*Oryctolagus* system provides one of the best data sets for modeling the evolution of a virus–host interaction, and as such, there have been several models developed to predict the changes in virus virulence associated with myxomatosis. Most of these are based on May–Anderson epidemiological models and have treated rabbit genetic resistance as constant (39). These models have predicted that grade IV viruses should predominate in the field, rather than the observed grade III. By using the limited data available on the development of genetic resistance in

**TABLE 6.** *The virulence of strains of myxoma virus recovered between 1959 and 1981 from wild rabbits in two different parts of Victoria where the genetic resistance differed substantially*

| Virulence grade<br>Case–fatality rate (%)<br>Mean survival times (d) | I<br>>99<br><13 | II<br>95–99<br>14–16 | III<br>70–95<br>17–28 | IV<br>50–70<br>29–50 | V<br><50 | Number of<br>samples |
|---|---|---|---|---|---|---|
| Mallee region (high genetic resistance) | | | | | | |
| 1959–63 | 0 | 4.3 | 57.1 | 34.3 | 4.3 | 70 |
| 1964–66 | 2.0 | 0 | 64.7 | 31.3 | 2.0 | 51 |
| 1967–69 | 0 | 0 | 68.1 | 31.9 | 0 | 31 |
| 1970–74 | 1.0 | 6.9 | 77.5 | 14.7 | 0 | 102 |
| 1975–81 | 3.0 | 5.8 | 67.8 | 23.4 | 0 | 121 |
| Elsewhere in Victoria (lower genetic resistance) | | | | | | |
| 1959–63 | 2.1 | 12.4 | 61.2 | 19.5 | 4.7 | 379 |
| 1964–66 | 0.4 | 0.4 | 63.5 | 34.5 | 1.2 | 255 |
| 1967–69 | 0 | 0 | 61.6 | 36.4 | 2.0 | 198 |
| 1970–74 | 0 | 1.4 | 69.4 | 29.2 | 0 | 72 |
| 1975–81 | 0 | 0 | 65.8 | 34.2 | 0 | 91 |

(From Fenner and Ross, ref. 2.)

rabbits, Dwyer et al. (40) predicted that increasing host resistance should lead to selection for increased virulence of virus strains and that the observed levels of genetic resistance in Australia would lead to the selection of grade III strains. This was based on the assumption that the decline in case mortality was similar for all strains of virus; however, on examining data on resistance and virus grades for Britain, they concluded that this approach may not be valid. For further predictive models to be developed, more data on the levels of transmission of field strains in genetically resistant rabbits are required.

## Myxomatosis in *Oryctolagus cuniculus* in Europe

Myxomatosis was introduced into Europe by a private landholder who wished to control rabbits on his estate near Paris. He used the Lausanne strain (see Fig. 5A), and within a few years myxomatosis had spread throughout Europe, although inoculation campaigns such as those promoted in Australia were illegal. The changes in the virulence of the virus occurred more slowly than in Australia. However, by 1962 grade III strains had become the preponderant strains; although the numbers of strains of grade I virulence fell greatly, viruses of grade II virulence remained much more common than they had in Australia (Table 7).

In England these changes in virus virulence were eventually followed by changes in rabbit resistance, at a much slower rate than was found in Australia. The first definite evidence of resistance was found in 1970, in several areas of Britain (Table 8). Later, in one area where rabbits were tested (Wiltshire) rabbits attained a very high level of resistance. The regional differences in resistance in Britain probably reflect the greater genetic diversity of rabbits in that country.

**TABLE 7.** *The virulence of strains of myxoma virus recovered from the field in Great Britain and France between 1953 and 1981 calculated on the basis of survival times (expressed as percentages)*

| Virulence grade<br>Case–fatality rate (%)<br>Mean survival times (d) | I<br>>99<br><13 | II<br>95–99<br>14–16 | III<br>70–95<br>17–28 | IV<br>50–70<br>29–50 | V<br><50 | Number of<br>samples |
|---|---|---|---|---|---|---|
| Britain | | | | | | |
| 1953 | 100[a] | | | | | |
| 1962 | 4.1 | 17.6 | 63.6 | 14.0 | 0.9 | 222 |
| 1975 | 1.6 | 25.8 | 66.4 | 5.5 | 0.8 | 128 |
| 1981 | 0 | 35.8 | 62.6 | 1.6 | 0 | 123 |
| France | | | | | | |
| 1953 | 100[a] | | | | | |
| 1962 | 11 | 19.3 | 55.4 | 13.5 | 0.8 | |
| 1968 | 2.0 | 4.1 | 35.1 | 58.8 | 4.3 | |

(From Fenner and Ross, ref. 2.)
[a]Based on very high mortality rates in the initial outbreaks.

**TABLE 8.** *Genetic resistance to myxomatosis in wild rabbits in Britain*

| Location | Year | Number of rabbits | Virulence of challenge virus | Case–fatality rate (%) | Mean survival time of fatal cases |
|----------|------|-------------------|------------------------------|------------------------|-----------------------------------|
| Norfolk | 1966 | 41 | Grade III | 90 | 26.7 |
| | 1967 | 34 | Grade III | 94 | 23.2 |
| | 1968 | 71 | Grade III | 86 | 25.5 |
| | 1969 | 74 | Grade III | 84 | 25.9 |
| | 1970 | 27 | Grade III | 59 | 24.8 |
| | 1974 | 15 | Grade III | 13 | |
| | 1974 | 11 | Grade I | 100 | 14.1 |
| | 1975 | 11 | Grade I | 100 | 17.1 |
| | 1976 | 63 | Grade III | 21 | 35.9 |
| Wiltshire | 1978 | 71 | Grade III | 45 | 28.6 |
| | 1979 | 53 | Grade II | 45 | 29.4 |
| | 1980 | 44 | Grade I | 56 | 22.1 |

(From Fenner and Ross, ref. 2.)

## SUMMARY

Little information is available on the evolution of the poxviruses of vertebrates, but biogeographic evidence suggests that they evolved with their reservoir hosts over geologic time extending back to the appearance of reptiles. Viruses of the genus *Leporipoxvirus* probably evolved in North America, as parasites of leporids.

Myxoma virus occurs naturally as a benign fibroma in American rabbits, transmitted mechanically from one animal to another by biting arthropods. When the European rabbit is infected with virus from these animals, an extremely severe and almost invariably fatal disease ensues. This led to the deliberate introduction of myxoma virus into Australia to control wild European rabbits, which were a major agricultural pest. Changes in the virulence of the virus, the genetic resistance of host rabbits, and the interactions between virus and rabbit provided a unique natural experiment of evolution in action and an opportunity to observe these rapid changes. Because of advantages in transmission during periods of low mosquito numbers, there was first a selection for less virulent, but not highly attenuated, viruses that would produce infectious lesions lasting for many days. The original highly virulent virus killed rabbits within 3 to 4 days of their lesions becoming infectious, whereas highly attenuated strains produced too few viruses in the skin, and the rabbits recovered quickly. The change from less than 1% recovery to 10% recovery that followed the change in the virulence of the dominant viruses provided the opportunity for the selection of resistance among the rabbits. Subsequently, there was coevolution of virulence and resistance such that, in regions where rabbit resistance was high, there was selection for viruses of higher virulence, because they were more efficiently transmitted. A similar pattern was seen in Britain, on a somewhat slower time scale.

## ACKNOWLEDGMENTS

We are grateful to Dr. R. Blasco and Dr. J. Ross for providing us with material before its publication, and to Dr. D. C. Regnery for updating the information on myxomatosis in the Americas.

## REFERENCES

1. Fenner F, Ratcliffe FN. *Myxomatosis.* Cambridge: Cambridge University Press; 1965.
2. Fenner F, Ross J. Myxomatosis. In: Thompson HV, King C, eds. *The European rabbit: The history and biology of a successful colonizer.* Oxford: Oxford University Press: 205–239, 1993.
3. Blasco R. Evolution of poxviruses and African swine fever virus. In: Gibbs AJ, Calisher CH, Garcia-Arenal F, eds. *Molecular evolution of viruses.* Cambridge: Cambridge University Press; [in press].
4. Moss B. Molecular biology of poxviruses. In: Binns MM, Smith GL, eds. *Recombinant poxviruses.* Boca Raton: CRC Press; 1992:45–80.
5. Fenner F, Wittek R, Dumbell KR. *The orthopoxviruses.* San Diego: Academic Press; 1989.
6. Robinson AJ, Lyttle DJ. Parapoxviruses: their biology and potential as recombinant vaccines. In: Binns MM, Smith GL, eds. *Recombinant poxviruses.* Boca Raton: CRC Press; 1992:285–327.
7. Robinson AJ, Barns G, Fraser K, Carpenter E, Mercer AA. Conservation and variation in orf virus genomes. *Virology* 1987;157:13–23.
8. Upton C, Delange AM, McFadden G. Tumorigenic poxviruses: genomic organization and DNA sequence of the telomeric region of the Shope fibroma virus genome. *Virology* 1987;160:20–30.
9. Esposito JJ, Knight JC. Orthopoxvirus DNA: a comparison of restriction profiles and maps. *Virology* 1985;143:230–251.
10. Knight JC, Goldsmith CS, Tamin A, Regnery RL, Regnery DC, Esposito JJ. Further analysis of the orthopoxviruses, volepox virus and raccoon poxvirus. *Virology* 1992;190: 423–433.
11. Gassmann U, Wyler R, Wittek R. Analysis of parapoxvirus genomes. *Arch Virol* 1985;83:17–31.
12. Gershon PD, Black DN. A comparison of the genomes of capripoxvirus isolates of sheep, goats and cattle. *Virology* 1988;164:341–349.
13. DeLange AM, Macauley C, Block W, Mueller T, McFadden G. Tumorigenic poxviruses: construction of the composite physical map of the Shope fibroma virus genome. *J. Virol.* 1984;50:408–416.
14. Russell RJ, Robbins SJ. Cloning and molecular characterization of the myxoma virus genome. *Virology* 1989;170:147–159.
15. Coupar BEH, Teo T, Boyle DB. Restriction endonuclease mapping of the fowlpox genome. *Virology* 1990;179:159–167.
16. Knight JC, Novembre FJ, Brown DR, Goldsmith DS, Esposito JJ. Studies on tanapox virus. *Virology* 1989;172:116–124.
17. Bugert J, Rosen-Wölff A, Darai G. Genomic characterization of molluscum contagiosum virus type 1: identification of the repetitive DNA sequences in the viral genome. *Virus Genes* 1989;3:159–173.
18. Fenner F. Genetic studies with mammalian poxviruses. II. Recombination between two strains of vaccinia virus in single HeLa cells. *Virology* 1959;8:499–507.
19. Block W, Upton C, McFadden G. Tumorigenic poxviruses: genomic organization of malignant rabbit virus, a recombinant between Shope fibroma virus and myxoma virus. *Virology* 1985;140: 113–124.
20. Upton C, Macen JL, Maranchuk RA, DeLange AM, McFadden G. Tumorigenic poxviruses: fine analysis of the recombination junctions in malignant rabbit fibroma virus, a recombinant between Shope fibroma virus and myxoma virus. *Virology* 1988;166:229–239.
21. Berkowitz EM, Pogo BG-T. Molecular characterization of two strains of Shope fibroma virus. *Virology* 1985;142:437–440.
22. Goebel SJ, Johnson GP, Perkus ME, Davis SW, Winslow JP, Paoletti E. The complete DNA sequence of vaccinia virus. *Virology* 1990;179:247–266.
23. Joklik WK, Pickup DJ, Patel DD, Moody MD. Virulence genes of poxviruses and reoviruses. *Vaccine* 1988;6:123–128.

24. Buenviaje GN, Ladds PW, Melville L. Poxvirus infection in two crocodiles. *Aust Vet J* 1992;69:15–16.
25. Karstad L, Thorsen J, Davies G, Kaminjolo JS. Poxvirus fibromas on African hares. *J Wildl Dis* 1977;13:245–247.
26. Fox RR, Taxonomy and genetics. In: Weisbroth SH, Flatt RE, Krause AL, eds. *The biology of the laboratory rabbit*. New York: Academic Press; 1974:1–19.
27. Regnery DC. The epidemic potential of Brazilian myxoma virus (Lausanne strain) for three species of North American cottontails. *Am J Epidemiol* 1971;94:514–519.
28. Regnery DC, Marshall ID. Studies in the epidemiology of myxomatosis in California. IV. The susceptibility of six leporid species to Californian myxoma virus and the relative infectivity of their tumors for mosquitoes. *Am J Epidemiol* 1971;94:508–513.
29. Regnery RL. Preliminary studies on an unusual poxvirus of the western grey squirrel (*Sciurus griseus griseus*) of North America. *Intervirology* 1975;5:364–366.
30. Aragão H de B. O virus do mixoma no coelho do mato (*Sylvilagus minensis*), sua transmissão pelos *Aedes scapularis* e *aegypti*. *Mem Inst Oswaldo Cruz* 1943;38:93–99.
31. Marshall ID. Myxomatosis investigations carried out in Central and South America. Report to the Australian Wool Research Fund Committee, 1961. Updated May 1992 by Regnery DC.
32. Marshall ID, Regnery DC, Grodhaus G. Studies in the epidemiology of myxomatosis in California. I. Observations on two outbreaks of myxomatosis in coastal California and the recovery of myxoma virus from a brush rabbit (*Sylvilagus bachmani*). *Am J Hyg* 1963;77:195–204.
33. Fenner F. Viruses of the myxoma–fibroma subgroup of the poxviruses. II. Comparison of soluble antigens by gel diffusion tests and a general discussion of the subgroup. *Aust J Exp Biol Med Sci* 1965;43:143–156.
34. Pasteur L. Sur la destruction des lapins en Australie et dans la Nouvelle-Zélande, Paris; 1988.
35. Myers K, Marshall ID, Fenner F. Studies on the epidemiology of infectious myxomatosis of rabbits. III. Observations on two succeeding epizootics in Australian wild rabbits on the Riverine plain of south-eastern Australia. *J Hyg Camb* 1954;52:337–360.
36. Hurst EW. Myxoma and the Shope fibroma. II. The effect of intracerebral passage on the myxoma virus. *Br J Exp Pathol* 1937;18:15–22.
37. Fenner F, Poole WE, Marshall ID, Dyce AL. Studies in the epidemiology of infectious myxomatosis of rabbits. VI. The experimental introduction of the European strain of myxoma virus into Australian wild rabbit populations. *J Hyg Camb* 1957;55:192–206.
38. Shepherd RCH, Edmonds JW. Myxomatosis: the transmission of a highly virulent strain of myxoma virus by the European rabbit flea *Spilopsyllus cuniculi* (Dale) in the Mallee region of Victoria. *J Hyg Camb* 1977;79:405–409.
39. Anderson RM, May RM. *Infectious diseases of humans. Dynamics and control*. Oxford: Oxford University Press; 1991.
40. Dwyer G, Levin SA, Buttel L. A simulation model of the population dynamics and evolution of myxomatosis. *Ecol Monogr* 1990;60:423–447.

The Evolutionary Biology of Viruses,
edited by Stephen S. Morse.
Raven Press, Ltd., New York © 1994.

# 14

# Evolution of Mosquito-Borne Viruses

Thomas W. Scott, *Scott C. Weaver, and Varuni L. Mallampalli

*Department of Entomology, University of Maryland, College Park,
Maryland 20742; *Department of Biology, University of California at San Diego,
La Jolla, California 92093*

Mosquito-borne viruses are two-host parasites, with life cycles that, except in transovarial transmission among insect hosts, require transmission between mosquito vectors and vertebrate hosts. This kind of transmission cycle is characteristic of numerous viruses that vary widely in their virological, host, and ecological characteristics. Their pattern of transmission is the common denominator of what, otherwise, could be considered a taxonomically diverse group of viruses.

Certain viruses that persist in a two-host transmission cycle—yellow fever, dengue, Rift Valley fever, and the equine encephalitides—have left important marks on history because they caused extensive epidemic and epizootic disease; consequently, pathogens of humans and domestic animals have received considerable study. Most investigations have focused on the epidemiology and natural history of virus transmission, prevention of disease, and virus structure and replication. Only recently have detailed studies begun to assimilate information that can be used to untangle the history and mechanisms of evolution for this important group of viruses.

Viruses with RNA genomes are of special concern in evolutionary studies because of their great capacity for change and their exploitation of new opportunities. They are among the most mutable of all parasites (1), an attribute that can lead to considerable variation among virus populations. The RNA viruses, therefore, can quickly exploit changes in their environment or the behavior of their hosts (2), resulting in continuous and unpredictable evolution (3). When one considers the potential for variation by RNA viruses, it is remarkable that few new diseases emerge. Apparently, genetic, structural, and environmental factors are important constraints in RNA virus evolution (1,4).

In this chapter we review current thoughts on the evolution of RNA viruses in three antigenic complexes of the genus *Alphavirus*. A growing body of knowledge on the evolution and population biology of alphaviruses makes them an appropriate choice for a synthesis of evolutionary information on mosquito-borne viruses. We believe that concepts discussed here will be applicable to the study of other arthropod-borne viruses (arboviruses).

Alphaviruses are classified serologically into seven complexes (5), which appear to have been derived from a common ancestor (6). Each complex contains at least one species, and some species contain several subtypes or varieties. We chose to examine New World viruses that are in the eastern equine encephalomyelitis (EEE), western equine encephalomyelitis (WEE), and Venezuelan equine encephalomyelitis (VEE) complexes (Table 1).

**TABLE 1.** *Classification, distribution, and transmission characteristics of New World equine encephalomyelitis viruses*

| Virus | Subtype | Variety | Distribution | Primary vector | Primary vertebrate host |
|---|---|---|---|---|---|
| EEE | | North America | Eastern North America | *Cs. melanura*[a] | Songbirds |
| | | | Caribbean | ?*Cx. taeniopus* | ?Birds |
| | | South America | Central and South America | *Cx. (Mel.)* sp. | Small mammals and birds |
| WEE | | | Western North America | *Cx. tarsalis* | Songbirds and small mammals |
| | R-43738 | | South Dakota | ? | ? |
| | Ag80-646 | | Argentina | ?*Cx. ocossa* | ? |
| | BeAr 102091 | | Brazil | ?*Cx. portesi* | ? |
| VEE | I | AB | South, Central, and North America | Numerous mosquitoes | Equines, other mammals and birds |
| | I | C | South, Central, and North America | Numerous mosquitoes | Equines, other mammals and birds |
| | I | D | Ecuador, Panama, Columbia, Venezuela | *Cx. ocossa* *Cx. panocossa* | Rodents and birds |
| | I | E | Central Am. | *Cx. taeniopus* | Rodents and birds |
| | I | F | Brazil | ? | ? |
| | II | (Everglades) | Southern Florida | *Cx. cedecei* | Rodents |
| | III | A (Mucambo) | South America | *Cx. portesi* | Rodents and birds |
| | III | B (Tonate) | South America | ? | Birds |
| | III | C | Peru | ? | ? |
| | IV | (Pixuna) | Brazil | ? | ? |
| | V | (Cabassou) | French Guiana | ? | ? |
| | VI | | Argentina | ? | ? |

[a]*Cs., Culiseta; Cx., Culex.*
(Data from Scott and Weaver, ref 9; Weaver et al, ref 10; Walton and Grason, ref 14; Reisen and Monath, ref 29; Calisher et al, ref 30.)

## ALPHAVIRUS REPLICATION

Alphavirus replication is reviewed in detail by Strauss and Strauss (7) and Schlesinger and Schlesinger (8). Scott and Weaver (9) and Weaver et al. (10) summarized information pertaining to replication in mosquito vectors.

Alphaviruses consist of an icosahedral nucleocapsid that is surrounded by a lipid bilayer, into which are inserted two membrane glycoproteins. Enveloped virions are approximately 50 nm in diameter. Their genomes consist of a single unsegmented strand of RNA, with plus polarity, and they are approximately 11,000 to 12,000 nucleotides in length.

Replication takes place within the cytoplasm of host cells and is completed when virions bud through the plasma membrane. Entry into vertebrate cells is believed to occur through receptor-mediated endocytosis, although the molecular details of this process have not been defined, and most information on the topic has come from cultured vertebrate cells, rather than intact animals or insect cells. In mammalian and mosquito cells (in vitro), Sindbis (SIN) virus uses the high-affinity laminin receptor for entry into cells (11). Other protein receptors have been identified for SIN virus in mouse neural (12) and chicken cells (13). Alphaviruses may be able to infect a wide range of host species—mammals, birds, and mosquitoes—and tissues in those hosts, by using the laminin receptor, which is highly conserved among vertebrates and mosquitoes, as well as multiple receptors (11).

Two alphavirus RNA species are found in infected cells. A genomic RNA functions as mRNA for the synthesis of four nonstructural proteins (nsP1–nsP4), and is the template for minus-stranded RNA synthesis, from which polymerization of plus-stranded progeny genomic RNA is initiated. A subgenomic 26S RNA, which is identical with the 3′ third of the genomic RNA, encodes the three structural proteins (capsid, E1, and E2). During translation, the capsid protein is cleaved in the cytoplasm and interacts with progeny genomic RNA to form nucleocapsids. The remaining structural polyprotein is cleaved, processed, and glycosylated in the rough endoplasmic reticulum and Golgi complex, followed by insertion of glycoproteins E1 and E2 into the plasma membrane through the secretory pathway. Nucleocapsids bind to a cytoplasmic domain of the inserted E2 glycoproteins. Through a series of binding events, the nucleocapsid pushes through the plasma membrane, is covered with a bilayer membrane, and is released from the cell.

## TRANSMISSION CYCLES

Alphaviruses are biologically transmitted by mosquitoes among vertebrate hosts (see Table 1). Only during times of epidemic or epizootic VEE virus activity do domestic animals—equines—contribute significantly to virus amplification (14). Humans are deadend hosts for most alphaviruses because they do not develop a viremia of sufficient magnitude to infect mosquito vectors. Horizontal transmission among vertebrates has been reported only for infected and caged gallinaceous birds during avian epizootics (15). Vertical transmission of alphaviruses from female mosquitoes to their offspring has not been demonstrated (9,16). Although transplacental infections have been reported for VEE virus to humans, horses, and mice, and for WEE virus to humans, this process is not believed to contribute to virus amplification (15). Within this system of obligate mosquito–vertebrate exchange, the antigenic and genetic characteristics of New World alphaviruses display an interesting pattern of conservation in North America and diversity in Central and South America (10).

Details concerning infection, replication, and effects on host biology of alphaviruses in vertebrates (15,17–19) and mosquitoes (9,20–26) are discussed in these reviews and recent publications.

## Eastern Equine Encephalomyelitis Virus

Eastern equine encephalomyelitis virus is the only species in the EEE complex; it consists of two serologic varieties (see Table 1; 27). The North American variety is transmitted in eastern North America and in the Caribbean. The South American variety has been recovered from Central and South America, and rarely from the Caribbean (9,28).

Enzootic transmission cycles have been most thoroughly described in North America (see Table 1), where the mosquito *Culiseta melanura* transmits virus among songbirds in freshwater swamps in the eastern part of the continent. A variety of other mosquito species are suspected as vectors to humans and horses during epidemics. The mechanism for EEE virus overwintering in temperate North America is unknown. In Central and South America mosquitoes in the subgenus *Melanoconion* of the genus *Culex* are regarded as the most important vectors and small rodents the principal hosts, although birds may be involved in some locations. Transmission in the Caribbean is not well described.

## Western Equine Encephalomyelitis Virus

Western equine encephalomyelitis virus is one of six species and six subtypes in the WEE complex (see Table 1; 29). The WEE virus isolates from North and South America have not been thoroughly studied for antigenic variation. Available information suggests that WEE virus populations may be composed of subpopulations that vary antigenically and have different virulence characteristics. Three distinct strains from South Dakota (R-43738), Argentina (AG80-646), and Brazil (BeAr 102091) are classified as subtypes (30). Although closely related, WEE viruses are also serologically distinct from Highlands J (HJ) virus, which is transmitted by *Cs. melanura* among songbirds in eastern North America in a transmission cycle similar to that of EEE virus.

Although transmission of WEE virus is well characterized in North America, little is known about virus maintenance in Central and South America (see Table 1). Reisen and Monath (29) suggest that in Argentina, subtype AG80-646 may be transmitted by *Cx. (Mel.) ocossa* mosquitoes among birds or small mammals and a variety of mammal-feeding mosquitoes serve as epidemic vectors. A recent study by Aviles et al. (31) incriminated *Aedes albifasciatus* as a vector of WEE virus in Argentina. They propose that this mosquito transmits virus among European hares in an *Aedes*–Leporidae–WEE virus cycle that is analogous to the *Ae. melanimon*–WEE virus transmission cycle reported for California (32,33). In Brazil, *Cx. (Mel.) portesi* may transmit WEE strains, such as subtype BeAr 102091, among small mammal or avian hosts (30).

In agroecosystems of western North America, *Cx. tarsalis* is the principal vector, and songbirds are the vertebrate hosts (29). Small mammals and other mosquitoes, such as *Ae. melanimon* and *Ae. dorsalis*, that blood-feed on them may be involved in the Northwest and in arid southwestern locations. The mechanism for virus overwintering in temperate North America is unknown.

### Venezuelan Equine Encephalomyelitis Virus

Venezuelan equine encephalomyelitis virus is the only species in the VEE complex, which consists of six subtypes and many varieties (14; see Table 1). Varieties IAB and IC are referred to as epizootic viruses because they have been isolated only during equine epizootics. Varieties ID, IE, IF, and subtypes II to VI are referred to as enzootic viruses because they typically circulate within sylvatic transmission cycles, without causing disease in humans or domestic animals. The geographic distribution of these viruses ranges from the southern United States, through Central America and South America, and into northern Argentina.

Epizootic varieties of VEE virus are transmitted by numerous mosquitoes that feed on large animals (see Table 1). Because equines can respond to infection with a high-titered viremia, a single infected equine can infect numerous mosquitoes. When large numbers of susceptible equines and mosquitoes are present epizootic virus can move swiftly through an area in a wavelike fashion. After a critical level of herd immunity is established in equine populations, transmission ceases and epizootic viruses cannot be isolated. The source of these periodic outbreaks is unknown.

Enzootic VEE complex viruses seem to persist year-round in a vector–host transmission cycle or in estivating mosquitoes. Mosquito–host associations for these viruses are more specific than those of the epizootic varieties. Suspected vectors vary with the virus subtype or variety and location, but *Melanoconion* mosquitoes are consistently identified as the primary vectors, and small rodents—in some places birds—are the principal vertebrate hosts.

## PHYLOGENETICS AND THE STUDY OF ARBOVIRUS EVOLUTION

Phylogenetic analyses using nucleotide and amino acid sequence data have made important contributions to the understanding of virus evolution. Recent studies have applied these methods to mosquito-borne alphaviruses (6,34) and flaviviruses (35–37).

## Phylogenetic Concepts

Phylogenetic methods are used to generate hypotheses of ancestor–descendant relationships among taxa. Relationships are reconstructed using homologous characters, such as nucleotides. Shared, derived characters provide evidence of common ancestry (38,39). The parsimony method is used to select, as the most likely hypotheses of evolutionary history, those trees with the minimum number of evolutionary steps (39). A group comprised of a hypothetical ancestor and all of its descendants is defined as a *monophyletic group*. Characters show *homoplasy* when they are shared by two taxa, but do not have a direct common ancestor. This could be due to reversal (reversion), convergence, or parallel evolution.

The inferred phylogenies for parasites—arboviruses—and their hosts—mosquito vectors—can provide robust estimates of genealogical relationships and, thereby, form a basis for studying several aspects of host and parasite biology (40). For example, concordance between independently derived phylogenetic trees for mosquito vectors and viruses is information consistent with the hypothesis that there has been cospeciation. In practice, concordance between phylogenetic trees is rarely observed, especially in host–parasite systems (41,42). Incongruence among trees could be due to (a) a host having more than one parasite that may or may not have an evolutionary history with that host, (b) one parasite may be present in several hosts, (c) critical taxa could be missing from the study sample—this could be due to incomplete sampling or extinction (43)—or (d) the parasites diversified by switching from one host to another (41).

Arbovirus phylogenetic studies may also be useful if we wish to determine what virus or host characteristics are associated with high diversity in some virus groups and a lack of diversity in others (Wiegmann B, Mitter C, personal communication). One way to address this kind of problem is the experimental approach, where one variable, such as the structure of a glycoprotein, is manipulated while other variables are held constant. By comparing diversity for manipulated viruses with diversity among a control group, the researcher can determine the possible effect of the glycoprotein on virus diversification. Often, this approach is not possible, for example, when studying virus transmission in the jungles of tropical America. An alternative strategy is a statistical approach that includes phylogenetic analyses, for which one examines many independent cases of high diversity among viruses and tries to identify trends, such as rapid replication, that are repeatedly associated with high virus diversity. Well-defined phylogenies are critical for this approach because traits being examined for an association with diversity must have independent origins. Related taxa that have inherited a trait from the same ancestor, represent only one data point in this kind of analysis.

## Alphavirus Diversity

Evolutionary relationships among alphaviruses were first determined based on antigenic similarities (44). Later, Wenger et al. (45) used genetic data to examine the evolution of alphaviruses. They employed RNA–RNA hybridization techniques to determine that Semliki Forest (SF), Sindbis (SIN), Chickungunya, and O'Nyong-nyong (ONN) viruses shared RNA sequence homologies. Bell et al. (46) subsequently examined the $NH_2$-terminal amino acid sequences of E1 and E2 glycoproteins of eight alphaviruses. They concluded that the alphaviruses they studied are closely related and appear to have descended from a common ancestor.

Levinson et al. (6) used sequences for the structural proteins E1, E2, and capsid—deduced from RNA nucleotide sequences—to examine phylogenetic relationships among seven New and Old World alphaviruses. Their analysis generated unrooted trees in which SIN, WEE, VEE, and EEE viruses formed one cluster and Ross River, SF, and ONN viruses formed another, separate group. Shifts in the placement of WEE virus in the capsid versus E1 and E2 trees supported the conclusion that WEE virus was derived from a recombination between EEE virus and a SIN-like virus.

Most recently, Weaver (47) used homologous nsP4 amino acid sequences among alphaviruses and viruses in the alphavirus-like superfamily to construct rooted trees of alphavirus evolution. He concluded that alphaviruses are monophyletic and that New and Old World alphaviruses fall into separate clusters. These data, and those mentioned earlier, support the notions that alphaviruses evolved from a common ancestor, there have been important geographic influences on alphavirus evolution, and VEE and EEE viruses are closely related. Moreover, similarities in amino acid sequences, three-dimensional capsid protein structure, 5' RNA cap structures, and the production of subgenomic mRNAs are consistent with the hypothesis that plus-stranded RNA viruses are in general monophyletic (48).

### *Western Equine Encephalomyelitis Virus*

Oligonucleotide fingerprinting analyses of WEE virus strains isolated over a 35-year period from North and South America indicated remarkable genetic conservation of these viruses. Isolates shared 75% to 100% of their T1-resistant oligonucleotides (49), which translated into an estimated substitution rate per nucleotide per year of $3 \times 10^{-4}$ (10). This conservation among viruses from North and South America suggests that (a) regular intercontinental exchange of viruses occurs, (b) viruses in both continents are under similar selective pressure for genome conservation, and (c) extant viruses diverged recently from a common ancestor.

### *Eastern Equine Encephalomyelitis Virus*

Antigenic and genetic characteristics of North American strains of EEE viruses have been shown to be remarkably conserved in recent RNA oligonucleotide fingerprinting and monoclonal antibody studies (34,50). Nucleotide sequence data supported this conclusion and indicated a substitution rate per nucleotide per year of $1.4 \times 10^{-4}$ (51). Most substitutions were synonymous, suggesting that there is selective pressure for conservation of structural proteins. The lack of genetic distinction among viruses isolated from different geographic locations during different years suggests that North American viruses undergo efficient and regular dispersal among their foci of transmission. Serological (27), oligonucleotide fingerprinting (34), and nucleotide sequence data (51,52) suggest that EEE viruses from Central and South America are distinct from, and are more diverse than, their North American relatives. In Central and South America, virus dispersal may be more restricted than in temperate America.

Recent phylogenetic analyses of alphaviruses in the EEE and VEE complexes, with nucleotide sequence data, have provided insights into mechanisms of evolution and diversification processes of alphaviruses (52,53). All EEE virus isolates yet examined form two monophyletic groups: one includes South American variety isolates and the other includes North American viruses (Fig. 1). The North American variety isolates are from North America and the Caribbean. Nested within the South American variety viruses there are two more monophyletic groups. One includes viruses from the Amazon basin in Brazil and Peru and the other includes strains from a wide geographic range in South and Central America extending

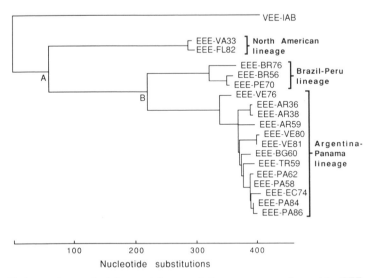

**FIG. 1.** Phylogenetic tree depicting evolutionary relations among members of the EEE complex viruses derived from nucleotide sequences. Homologous sequences of the Trinidad donkey strain of VEE virus were used as an outgroup to root the tree. Virus strains are listed by the abbreviation of the country or state (USA) where they were isolated, followed by the year of isolation. The scale *below* indicates numbers of nucleotide substitutions represented in branch lengths. (From Weaver et al., ref. 52.)

north, east, and south of the Amazon basin from Panama to Argentina. Therefore, as was reported previously for VEE virus (54), geographic factors are correlated with variation among EEE virus isolates.

Within each EEE virus monophyletic group, some isolates from the same time period are clustered, based on location of isolation. This suggests that, over short time periods, evolution may occur within some geographic regions. However, over longer time periods and when trees for North American viruses are rooted with the oldest isolate, relationships are based on year of isolation, rather than location (34,52). This suggests that, within a monophyletic group, dispersal of EEE virus occurs periodically among all transmission foci, rendering the virus population a genetic mixture when viewed over several decades. Regionally independent evolution, therefore, is obscured when (a) periodic extinction is followed by reintroduction from other locations or (b) regionally evolved genotypes are displaced by more fit viruses, which represent the consensus genotype of the monophyletic group (52).

In North America, a single EEE virus group appears to have diverged into two closely related monophyletic groups during the early 1970s (52). These virus populations were sympatric and may have been competing in nature through 1990. Although the time when these groups diverged cannot be determined with certainty, the estimated time of their divergence was accompanied by a tenfold rise in the rate of evolution, from approximately $4 \times 10^{-5}$ substitutions per nucleotide per year to about $4 \times 10^{-4}$. This variation in the EEE virus evolutionary rate, over a relatively short time period, contrasts with the constant rate of *NS* gene evolution in influenza A virus reported by Buonagurio et al. (55) and the molecular clock predicted by the neutral theory of molecular evolution (56).

The factors responsible for the increase in the North American EEE virus evolutionary rate are unclear. However, divergence of the two monophyletic groups may have been facilitated by genetic drift, associated with reduced population sizes, within isolated transmission foci, such as the upstate New York transmission focus (52). If isolated populations undergo reduced sizes during periods without virus immigration—perhaps during the winter when mosquitoes do not transmit virus—relatively rapid evolution may ensue owing to founder effects and genetic drift. This could result in genetic divergence of the isolated virus population. This scenario is similar to peripatric speciation events proposed to explain the punctuated equilibrium pattern of evolution reported for some higher eukaryotes (57).

Diversification of the EEE complex probably occurred relatively recently, during the past two millennia. The time of divergence of the North and South American monophyletic groups has been estimated at 1,000 years ago (52). North American (*Cs. melanura*) and South-Central American [*Cx. (Mel.)* sp.] enzootic mosquito vectors are members of different genera, and the closely related VEE viruses are also transmitted by *Cx. (Mel.)* sp. mosquitoes in tropical and subtropical America (14). Given the phylogenies for alphaviruses (47), the EEE complex (52), and the VEE complex (58), the most parsimonious scenario for EEE virus diversification is a single introduction of an ancestor of the extant North American monophyletic virus group from tropical America into temperate North America. Other scenarios would require multiple introductions and extinctions. Given the current mosquito–virus associations, introduction of EEE virus from Latin America into North America would have included switching hosts from a *Cx. (Mel.)* sp. to *Cs. melanura* and, perhaps, from small mammals to birds.

Phylogenetic studies with EEE viruses suggest that many viruses may have gone extinct in the past (52). Frequent extinction of mosquito-borne viruses would not be surprising, considering the complex series of events required to maintain transmission cycles.

### *Venezuelan Equine Encephalomyelitis Virus*

Antigenic and geographic variation among VEE complex viruses was first established serologically with polyclonal sera by Young and Johnson (54). Subsequent monoclonal antibody-based studies supported and refined the analysis of subtypes and varieties in this group. Oligonucleotide fingerprinting indicated that there is considerable diversity in the RNA nucleotide sequence among VEE virus subtypes and varieties (59–61). Within a subtype, isolates from different geographic locations showed less than 50% similarity. When isolates were from the same general location, however, their genetic similarity was high—over 85% of the oligonucleotides were identical. An estimate of the divergence rate for variety ID VEE viruses is $5 \times 10^{-4}$ substitutions per nucleotide per year (10).

Two serologically distinct VEE viruses, IC and ID, from Colombia are over 85% similar in T1-resistant oligonucleotides (61). This result is consistent with the hypothesis that IC viruses in Colombia were derived from an ID-like ancestor. In total, data on variation in VEE viruses suggest that the pattern of evolution for these viruses is diversification among numerous geographically discrete populations, or foci of transmission, with little exchange between different virus populations (foci).

Similar to EEE virus, evolution of VEE complex viruses has been examined using RNA nucleotide sequence data from limited portions of nsP4, E1, and 3'-untranslated regions of virion genomes (58). The VEE complex was found to be monophyletic (Fig. 2). Some serological VEE varieties, such as II and III, comprised monophyletic groups that were nested within the complex.

Subtype II (Everglades virus) and variety ID viruses formed a monophyletic group that included all epizootic variety IAB and IC VEE isolates. Weaver et al. (58) suggested that subtype II virus diverged from the ID South American ancestor and colonized Florida during the past 100 to 150 years, followed by two or more emergences of epizootic IAB and IC viruses in South or Central America. Variety IAB appears to have been derived from ID or an ID ancestor once during the early part of this century, whereas variety IC appears to have been derived from a ID-like virus on two separate occasions during the same period. An estimate of approximately 1,400 years was calculated for the time of divergence of EEE and VEE viruses from a common alphavirus ancestor.

Results of these analyses suggest that the source of epizootic VEE viruses was the

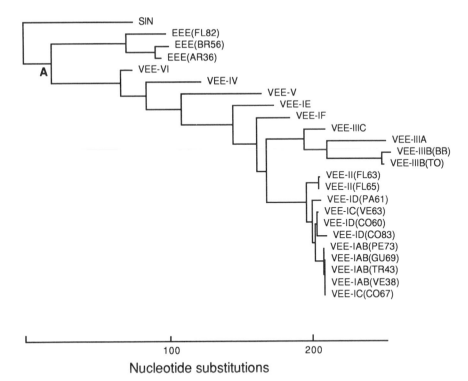

Nucleotide substitutions

**FIG. 2.** Phylogenetic tree depicting evolutionary relationships among members of the VEE complex viruses derived from nucleotide sequences. Homologous sequences of SIN and EEE viruses were used as outgroups to root the tree. The scale *below* indicates the number of nucleotide substitutions separating viruses from ancestral nodes. Virus strains are listed by the abbreviation of the country or state (USA) where they were isolated, followed by the year of isolation. Node *A* indicates the hypothetical ancestor for EEE and VEE complex viruses. (From Weaver et al., ref. 58.)

variety ID enzootic virus in northern South America and Panama. Even if variety IAB and IC viruses are now extinct, the suggestion that there have been multiple emergences of epizootic viruses from an enzootic virus suggests that other epizootic VEE viruses could evolve again in the future.

Evolutionary relationships of the VEE complex viruses suggest that, similar to EEE virus, host switching has accompanied virus diversification. Introduction of subtype II VEE virus might have occurred by a migratory bird, because some birds that migrate between South and North America are susceptible to VEE virus infection and develop viremias sufficient to infect mosquito vectors (62). After its introduction, the subtype II ancestor probably switched to a different mosquito host because *Cx. (Mel.) cedecei*, the subtype II vector in Florida, is not known to exist in South or Central America (63). Similarly, the Bijou Bridge strain of Tonate virus appears to be the descendant of a VEE subtype III-like virus that was introduced from South America into western North America and then switched from a mosquito to a cimicid (bed bug) vector (64).

## FACTORS THAT REGULATE ALPHAVIRUS EVOLUTION

### Ecological and Evolutionary Concepts

Basic concepts summarized by Price (65) for the evolution of parasites provide a framework for understanding the genesis of New World alphaviruses. Through a combination of genetics and population biology, these concepts succinctly summarize essential features of the parasitic lifestyle and suggest processes that may be important in alphavirus diversification (10).

### *Ecological Concepts*

#### *Parasites Are Often Adapted for Taking Advantage of Small, Disjointed Environments*

The spatial and temporal distributions of individual alphavirus hosts or populations of hosts are rarely identical. Infected mosquito and vertebrate hosts are often clumped in time and space within geographically distinct or isolated foci of transmission. We, therefore, propose that alphaviruses exist in a coarse-grained environment with significant distances of inhospitable space between hosts and transmission foci.

For parasites, in general, the risks associated with infecting a new host are often overcome by high fecundity. In fact, parasites are traditionally characterized by an extremely high reproductive rate (66). Alphaviruses fit this pattern by being asexual and having a short generation time (67–69; Weaver SC, Scott TW, unpublished data); they are able to quickly replicate from a small inoculum to high titers (9). This can occur within the cells of a single host, or a single infected mosquito can infect several vertebrates, and a single viremic vertebrate can infect numerous mosquitoes.

Efficient dispersal increases the opportunities for a virus, or its progeny, to locate and infect a new host. Alphaviruses are reliant on arthropod and vertebrate hosts as vehicles for movement (9). Without dispersal, viruses within a focus might tend to become isolated, opportunities for colonizing new sites would diminish and, if the population size was small, the probability of extinction could increase.

#### *Parasites Are Often Specialized for Exploiting Resources That Are Critical for Their Survival and Reproduction*

Because of the potential for a virus to infect new hosts, or for a host to change in some small way that can dramatically influence virus survival, the natural history of alphaviruses is consistent with Price's (65) prediction that hosts constitute a critical resource for parasite survival and reproduction. We, therefore, predict that host associations are important correlates in the geographic and genetic variation of extant alphaviruses.

Host specificity could increase the probability of virus persistence. Ideally, mosquito and vertebrate hosts would have the following characteristics: be abundant, remain infectious for a long time, and be in frequent contact with each other. The ability of certain viruses to develop and maintain finely tuned associations with certain hosts may confer a selective advantage by increasing their relative efficiency of transmission.

### *Because Survival Within a Host Is Brief, Parasites Exist in a Nonequilibrium Condition and Take Great Risks in Colonizing a New Host; the Chances of Failure and Extinction Are High*

Spatially and temporally, hosts for alphaviruses tend to exist in patches (transmission foci) where the potential for virus infection of a susceptible host and, thus, survival of a virus genotype through transmission, is relatively high. Beyond the bounds of those host patches, the probability of infecting a susceptible host is low. The ephemeral aspects of survival within a host or patch, and the presence of discrete patches themselves, highlight risks associated with the parasitic lifestyle of alphaviruses.

When multiple hosts are required for parasite persistence, as they are with alphaviruses, risks associated with virus survival increase. It becomes increasingly unlikely that both the hosts, which have different survival requirements, and the environmental conditions that support transmission will overlap in time and space. Deviation from critical factors that support transmission may prevent colonization of a host or patch, and virus transmission may be blocked. If all else remains the same, the more complicated the transmission cycle, the greater the opportunity for failure and extinction.

The presence in the same place and time of all the factors necessary for transmission of an alphavirus is probably a rare and unpredictable event. We predict that extinction rates for dispersing viruses are high. Moreover, we speculate that chance is an important component for establishing a new focus of transmission, and most colonization attempts, whether they be to new geographic regions, vectors, or vertebrates, are doomed to failure. Founder effects or genetic drift may, therefore, play an important role in alphavirus diversification (10). Specific information on dispersal among natural alphavirus populations is needed to test these hypotheses.

### *Evolutionary Concepts*

#### *Rates of Parasite Evolution and Diversification Can Be High*

The three ecological concepts discussed in the foregoing section explain constraints on alphavirus evolution that may promote the partitioning of virus populations. In combination with the rapid reproductive rates and high mutation frequencies of alphaviruses (10,51), these constraints have the potential to support swift diversification of viruses through chance or when they are exposed to different selective pressures. The result could be the evolution of distinct geographic strains of virus.

The remarkable diversity of viruses in the VEE complex (six subtypes and eight varieties) is consistent with the notion that small founding populations establish new transmission foci and that chance in the form of genetic drift is a significant factor in New World alphavirus evolution. Similarly, recent studies with EEE virus indicate that it is more diverse in tropical America than was previously recognized, and that the two main South American monophyletic groups of that virus diversified as recently as about 450 years ago (52).

It has been suggested that constraints associated with a two-host transmission cycle may be an important restriction on the evolutionary options available to alphaviruses (see section on mutation frequencies; 70,71). However, compared with that of their hosts, the short replication time [alphavirus = 3–5 hr at 30–37°C in vertebrate cells and 6–12 hr at 26–34.5°C in mosquito cells (67–69; Weaver SC, Scott TW, unpublished data), mosquito = 1 month (9); and bird = 1 year], high mutation rate [RNA virus, $10^{-3}$ to $10^{-5}$ (72) and DNA in mosquito or bird, $10^{-7}$ to $10^{-10}$ (73) per nucleotide per year], and lack of evidence for frequent exchange of virus genetic information (recombination or sex), alphaviruses are well suited to quickly generate genetic variation.

### The Role of Adaptive Radiation in Parasite Evolution

The role of adaptive radiation in parasite evolution is dependent on (a) host diversity, (b) host population size and geographic distribution, and (c) selective pressure for coevolution. Alphaviruses, like many other parasites, are able to evolve quickly because they can colonize small, isolated environments; can exploit critical resources; and have evolved to exist in conditions of nonequilibrium. In the following sections we discuss three evolutionary factors that can influence the availability of niches to alphaviruses and affect whether viruses will flourish in a new setting.

*Host Diversity.* Parasite colonization of new hosts, and diversification through adaptive radiation, tends to be most extensive when hosts are most diverse (41). The New World alphaviruses follow this pattern. In tropical America, currently recognized mosquito vectors are classified in the remarkably diverse subgenus *Melanoconion*, which contains more than 150 species (10). In temperate North America, however, EEE virus is transmitted enzootically by one species of mosquito that is placed in a relatively species-poor genus (9). Thus, in tropical America, VEE complex viruses are diverse, like their species-rich *Melanoconion* vectors and, in temperate America, EEE virus is remarkably homogeneous, like its species-poor *Culiseta* vector.

The WEE complex viruses appear to have diversified in North America by adapting to nonmosquito, insect vectors. Fort Morgan virus is transmitted by the swallow bug, *Oeciacus vicarius* (74) and appears to have little or no ability to replicate in mosquitoes (75). Buggy Creek virus has similarly been isolated from the swallow bug (30). This information on VEE, EEE, and WEE complex viruses suggest that the availability of arthropod hosts, which vary in some undefined critical way, has been an important influence in alphavirus diversification.

Alphavirus association with vertebrate hosts does not appear to be as specific as it is with vectors. For example, in North America the enzootic vertebrate hosts of EEE virus are a wide variety of songbirds (9). In South and Central America, VEE virus transmission cycles include several small-mammal species (14). Because vertebrates become immune after a brief viremia, retaining the ability to replicate in a variety of vertebrate species may increase the probability that a virus will locate a susceptible vertebrate host and, thereby, continue the risky business of transmission. Both passerine birds and forest-dwelling rodents are small and frequently behaviorally defensive to mosquito vectors (76), are often abundant, and have a rapid population turnover. Susceptible individuals would regularly be introduced into their populations (15).

Moreover, it may be unlikely that alphaviruses can diversify by avoiding antiviral immunity in previously infected vertebrate hosts. Changes in glycoprotein structure associated with escape from neutralizing antibody could often be lethal, if, as some studies suggest, such changes interfered with virus adsorption to susceptible cells (71).

*Host Population Size and Distribution.* Patterns of alphavirus distribution and diversity appear to be influenced by the dispersal capabilities of their vertebrate hosts. The close association of VEE complex viruses with sedentary forest-dwelling rodents is consistent with the geographically discrete distributions of different VEE virus subtypes and varieties. In North America, where migratory songbirds are vertebrate hosts, molecular investigations suggest that EEE virus has existed as a large, genetically mixed population for almost 60 years (34,51).

Long-range movement of alphavirus-infected insect vectors in air currents has been postulated (77). Migration of infected mosquitoes might also contribute to local mixing of virus populations. A recent study by Tabachnick (78) suggests that gene flow is limited among *Culicoides variipennis* gnats that transmit bluetongue virus. Local gnat populations were not panmictic, and colonization of temporary breeding sites or mixing among permanent sites was limited to a relatively small number of insects.

Such seemingly rare events as wind-borne dispersal of insects or migration of infected mosquitoes could be important for occasional dispersal of alphaviruses. However, we suspect that regular, migratory movements or postbreeding random wanderings by infected birds—and possibly bats—are the principal mechanism for mixing of genetically different viruses among broadly distributed and genetically homogeneous alphavirus populations.

Regular and efficient movement of alphaviruses among foci of transmission is important because it would increase the effective population size of a virus monophyletic group and, thereby, reduce selective pressure needed to maintain genetic homogeneity. Occasional or inefficient dispersal, however, might increase virus diversification by increasing the probability of isolation, reducing the effective population size, and reducing the relative effects of similar selective pressures (79,80).

*Coevolution.* An extension of the assumption that parasites tend to exist in conditions of nonequilibrium and exploit critical host resources, is the idea that parasites and hosts have coevolved (41,65). The concept of *coevolution* was first proposed by Ehrlich and Raven (81), based on their study of butterflies and plants. Later, the term was modified into a succinct definition of reciprocal change between two species by Janzen (82). A recent definition by Futuyma and Slatkin (83) is a relaxation of Janzen's strict sense and states that "the study of coevolution is the analysis of reciprocal genetic changes that might be expected to occur in two or more ecological interacting species and the analysis of whether the expected changes are actually realized."

Most discussions of arbovirus coevolution have focused on virus–vector, rather than virus–vertebrate interactions because of the often-specialized association between a virus and its arthropod host. Recently, Eldridge (84) suggested that New World bunyaviruses may have coevolved with their *Aedes* mosquito hosts.

The nature of specialization in the vector–virus relation could affect the prospects for adaptive radiation and parallel diversification. Well-established specificity of interaction, which has been only partially characterized, between VEE complex viruses and certain *Melanoconion* vectors is superficially consistent with the notion of virus–vector coevolution (10). In their phylogenetic analysis of VEE complex viruses, Weaver et al. (58) argue that alphaviruses have evolved too rapidly to have diverged in concert with their mosquito vectors. They assert that eukaryotic organisms, in this case mosquitoes, tend to evolve much more slowly than the rates they reported for alphaviruses. Phylogenetic analyses of *Melanoconion* vectors, including nucleotide sequence data that can be used to calibrate rates of mosquito evolution, are needed to test this hypothesis.

A model that arbovirologists could follow for studying mosquito–virus coevolution is the approach taken by Farrell and Mitter (85) for investigating parallel diversification of plants and herbivorous beetles that eat the plants' leaves. They began by using contemporary phylogenetic methods to reconstruct the evolutionary histories of insects. The insect phylogeny was then compared with a published, independently derived plant phylogeny for concordance of patterns in diversification and to predict the constraints that might select against locally adapted traits. Finally, extant communities were used to determine if unknown host associations could be correctly predicted, based on the hypothesized phylogenies. A research approach such as this, which combines phylogenetics with comparative analyses of field populations, would be a significant contribution to an improved understanding of arbovirus evolution.

Rigorous studies of coevolution have not been carried out with arboviruses and, until they are, it is premature to conclude that the viruses and their hosts have coevolved (86). Differences in replication time between viruses and their hosts might result in the virus simply following or tracking its host without reciprocal host change. Constraints associated with host specificity might be more likely to lead to coevolution when a virus is transmitted among a single host species (vertical or transovarial transmission) than when two hosts are required for transmission (horizontal transmission), as for alphaviruses.

Coevolution is not inconsistent with a system in which host-switching is an important mechanism of diversification. For example, coevolution between EEE virus and its arthropod or vertebrate hosts may have taken place after its presumed introduction from tropical South or Central America to temperate North America; after the virus had switched to new hosts. During alphavirus evolution, coevolution may have occurred periodically, without there being an overall pattern of codiversification.

*Parasite Speciation Does Not Necessarily Require Geographic Isolation*

Much of the foregoing discussion focuses on evolution through geographic isolation and the subsequent establishment of unique virus strains through random events, such as genetic drift, or change owing to different selective pressures. Geographically separate patches that contain only one subtype or variety of VEE virus are consistent with a mechanism of colonization, followed by drift or adaptive radiation. However, because alphaviruses reproduce asexually, allopatry is not necessary for their diversification. Sympatric diversification could occur if colonization of a new, ecologically distinct host was accompanied by minor modifications in a virus that is already quite specialized for a particular host. For example, subtle changes in the virus genome that influence host susceptibility might lead to temporal or ecological isolation of viruses. The testing of this kind of mechanism is not necessarily a simple task. A virus could diversify after switching to a sympatric host, then displace the ancestral or parental virus and, in the process, remove all signs of their former sympatric existence.

### Alphavirus Virulence for Mosquito Vectors

Long-standing views about the benign nature of mosquito infections by arboviruses have been challenged by recent reports of pathology in alphavirus-infected mosquito vectors. Weaver et al. (87) detected pathological changes in the abdominal midgut epithelial cells of *Cs. melanura* 2 to 5 days after oral infection with EEE virus. Some cells were sloughed into the midgut lumen, other cells degenerated in situ. Pathological changes were detected only after mosquitoes imbibed moderate- or high-titer EEE virus blood meals.

Similar cytopathic effects accompanied infection of *Cx. tarsalis* with WEE virus (88). In addition to the lesions just described, WEE virus infections resulted in dramatic midgut disruptions, including chicken blood cells and luminal bacteria within necrotic regions of the epithelium. These pathological changes were also dependent on the amount of virus imbibed. Results of these two studies suggest that alphaviruses, in general, may adversely affect their mosquito vectors in nature.

These observations of alphavirus-associated lesions in enzootic mosquito vectors are noteworthy because they challenge long-standing dogma concerning the nature of the interaction between mosquitoes and the arboviruses that they transmit. Traditional views in arbovirology assert that viruses often cause severe pathology and disease in vertebrates, but have no noticeable effect on their arthropod vectors (17,20,89,90).

This view is similar to a traditional explanation for the evolution of host–parasite relationships in general, which states that, over time, virulent parasites will become attenuated, hosts will become increasingly resistant, or attenuation and resistance will occur together as the result of coevolution between host and parasite (91). "Well-adapted" host–parasite relationships are benign; if they are not, it is only because the host and parasite have only recently come together. Given enough time, a benign, commensal relationship will evolve.

More recent theoretical analyses challenge this conclusion by proposing that, under certain circumstances, virulence can confer a selective advantage and, therefore, parasite attenuation would not be expected (91,92). The following model (91) for the reproductive rate of a parasite is used to explain how this could happen.

$$R_0 = \frac{\beta(\alpha,N)N}{\alpha + \mu + \nu(\alpha)}$$

Where $R_0$ is the basic reproductive rate of the parasite (mean number of secondary infections produced from one infection in a population where all hosts are susceptible), $\beta$ is the coefficient of transmission, $\alpha$ is the disease-induced host mortality rate, $N$ is the total host population size, $\mu$ is the disease-free host mortality rate, and $\nu$ is the host recovery rate.

If $\beta N$ or $\nu$ are not in some way linked to $\alpha$, then $R_0$ would be maximized as $\alpha$ is attenuated or as $\alpha$ approaches 0. Alternatively, if $\alpha$ has a functional relation to $\beta(\alpha)$ or $\nu(\alpha)$, then $R_0$ will change, dependent on the nature of that interaction. Because mosquitoes are infected for life with arboviruses, we focus our attention on $\beta(\alpha)$. If an increase in $\alpha$ increases $\beta$, natural selection would favor moderately virulent parasites over their less virulent cohorts. This would be especially true if $\alpha$ increases $\beta$ in a nonlinear fashion.

Earlier studies with mosquito-borne bunyaviruses and flaviviruses reported that mosquito infection with an arbovirus can reduce the blood-feeding success of adult female mosquitoes (93,94), survival of larvae and adult mosquitoes (95), the speed of larval development (96–98), and fecundity (95–97).

Results from laboratory studies we have conducted are consistent with the hypothesis that North American strains of EEE virus have detrimental effects on *Cs. melanura*, and that virus virulence has not been measurably attenuated during the past 57 years (Scott TW, unpublished data). When compared with control mosquitoes, EEE virus-infected *Cs. melanura* showed a reduction in survival and fecundity. The effect on survival did not become apparent until after approximately 3 weeks of incubation, which is a sufficient time period for *Cx. melanura* to transmit virus to an avian host (9,24,99,100). Virus infection did not appear to affect the developmental time for mosquito eggs or the ability of mosquitoes to find and feed on an avian host. The detected differences are important because they translate into an approximately 50% reduction in the net replacement rate (101) of infected mosquitoes. There were no measurable differences in detrimental effects when mosquitoes were infected with viruses isolated in 1933 versus 1990.

All available information indicates that *Cs. melanura* is necessary for the transmission and maintenance of EEE virus in North America (9). The EEE virus would perish, or at least its transmission dynamics would be notably altered, if *Cs. melanura* were removed from the continent. The association between EEE virus and *Cs. melanura* appears to be an example of an obligate host–virulent parasite interaction.

The evolutionary significance of these findings may be that a moderate level of virus virulence for the mosquito vector is associated with an increase in the efficiency of virus transmission and, therefore, an increase in virus fitness. More virulent mutant viruses may kill the mosquito before virus can be transmitted to an avian host, and less virulent mutants may not be transmitted as efficiently as viruses with moderate virulence characteristics. A selective advantage in favor of moderately virulent viruses may contribute to homogeneity among North American isolates of EEE virus (51). Frequent dispersal from regional foci by viremic birds is hypothesized to be an efficient mechanism for mixing EEE viruses throughout their North American range (10,51). Within such a large and fluid population, viruses with moderate virulence could regularly displace mutants.

## Molecular and Cellular Factors

### *Mutation Frequencies*

All RNA viruses that have been examined in detail have high polymerase error frequencies, on the order of $10^{-4}$ to $10^{-5}$. Relative to DNA-based organisms, these high mutation frequencies provide RNA viruses with the potential for rapid evolution under appropriate conditions (72,102,103). In fact, rates of evolution on the order of $10^{-3}$ substitutions per nucleotide per year have been estimated for genes of several RNA viruses (reviewed in 10).

Available information indicates that mosquito-borne alphaviruses and flaviviruses have evolved at rates approximately tenfold slower (about $10^{-4}$ substitutions per nucleotide per year) than many nonarthropod-borne RNA viruses (10). Experimental transmission studies indicate that the bunyavirus, La Crosse virus, is genetically stable during three laboratory transmission cycles–two vector–vector (transovarial) and one vector–vertebrate cycles (104). Field isolates of vesicular stomatitis virus, which is probably transmitted by sandflies and other biting flies, also exhibit a high degree of genetic stability within certain natural virus populations (Nichol S, personal communication).

The reason(s) for these slow rates of evolution in insect-borne RNA viruses is unknown. Several hypotheses (10,51,70) are reviewed in the following discussion.

The simplest explanation for slower evolution of mosquito-borne viruses is higher fidelity of RNA genome replication, such as a proofreading function. Measures of mutation frequencies are available for only one alphavirus, EEE virus. The T1-resistant RNA oligonucleotide fingerprinting studies of plaque-purified clones from unpassaged mosquito and bird isolates, which represent genomes within naturally infected hosts, demonstrated that most EEE virus populations contain a consensus or majority genotype, along with genetic variants that differ by one to a few oligonucleotides (53). This level of heterogeneity is similar to that reported for a variety of nonarthropod-borne RNA viruses (72,102,103). Phenotypic variants (plaque size and temperature sensitivity) have also been identified in natural populations of the alphaviruses EEE (53) and Getah viruses, as well as the flavivirus, Japanese encephalitis virus (105).

These results suggest that alphaviruses, and probably flaviviruses, have mutation frequencies similar to those of faster-evolving, nonarthropod-borne RNA viruses. Lack of genetic and phenotypic diversity, therefore, seems an unlikely explanation for natural conservation of alphaviruses.

Another hypothesis for slow evolution of mosquito-borne viruses is that slow replication in poikilothermic mosquito vectors reduces the opportunity for RNA sequence change, because each replication cycle presumably takes longer in cold-blooded vectors than in warm-blooded vertebrates. Because of the role of exo-thermic arthropods in transmission, rates of change would be expected to be greater in the tropics than in temperate regions. This prediction is supported by recent findings for EEE virus nucleotide substitution rates; South American strains are changing about two to threefold faster than North American strains (51). The, as yet unknown, overwintering mechanisms for alphaviruses in temperate North America could also be an important constraint on virus change in that region.

The number of replication cycles completed in mosquito versus vertebrate hosts during natural arbovirus transmission cycles has not been determined. That kind of information is needed for a definitive analysis of this topic.

Persistent virus infection of mosquitoes might also lower rates of evolution. Because titers for infectious alphavirus are reduced after approximately 1 week of mosquito infection (20,99), persistent infections of many weeks or even months—during times when environmental conditions are unfavorable for transmission—could contribute to genetic stasis.

Alternatively, persistent infections of mosquito cells in vitro leads to the production of defective interfering (DI) virus (89), which can promote rapid evolution of RNA viruses (72,102,103). However, the ability of persistently infected *Cs. melanura* to transmit EEE virus to vertebrate hosts decreases with time (24); therefore, persistently infected mosquitoes may not be able to transmit viruses that have evolved by escaping the selective pressure of DI virus (10). In addition, natural barriers to virus dissemination within the mosquito may reduce multiplicities of infection and thereby, delay the production and effect of DI virus (10). More studies that examine the population genetics of mosquito-borne viruses during persistent infections of mosquitoes in vivo, as well as the ability of these mosquitoes to transmit virus, are needed to evaluate the effect of long-term mosquito infections on rates of alphavirus evolution.

A third hypothesis for alphavirus conservation is that the alternation of hosts during the transmission cycle constrains the evolution of these viruses (7,10,68, 70,71). Several lines of reasoning are consistent with this hypothesis. Studies of alphavirus replication indicate that there are different host cell factors involved when alphaviruses replicate in mosquito versus vertebrate cells. Therefore, interaction with these different sets of cellular factors may place two different sets of constraints on the variability of alphavirus proteins or RNA structure (10,51). Similarly, different host cell receptors and glycoprotein sorting signals may be required for infection and replication in vertebrate versus invertebrate hosts (51). Primary structure of the virus genome may be constrained by selection for optimal (possibly compromise) codon usage by isoaccepting tRNA species in mosquito versus vertebrate hosts. These considerations all suggest that mosquito-borne viruses may be subject to tight selective constraints imposed by their two-host life cycle.

### Adaptive Landscape Model

The hypothesis that a two-host life cycle can constrain the evolution of alphaviruses is facilitated conceptually using the adaptive landscape model, a popular metaphor in evolutionary biology (10, 106). In this model the landscape is a two-dimensional plot of different genotypes; in reality the field is $N$ dimensional and describes all possible genomes (Fig. 3). The vertical axis, the height of a peak, indicates the relative fitness of each genotype, which is defined by the horizontal axis.

For a single-host virus, a single adaptive landscape can be used to describe the fitness of different genotypes. Genotypes of high fitness occupy landscape peaks, whereas unfit genotypes occupy valleys (see Fig. 3A). Evolution by natural selection alone proceeds toward the top of the nearest peak. Selection will tend to be relatively more important than genetic drift in large populations and prevent genotypes from moving down the slope of an adaptive peak, across a valley of low fitness, and then up a new peak, even if the new peak has a higher fitness. The probability of moving across the landscape to a new peak increases when (a) the population is small, which increases the probability of changes in allele frequencies by chance; (b) variability in the population is high; (c) there are ridges between fitness peaks; (d) the environment changes, which changes the topography of the landscape; or (e) individuals with high fitness disperse to new environments. High mutation frequencies associated with RNA genome replication might promote movement across the landscape by continually producing different genotypes, some of which might lie near a new peak of higher fitness.

For a two-host virus with different fitness characteristics in each host, different

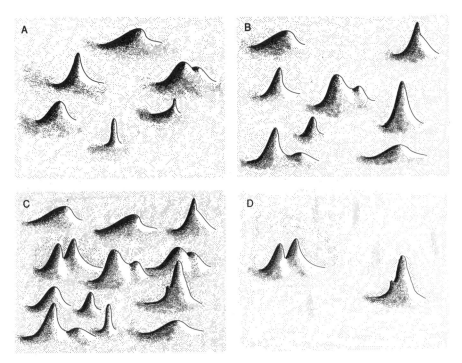

**FIG. 3.** Two-dimensional adaptive landscapes for the evolution of mosquito-borne viruses. *Horizontal axes* define different virus genotypes (actually, the field is *N*-dimensional and represents all possible genomes of the virus). The *vertical axis* indicates the relative fitness of any genotype defined by the horizontal axes. Landscapes depict fitness of different genotypes in a particular environment, high-fitness genotypes occupy peaks, and low-fitness genotypes occupy valleys. **A:** Landscape for virus fitness in a mosquito host. **B:** Landscape for virus fitness in a vertebrate host. **C:** Mosquito and vertebrate landscapes superimposed to show the topography of the landscape for alphavirus evolution. Transmission requires that viruses alternately infect mosquito and vertebrate hosts. **D:** Host landscape containing only mosquito and vertebrate peaks that overlap or coincide. Alphaviruses are restricted to the area of peak overlap where they can be transmitted between hosts, without having to consistently traverse a valley of low fitness between nonoverlapping peaks. (From Wright, ref 106.)

adaptive landscapes are predicted because each host can impose different genetic constraints on the virus (see Fig. 3A,B). Since an alphavirus must alternate between hosts, it must evolve within regions where the fitness peaks on vertebrate and invertebrate landscapes coincide or overlap sufficiently to avoid movement through a low-fitness valley during part of each transmission cycle (see Fig. 3C). If the two-host landscapes are independent and fitness peaks are rare relative to low-fitness valleys (i.e., most mutations are deleterious) overlapping peaks will be rare relative to peaks for either host by itself (see Fig. 3D). Therefore, the transition from one pair of overlapping peaks to another will involve movement through even wider low-fitness valleys than would be required on either of the single-host landscapes. This implies that reaching two coinciding higher-fitness peaks in a two-host life cycle is less likely than reaching a single higher-fitness peak in a one-host life cycle.

Application of this model to alphavirus biology leads to two predictions that can be tested with current technology. One prediction is that some virus genotypes will exhibit a high degree of fitness for infection of the vertebrate host, but not the mosquito, and vice versa. An example of a host-dependent alphavirus is the Sindbis virus mutant $SV_{ap15/21}$, which replicates well in mosquito cells, but grows poorly in chick or baby hamster kidney cells (107). Another prediction of the model is that alphaviruses will exhibit subtle genetic shifts during movement between partially overlapping vertebrate and mosquito fitness peaks (see Fig. 3D).

Important questions about the adaptive landscape model and alphavirus evolution center on defining the topography of the landscape upon which alphaviruses exist or upon which they are introduced. What are the frequencies of each genotype and what are their relative magnitudes of fitness (height of a peak) in the two different hosts? Are fitness differences large relative to genotype differences, so that a large genetic distance is needed to obtain meaningful differences in fitness? In other words, how rugged is the landscape, and how does this topography influence the prospects for change within a given virus population in a particular environment? If small genetic distances are associated with large differences in fitness, the landscape would be considered rugged. Conversely, if small differences in fitness are associated with large genetic differences, the landscape would be relatively smooth.

It is also important to note that the adaptive landscape model discussed in the foregoing assumes that virus genes responsible for vector infection are functionally independent from those that govern vertebrate infection. In a scenario with independent genes (a) overlap between peaks and overall fitness would be expected to be relatively low, and (b) the probability would be relatively high for host-restricted characteristics; that is, high fitness in one host and relatively low fitness in the other. If, however, genes that regulate host infection are interactive or additive, (a) overlap between peaks and overall fitness would be relatively high, and (b) the probability of host-restricted characteristics would decrease; most viruses would have relatively high fitness in both hosts.

### Recombination

Genetic reassortment in segmented RNA viruses, including mosquito-borne bunyaviruses, and recombination in unsegmented viruses, including mosquito-borne alphaviruses and flaviviruses, can provide another means of arbovirus diversification. Reassortment in bunyaviruses occurs at high frequency in vitro, and reassortment viruses have been detected in nature, as well as during experimental infections of mosquito vectors. Beaty et al. (71) suggested that reassortment may have contributed to the great diversity of mosquito-borne bunyaviruses.

Genetic recombination has also been described for several unsegmented animal RNA viruses including picornaviruses and coronaviruses, as well as for some RNA plant viruses (reviewed in 108). Although recombination has not been demonstrated during experimental infections with any mosquito-borne alphavirus, WEE virus is believed to have descended from a recombinant alphavirus that derived its nonstructural and capsid proteins from an EEE-like virus, and its envelope glycoproteins from a Sindbis-like virus. Presumably, this was the result of a copy-choice mechanism in which the viral polymerase jumped from one template to another during RNA replication (109).

The role of recombination in diversification of viruses in the genus *Alphavirus* does not appear to be widespread. Phylogenetic analyses of members of the EEE (52,53) and VEE complexes (58) have revealed no evidence of recombination. Although apparently rare, occasional recombination among phylogenetically distinct genotypes may still be an important event in alphavirus evolution. Moreover, recombination among alphaviruses may be more common than current studies suggest if it involves closely related viruses, in which it would be more difficult to detect. Beaty et al. (71) suggest that recombination of closely related viruses might help explain the evolution of the VEE virus complex. The VEE subtypes IAB, IC, and II share a conserved neutralizing epitope.

### Muller's Ratchet and Genetic Drift

Partitioning of alphaviruses into individual hosts within discrete transmission foci may enhance the probability for genetic bottlenecks and genetic drift, especially during times of the year when environmental factors do not favor transmission and relatively few hosts are infected. Because most mutations in mosquito-borne RNA viruses are probably delterious (110), forward mutations are expected to occur at a greater rate than back mutations or reversions (111,112), and recombination probably occurs too infrequently to correct mutations in nonsegmented alphavirus genomes (102), repeated reductions in population size (genetic bottlenecks) can lead to a dramatic reduction in fitness owing to Muller's ratchet (113,114).

Muller's theory predicts that, for asexual organisms, chance elimination of mutation-free individuals from finite populations can lead to an irreversible accumulation of delterious mutations and corresponding reductions in fitness. As the least-mutated individuals are successively eliminated, the mutation load increases. Genetic drift within small populations could minimize the effect of differences in fitness by reducing the relative effect of selection. High mutation rates, which characterize replication of RNA virus genomes (72,102), might expedite the ratchet process, because most progeny virions will have one or more mutations than their parent. Fitness reductions owing to Muller's ratchet have been reported during experimental studies of plaque-to-plaque passage of an RNA bacteriophage φ6 (112,115) and the RNA arbovirus, vesicular stomatitis virus (110). The effects of genetic drift, in these studies, were amplified by forcing each transfer population through a genetic bottleneck.

Anthropod-borne bunyaviruses and reoviruses can use reassortment of their segmented genomes to correct deleterious mutations. The cost associated with evolution of a segmented genome (114) may have prevented segmentation of alphavirus RNA. Therefore, assuming that recombination among single-strained RNA viruses is rare (102), alphaviruses and flaviviruses must avoid fitness losses from genetic drift and Muller's ratchet by some mechanism other than reassortment or recombination. This could include processes that maintain large effective population sizes (10,51), especially during transmission events; for example, (a) high-titered vertebrate viremias, (b) low-infection thresholds for mosquito and vertebrate hosts, (c) high-titered inocula in mosquito saliva, or (d) efficient virus dispersal. Large infecting virus populations would minimize the probability of genetic bottlenecks or founder effects resulting from infections initiated by a single or very few virions.

## Individual Versus Interdemic (Group) Selection

An issue of central concern in alphavirus evolution is why some virus genotypes are favored and become so prevalent that they are regularly isolated by field workers, whereas other genotypes are not detected in wild populations, leading to the conclusion that they have been selected against. We propose two general scenarios, that are not necessarily mutually exclusive, to explain this observation.

1. Reproductively successful virus genotypes outcompete less successful genotypes (individual selection).
2. Some virus *demes* (a local, genetically mixed group of viruses) cause the equilibrium of other demes to shift, through differential dispersal, toward their genotype (interdemic or group selection).

### *Individual Selection*

Recent reviews suggest that RNA virus populations, even clonal populations, consist of a mixture of genotypes clustered around one or more master sequences. These populations are referred to as *quasispecies* (72,73,116). Eigen and Biebricher (116) argue that selection acts on "the distribution of the quasispecies as a whole," rather than individual genotypes or phenotypes.

Keeping this in mind, one could consider selection of viruses at the level of the individual to include either direct or indirect competition among genotypes or quasispecies. Indirect competition could occur through fitness differences among mutant and master virus genotypes and among viruses that vary in their ability to be transmitted or to disperse. If most mutations are deleterious and negatively affect transmission, most mutants would have low fitness, compared with more transmissible genotypes. Important constraints in this scenario could be factors imposed on the virus itself, for example, secondary structure of RNA that is critical for replication of the virus genome. Viruses that are transmitted or disperse most efficiently might infect new hosts or establish new transmission foci more quickly and, thus, outrace less fit variants to numerically abundant status.

Direct competition could occur when two different viruses simultaneously infect the same host and one genotype interferes with replication of the other. The phenomenon of interference between different viruses infecting the same mosquito host has been demonstrated for alphaviruses (117), bunyaviruses (71), and flaviviruses (118). Interference was not detected when *Cx. tarsalis* were sequentially infected with EEE and WEE viruses (119). When monkeys were the experimental host, interference was demonstrated between Rift Valley fever and yellow fever viruses (120) as well as dengue and yellow fever viruses (118).

Isolations of two different arbovirus genotypes in the same naturally infected host has not been reported often (121,122). In fact, detecting simultaneous infection with serologically identical alphaviruses requires detailed studies of the virus genome, something that has only been done for EEE virus (53). Over a 3-day period two different monophyletic groups of EEE virus, based on nucleotide sequence data, were isolated from a sentinel bobwhite quail in Maryland. Of ten North American mosquito isolates, only one contained phenotypic—plaque size—variants of EEE virus.

Laboratory studies are needed to better understand temporal patterns in alphavirus competition. For example, if the competition process if protracted, coinfection of a single host with different virus genotypes may be detected simply by carrying out more studies designed to detect genetic variation in existing virus isolates. However, if the outcome of competition is decided during the very early stages of coinfection, competition may be common, but difficult to detect, because it occurs so quickly.

### Interdemic Selection

The adaptive landscape model discussed in the foregoing is part of Wright's shifting balance theory of evolution which explains how the combined effects of natural selection and genetic drift can move the genetic makeup of a population from one peak on an adaptive landscape to another (123,124). The combined effects of selection and genetic drift are able to do what selection by itself is unable to do: move the genetic composition of a population across the landscape to another peak. The theory describes how a trait can increase in frequency by drift in a small population and, then, after reaching a critical frequency, be swept to fixation by mass selection. Wright (123) proposed that evolution through adaptation is most likely in species with broad distributions, which are composed of many discrete demes, and have limited gene flow between such demes. Within this framework, evolution can occur rapidly if the environment changes; this would include the availability of new niches.

Although shifting balance is one of the most prominent theories in modern evolutionary biology, it is controversial. Certain critical phases of the theory seem to contradict each other if one assumes that all the major components of the process need to occur simultaneously in the same population. Initially, limited migration is required for drift to operate, but then considerable migration is needed to move genes from one deme to another (125).

Information is not currently available on the rates and mechanisms of alphavirus dispersal, colonization, and extinction to test Wright's theory for alphaviruses. These are topics within the discipline of arbovirus evolution that deserve more research emphasis and would add much to the study of evolution in general. A comparative analysis of the role of individual versus interdemic selection in alphavirus evolution would be an important contribution to arbovirology.

## CONCLUDING REMARKS

In this chapter we have presented a synthesis of information, from a variety of scientific disciplines, on the evolution of mosquito-borne alphaviruses. Our approach emphasizes contributions from ecological and evolutionary theory, molecular aspects of virus replication, population genetics, phylogenetics, and arbovirus natural history. We have identified questions or areas of research that we feel deserve additional research emphasis. Without continued research efforts on all aspects of arbovirus biology, including longitudinal studies of natural transmission cycles, subsequent discussions of arbovirus evolution will remain incomplete.

The following generalizations of alphavirus evolution can be used as a hypothetical basis for additional study. Homogeneity among North American alphaviruses appears to be strongly influenced by the capability of viruses to disperse efficiently, which results in large populations that are relatively resistant to the effects of chance events. Greater diversity among South American alphaviruses may be due to restricted dispersal by mammalian hosts, which increases the opportunities for establishing small founder populations and subsequent diversification by adaptive radiation or chance. Anthropod host diversity is positively associated with alphavirus diversity in the New World. Host switching appears to have been more common than codiversification during alphavirus evolution. The two-host alphavirus life cycle may place stringent selective constraints on variability in alphavirus populations.

Stochastic events are discussed repeatedly in this chapter as important factors in alphavirus evolution. The relative role of natural selection versus random events, such as genetic drift, is one of the most important unanswered questions in alphavirus evolutionary biology, as it is for the study of evolution in general (125,126). We conclude that because chance is important, the direction and specific events in alphavirus evolution will be unpredictable (72).

## ACKNOWLEDGMENTS

We wish to thank John J. Holland, Charles Mitter, George K. Roderick, and Walter J. Tabachnick for reviewing and offering helpful comments on earlier drafts on this manuscript. Work reported in this chapter was supported by National Institutes of Health grants AI-26787 and AI-22119 and the Maryland Agricultural Experiment Station.

## REFERENCES

1. Kilborne ED. New viral diseases: a real and potential problem without boundaries. *JAMA* 1990;264:68–70.
2. Krause RM. After AIDS: the risk of other plagues. *Cosmos* 1991;1:15–21.
3. Morse SS. Emerging viruses: defining the rules for viral traffic. *Perspect Biol Med* 1991;34:387–409.
4. Morse SS, Schluederberg A. Emerging viruses: the evolution of viruses and viral diseases. *J Infect Dis* 1990;162:1–7.
5. Calisher CH, Karabatsos N. Arbovirus serogroups definition and geographic distribution. In: Monath TP, ed. *The arboviruses: epidemiology and ecology*, vol 1. Boca Raton: CRC Press; 1988:19–57.
6. Levinson RS, Strauss JH, Strauss EG. Complete sequence of the genomic RNA of O'nyong-nyong virus and its use in the construction of alphavirus phylogenetic trees. *Virology* 1990;175:110–123.
7. Strauss EG, Strauss JH. Structure and replication of the alphavirus genome. In: Schlesinger S, Schlesinger M, eds. *The togaviruses and flaviviruses*. New York: Plenum Press; 1986:35–90.
8. Schlesinger S, Schlesinger MJ. Replication of Togaviridae and Flaviviridae. In: Fields BN, Knipe DM, eds. *Virology*, 2nd ed. New York: Raven Press; 1990:697–711.
9. Scott TW, Weaver SC. Eastern equine encephalomyelitis virus: epidemiology and evolution of mosquito transmission. *Adv Virus Res* 1989;37:277–328.
10. Weaver SC, Rico-Hesse R, Scott TW. Genetic diversity and slow rates of evolution in New World alphaviruses. *Curr Top Microbiol Immunol* 1992;176:99–117.
11. Wang K-S, Kuhn RJ, Strauss EG, Ou S, Strauss JH. High-affinity laminin receptor is a receptor for Sindbis virus in mammalian cells. *J Virol* 1992;66:4992–5001.
12. Ubol S, Griffin DE. Identification of a putative alphavirus receptor on mouse neural cells. *J Virol* 1991;65:6913–6921.
13. Wang K-S, Schmaljohn AL, Kuhn RJ, Strauss JH. Antiidiotype antibodies as probes for the Sindbis virus receptor. *Virology* 1991;181:694–702.
14. Walton TE, Grayson MA. Venezuelan equine encephalomyelitis. In: Monath TP, ed. *The arboviruses: epidemiology and ecology*, vol. 1. Boca Raton: CRC Press; 1988:59–85.
15. Scott TW. Vertebrate host ecology. In: Monath TP, ed. *The arboviruses: epidemiology and ecology*, vol 1. Boca Raton: CRC Press; 1988:257–280.
16. Rosen L. Overwintering mechanisms of mosquito-borne arboviruses in temperate climates. *Am J Trop Med Hyg* 1987;37:69S–70S.
17. Shope RE. Togaviruses. In: Fields BN, ed. *Virology*. New York: Raven Press; 1985:1055–1082.
18. Scott TW, Edman JD, Lorenz LH, Hubbard JL. Effects of disease on vertebrates ability behaviorally to repel host-seeking mosquitoes. In: Scott TP, Grumpstrup-Scott J, eds. *Proceedings of a symposium: the role of vector–host interactions in disease transmission*. Misc. Publ. 68, College Park, Maryland: Entomology Society America; 1988:9–17.
19. Scott TW, Lorenz LH, Edman JD. Effects of house sparrow age and arbovirus infection on attraction of mosquitoes. *J Med Entomol* 1990;27:856–863.
20. Hardy JL, Houk EJ, Kramer LD, Reeves WC. Intrinsic factors affecting vector competence of mosquitoes for arboviruses. *Annu Rev Entomol* 1983;28:229–262.
21. Hardy JL. Susceptibility and resistance of vector mosquitoes. In: Monath TP, ed. *The arboviruses: epidemiology and ecology*, vol 1. Boca Raton: CRC Press; 1988:87–126.
22. Scott TW, Olson JG, Lewis TE, Carpenter JW, Lorenz LH, Lembeck LA, Joseph SR, Pagac BB. A prospective field evaluation of an enzyme immunoassay: detection of eastern equine encephalomyelitis virus antigen in pools of *Culiseta melanura*. *J Am Mosq Control Assoc* 1987;3:412–417.
23. Weaver SC. Electron microscopic analysis of infection patterns for Venezuelan equine encephalomyelitis virus in the vector mosquito, *Culex (Melanoconion) taeniopus*. *Am J Trop Med Hyg* 1986;35:624–631.
24. Weaver SC, Scott TW, Lorenz LH. Patterns of eastern equine encephalomyelitis virus infection in *Culiseta melanura* (Diptera: Culicidae). *J Med Entomol* 1990;27:878–891.

25. Weaver SC, Scott TW, Lorenz LH, Repik PM. Detection of eastern equine encephalomyelitis virus deposition in *Culiseta melanura* following ingestion of radiolabeled virus in blood meals. *Am J Trop Med Hyg* 1991;44:250–259.
26. Weaver SC, Lorenz, LH, Scott TW. Distribution or western equine encephalolmyelitis virus in the alimentary tract of *Culex tarsalis* (Diptera: Culicidae) following natural and artificial blood meals. *J Med Entomol* 1993;30:391–397.
27. Casals J. Antigenic variants of eastern equine encephalitis virus. *J Exp Med* 1964;119:547–565.
28. Morris CD. Eastern equine encephalomyelitis. In: Monath TP, ed. *The arboviruses: epidemiology and ecology*, vol 3. Boca Raton: CRC Press; 1988:1–20.
29. Reisen WK, Monath TP. Western equine encephalitis. In: Monath TP, ed. *The arboviruses: epidemiology and ecology*, vol 5. Boca Raton: CRC Press; 1988:89–137.
30. Calisher CH. Karabatsos N, Lazuick JS, Monath TP, Wolff KL. Reevaluation of the western equine encephalitis antigenic complex of alphaviruses (family Togaviridae) as determined by neutralization test. *Am J Trop Med Hyg* 1988;38:447–452.
31. Aviles G, Sabattini MS, Mitchell CJ. Transmission of western equine encephalomyelitis virus by Argentine *Aedes albifasciatus* (Diptera: Culicidae). *J Med Entomol* 1992;29:850–853.
32. Hardy JL, Bruen JP. *Aedes melanimom* as a vector of WEE virus in California. *Proc Calif Mosq Control Assoc* 1974;42:36.
33. Reeves WC, Asman SM, Hardy JL, Milby MM, Reisen WK. *Epidemiology and control of mosquito-borne arboviruses in California, 1943–1987*. Sacramento: California Mosquito and Vector Control Association; 1990;504.
34. Weaver SC, Hagenbaugh A, Bellew LA, Calisher CH. Genetic characterization of an antigenic subtype of eastern equine encephalomyelitis virus. *Arch Virol* 1992;127:305–314.
35. Chen WR, Tesh RB, Rico-Hesse R. Genetic variation of Japanese encephalitis virus in nature. *J Gen Virol* 1990;71:2915–2922.
36. Mandl CW, Heinz FX, Stockl E, Kunz C. Genome sequence of tick-borne encephalitis virus (western subtype) and comparative analysis of nonstructural proteins with other flaviviruses. *Virology* 1989;173:291–301.
37. Rico-Hesse R. Molecular evolution and distribution of dengue viruses type 1 and 2 in nature. *Virology* 1990;174:479–493.
38. Hennig W. *Grundzuge einer theorie der phylogenetischen systematik*. Berlin: Deutscher Zentralverlag; 1950.
39. Hennig W. *Phylogenetic systematics*. Urbana: University of Illinois Press; 1966:263 pp.
40. Wanntorp HE, Brooks, DR, Nilsson T, Nylin S, Ronquist F, Stearns S, Wedell N. Phylogenetic approaches in ecology. *Oikos* 1990;57:119–132.
41. Mitter C, Brooks DR. Phylogenetic aspects of coevolution. In: Futuyma DJ, Slatkin M, eds. *Coevolution*. Sunderland, Mass: Sinauer Associates; 1983:65–98.
42. Hafner MS, Nadler SA. Cospeciation in host–parasite assemblages: comparative analysis of rates of evolution and timing of cospeciation events. *Syst Zool* 1990;39:192–204.
43. Page DMR. Temporal congruence and cladistic analysis of biogeography and cospeciation. *Syst Zool* 1990;39:205–226.
44. Calisher CH, Shope RE, Brandt W, Casals J, Karabatsos N, Murphy FA, Tesh RB, Woebe ME. Proposed antigenic classification of registered arboviruses. I. Togaviridae, *Alphavirus*. *Intervirology* 1980;14:229–232.
45. Wengler G, Wengler G, Filipe AR. A study of nucleotide sequence homology between the nucleic acids of different alphaviruses. *Virology* 1977;78:124–134.
46. Bell JR, Kinney RM, Trent DW, Strauss EG, Strauss JH. An evolutionary tree relating eight alphaviruses based on amino-terminal sequences of their glycoproteins. *Proc Natl Acad Sci USA* 1984; 81:4702–4706.
47. Weaver SC. Alphaviruses. In: Calisher CH, Gibbs AJ, Garcia-Arenel F, eds. *Molecular evolution of viruses*. Cambridge: Cambridge University Press; 1993.
48. Goldbach R, Wellink J. Evolution of plus-strand RNA viruses. *Intervirology* 1988;29:260–267.
49. Trent DW, Grant JA. A comparison of New World alphaviruses in the western equine encephalitis virus complex by immunochemical and oligonucleotide fingerprint techniques. *J Gen Virol* 1980; 47:261–282.
50. Roehrig JT, Hunt AR, Chang G-J, Sheik B, Bolin RA, Tsai TF, Trent DW. Identification of monoclonal antibodies capable of differentiating antigenic varieties of eastern equine encephalitis viruses. *Am J Trop Med Hyg* 1990;42:394–398.
51. Weaver SC, Scott TW, Ricco-Hesse R. Molecular evolution of eastern equine encephalomyelitis virus in North America. *Virology* 1991;182:774–784.

52. Weaver SC, Hagenbaugh A, Bellew LA, Gousset L, Mallampalli V, Scott TW. Evolution of alphaviruses in the eastern equine encephalomyelitis complex. *J Virology* [submitted].
53. Weaver SC, Bellew LA, Gousset LA, Repik PM, Scott TW, Holland JJ. Diversity within natural populations of eastern equine encephalomyelitis virus. *Urology* [in press].
54. Young NA, Johnson KM. Antigenic variants of Venezuelan equine encephalitis virus: their geographic distribution and epidemiologic significance. *Am J Epidemiol* 1969;89:286–307.
55. Buonagurio DA, Nakada S, Parvin J, Krystal M, Palese P, Fitch WM. Evolution of human influenza A viruses over 50 years: rapid, uniform rate of change in *NS* gene. *Science* 1986;232:980–982.
56. Kimura M. *The neutral theory of evolution*. London: Cambridge University Press; 1983.
57. Eldredge N, Gould SJ. Punctuated equilibria: an alternative to phyletic gradualism. In: Schopf TJM, ed. *Models in paleobiology*. New York: Freeman, Cooper & Co.; 1972:82–115.
58. Weaver SC, Bellew LA, Rico-Hesse R. Phylogenetic analysis of alphaviruses in the Venezuelan equine encephalitis complex and identification of the source of epizootic viruses. *Virology* 1992; 191:282–290.
59. Trent DW, Clewey JP, France JK, Bishop DHL. Immunochemical and oligonucleotide fingerprint analysis of Venezuelan equine encephalomyelitis complex viruses. *J Gen Virol* 1979;43:365–381.
60. Kinney RM, Trent DW, France JK. Comparative immunological and biochemical analysis of viruses in the Venezuelan equine encephalitis complex. *J Gen Virol* 1983;64:135–147.
61. Ricco-Hesse R, Roehrig JT, Trent DW, Dickerman RW. Genetic variation of Venezuelan equine encephalitis virus strains of the ID variety in Columbia. *Am J Trop Med Hyg* 1988;38:195–204.
62. Dickerman RW, Bonacorsa CM, Scherer WF. Viremia in young herons and ibis infected with Venezuelan encephalitis virus. *Am J Epidemiol* 1976;104:678–683.
63. Cupp EW, Kreutzer RD, Weaver SC. The biosystematics of *Culex (Melanoconion) taeniopus sensu lato* in relation to Venezuelan equine encephalomyelitis. *Mosq Syst* 1989;21:216–221.
64. Monath TP, Lazuick JS, Cropp CB, Rush WA, Calisher CH, Kinney RM, Trent DW, Kemp GE, Bowen GS, Francy DB. Recovery of tonate virus ("Bijou Bridge" strain), a member of the Venezuelan equine encephalomyelitis virus complex, from cliff swallow nest bugs (*Oeciacus vicarius*) and nestling birds in North America. *Am J Trop Med Hyg* 1980;29:969–983.
65. Price PW. *Evolutionary biology of parasites*. Princeton: Princeton University Press; 1980:176.
66. Tinsley RC. Host behavior and opportunism in parasitic life cycles. In: Barnard CJ, Behnke JM, eds. *Parasitism and host behavior*. London: Taylor & Francis; 1990:158–192.
67. Gieldman JB, Smith JF, Brown DT. Morphogenesis of Sindbis virus in cultured *Aedes albopictus* cells. *J Virol* 1975;16:913–926.
68. Strauss EG, Birdwell CR, Lenches EM, Staples SE, Strauss JH. Mutants of Sindbis virus. II. Characterization of a maturation-defective mutants *ts*103. *Virology* 1977;82:122–149.
69. Kowal KJ, Stollar V. Temperature-sensitive host-dependent mutants of Sindbis virus. *Virology* 1981;114:140–148.
70. Strauss JH, Strauss EG. Evolution of RNA viruses. *Annu Rev Microbiol* 1988;42:657–683.
71. Beaty BJ, Trent DW, Roehrig JT. Virus variation and evolution. In: Monath TP, ed. *The arboviruses: epidemiology and ecology*, vol. 1. Boca Raton: CRC Press; 1988:59–85.
72. Holland JJ, De La Torre JC, Steinhauer DA. RNA virus populations as quasispecies. *Curr Top Microbiol Immunol* 1992;176:1–20.
73. Domingo E, Holland JJ. High error rates, population equilibrium and evolution of RNA replication systems. In: Domingo E, Holland JJ, Alquist P, eds. Boca Raton, FL: CRC Press.
74. Rush WA, Francy DB, Smith GC, Cropp CB. Transmission of an arbovirus by a member of the family Cimicidae. *Ann Entomol Soc Am* 1980;73:315–318.
75. Calisher CH, Monath TP, Muth DJ, Lazuick JS, Trent DW, Francy DB, Kemp GE, Chandler FW. Characterization of Fort Morgan virus, an alphavirus of the western equine encephalitis virus complex in a unusual ecosystem. *Am J Trop Med Hyg* 1980;29:1428–1440.
76. Edman JD, Scott TW. Host defensive behavior and the feeding success of mosquitoes. *Insect Sci Appl* 1987;8:617–622.
77. Sellers RF. Eastern equine encephalitis in Quebec and Connecticut, 1972: introduction by infected mosquitoes on the wind? *Can J Vet Res* 1989;53:76–79.
78. Tabachnick WJ. Microgeographic and temporal genetic variation in populations of the bluetongue virus vector *Culicoides variipennis* (Diptera: Ceratopogonidae). *J Med Entomol* 1992;29:384–394.
79. Slatkin M. Gene flow in natural populations. *Annu Rev Ecol Syst* 1985;16:393–430.
80. Slatkin M. Gene flow and the geographic structure of natural populations. *Science* 1987;236:787–792.

81. Ehrlich PR, Raven PH. Butterflies and plants: a study in coevolution. *Evolution* 1964;18:586–608.
82. Janzen DH. When is it coevolution? *Evolution* 1990;34:611–612.
83. Futuyma DJ, Slatkin M. The study of coevolution. In: Futuyma DJ, Slatkin M, eds. *Coevolution.* Sunderland, Mass: Sinauer Associates; 1983:456–464.
84. Eldridge B. Evolutionary relationships among California serogroup viruses (Bunyaviridae) and *Aedes* mosquitoes (Diptera: Culicidae). *J Med Entomol* 1990;27:738–749.
85. Farrell B, Mitter C. Phylogenesis of insect/plant interactions: have *Phyllobrotica* leaf beetles (chrysomelidae) and the lamiales diversified in parallel? *Evolution* 1990;44:1389–1403.
86. Tabachnick WJ. A reappraisal of the consequences of evolutionary relationships among California serogroup viruses (Bunyaviridae) and *Aedes* mosquitoes (Diptera: Culicidae). *J Med Entomol* 1991;28:297–298.
87. Weaver SW, Scott TW, Lorenz LH, Lerdthusnee K, Romoser WS. Togavirus-associated pathologic changes in the midgut of a natural mosquito vector. *J Virol* 1988;62:2083–2090.
88. Weaver SC, Lorenz LH, Scott TW. Pathologic changes in the midgut of *Culex tarsalis* following infection with western equine encephalomyelitis virus. *Am J Trop Med Hyg* 1992;47:691–701.
89. Brown DT, Condreay LD. Replication of alphaviruses in mosquito cells. In: Schlesinger S, Schlesinger M. eds. *The togaviruses and flaviviruses.* New York: Plenum Press; 1986:171–207.
90. DeFoliart GR, Grimstad PR, Watts DM. Advances in mosquito-borne arbovirus/vector research. *Annu Rev Entomol* 1987;32:479–505.
91. Anderson RM, May RM. *Infectious diseases of humans: dynamics and control.* Oxford: Oxford University Press; 1991:656.
92. Levin BR, Allison AC, Bremermann HJ, Cavalli-Storza LL, Clarke BC, Frentzel-Beyme R, Hamilton WD, Levin SA, May RM, Thieme HR. Evolution of parasites and hosts. In: Anderson RC, May RM, eds. *Population biology of infectious diseases.* New York: Springer-Verlag; 1982:213–243.
93. Grimstad PR, Ross QE, Graig GB Jr. *Aedes triseriatus* (Diptera: Culicidae) and La Crosse virus. II. Modification of mosquito feeding behavior by virus infection. *J Med Entomol* 1980;17:1–7.
94. Turell MJ, Gargan TP, Bailey CL. *Culex pipiens* (Diptera: Culicidae) morbidity and mortality associated with Rift Valley fever virus infection. *J Med Entomol* 1985;22:332–337.
95. Faran ME, Turell MJ, Romoser WS, Routier RG, Gibbs PH, Cannon TL, Bailey CL. Reduced survival of adult *Culex pipiens* infected with Rift Valley fever virus. *Am J Trop Med Hyg* 1987;37:403–409.
96. Beaty BJ, Tesh RB, Aitken THG. Transovarial transmission of yellow fever virus in *Stegomyia* mosquitoes. *Am J Trop Med Hyg* 1980;29:125–132.
97. Tesh RB. Experimental studies on the transovarial transmission of Kunjin and San Angelo viruses in mosquitoes. *Am J Trop Med Hyg* 1980;29:657–666.
98. Turell MJ, Reeves WC, Hardy JL. Transovarial and trans-stadial transmission of California encephalitis virus in *Aedes dorsalis* and *Aedes melanimom. Am J Trop Med Hyg* 1982;31:1021–1029.
99. Scott TW, Hildreth SW, Beaty BJ. The distribution and development of eastern equine encephalitis virus in its enzootic mosquito vector, *Culiseta melanura. Am J Trop Med Hyg* 1983;33:300–310.
100. Scott TW, Burrage TG. Rapid infection of salivary glands in *Culiseta melanura* with eastern equine encephalitis virus: an electron microscopy study. *Am J Trop Med Hyg* 1984;33:961–964.
101. Price PW. *Insect ecology.* New York: John Wiley & Sons; 1984; 510.
102. Holland JJ, Spindler K, Horodyski F, Grabau E, Nichol S, VandePol S. Rapid evolution of RNA genomes. *Science* 1982;215:1577–1585.
103. Steinhauer DA, Holland JJ. Rapid evolution of RNA viruses. *Annu Rev Microbiol* 1987;41:409–433.
104. Baldridge GD, Beaty BJ, Hewlett MJ. Genomic stability of La Cross virus during vertical and horizontal transmission. *Arch Virol* 1989;108:89–99.
105. Igarashi A, Sasao F, Fukai K, Buei K, Ueba N, Yoshida M. Mutants of Getah and Japanese encephalitis viruses isolated from field-caught *Culex tritaeniorynchus* using *Aedes albopictus* clone C6/36 cells. *Ann Virol* 1981;132:235–245.
106. Wright S. The roles of mutation, inbreeding, crossbreeding and selection in evolution. *Proceedings of the Sixth International Congress of Genetics* 1932;1:356–366.
107. Durbin RK, Stollar V. A mutant of Sindbis virus with a host-dependant defect in maturation associated with hyperglycosylation of E2. *Virology* 1984;135:331–344.
108. Lai MMC. Genetic recombination in RNA viruses. *Curr Top Microbiol Immunol* 1992;176:21–32.
109. Hahn CS, Lustig S, Strauss EG, Strauss JH. Western equine encephalitis virus is a recombinant virus. *Proc Natl Acad Sci USA* 1988;85:5997–6001.

110. Duarte E, Clarke D, Moya A, Domingo E, Holland J. Rapid fitness losses in mammalian RNA virus clones due to Muller's ratchet. *Proc Natl Acad Sci USA* 1992;89:6015–6019.

111. Haigh J. The accumulation of deleterious genes in a population—Muller's ratchet. *Theor Popul Biol* 1978;14:251–267.

112. Chao L. Fitness of RNA decreased by Muller's ratchet. *Nature* 1991;348:454–455.

113. Muller HJ. The relation of recombination to mutational advance. *Mutat Res* 1964;1:2–9.

114. Chao L. Evolution of sex in RNA viruses. *Trends Ecol Evol* 1992;7:147–151.

115. Chao L, Tran T, Matthews C. Muller's ratchet and the advantage of sex in the RNA virus φ6. *Evolution* 1992;46:289–299.

116. Eigen M, Biebricher DK. Sequence space and quasispecies distribution. In: Domingo E, Holland JJ, Ahlquist P, eds. *RNA genetics*, vol 3. Boca Raton: CRC Press, 1988:211–245.

117. Davey MW, Mahon RJ, Gibbs AJ. Togavirus interference in *Culex annulirostris* mosquitoes. *J Gen Virol* 1979;42:641–643.

118. Sabin AB. Research on dengue during World War II. *Am J Trop Med Hyg* 1952;1:30–50.

119. Chamberlain RW, Sudia WD. Dual infection of eastern and western equine encephalitis viruses in *Culex tarsalis*. *J Infect Dis* 1957;101:233–236.

120. Findlay GM, MacCallum FO. An interference phenomenon in relation to yellow fever and other viruses. *J Pathol Bacteriol* 1937;44:405–424.

121. Gubler DJ, Kuno G, Sather GE, Waterman SH. A case of natural concurrent infection with two dengue viruses. *Am J Trop Med Hyg* 1985;34:170–173.

122. Gubler DJ. Dengue. In: Monath TP, ed. *The arboviruses: epidemiology and ecology*, vol 2. Boca Raton: CRC Press; 1988:223–260.

123. Wright S. *Evolution and the genetics of populations*, vol. 3. *Experimental results and evolutionary deductions*. Chicago: University of Chicago Press; 1977.

124. Wright S. Genic and organismic selection. *Evolution* 1980;34:825–843.

125. Wade MJ, Goodnight CJ. Wright's shifting balance theory: an experimental study. *Science* 1991; 253:1015–1018.

126. Dobzhansky T, Ayala FJ, Stebbins GL, Valentine JW. *Evolution*. San Francisco: WH Freeman; 1977:164.

*The Evolutionary Biology of Viruses*,
edited by Stephen S. Morse.
Raven Press, Ltd., New York © 1994.

15

# The Viruses of the Future?
# Emerging Viruses and Evolution

## Stephen S. Morse

*The Rockefeller University, New York, New York 10021*

The historiography of epidemic disease is one of the last refuges of the concept of special creationism.

Joshua Lederberg (1)

The chapters in this book have examined the evolution of viruses from a number of different perspectives. It is my self-imposed task here to try to glimpse into the future and discuss what we might expect from viruses. I might add, it is thankless. Predicting the future is a rash venture, usually unscientific, and undertaken at great peril. Most attempts serve mostly to embarrass the predictors, except (or, perhaps, especially) when they become self-fulfilling prophecies. Certainly, the acquired immunodeficiency syndrome (AIDS) epidemic was not predictable by extrapolation from diseases known at the time AIDS first began to gain momentum in the 1970s. (I will return to this question shortly.) However, if we take it as axiomatic that the processes (evolutionary and other) that went on in the past are going on today and, therefore, that history can be used as a guide, we can try to infer what likely processes and mechanisms could be involved and where they might lead. If viral evolution is also often rapid, then we may also have some hope that recorded history, which is usually insufficient to follow most geologic or evolutionary processes, may be a long enough timescale to afford us some helpful information about viruses (and throughout this chapter, *recent* will be used in the historical, not the geologic, sense).

We can begin by expecting the unexpected. The AIDS pandemic has many unusual features and has held many surprises (and, as Wain-Hobson argues, probably will have some further surprises in store), but at least one thing is certain: it has not been the first "new" epidemic to appear (2). Apparently sudden manifestations of outbreaks or epidemics of new infectious diseases have been noted since antiquity (3,4). The most notorious example was the bubonic plague, caused by the bacterium *Yersinia pestis*. In the 14th century, plague was estimated to have killed almost one-third of the European population (5). Smallpox (see Chapter 13 by Fenner and

*325*

**TABLE 1.** *Some examples of "emerging" viruses*

| Virus | Signs/symptoms | Distribution | Natural host |
|---|---|---|---|
| *Orthomyxoviridae* (RNA, 8 segments) | | | |
| **Influenza** | Respiratory | Worldwide (?from China) | Fowl, pigs |
| *Bunyaviridae* (RNA, 3 segments) | | | |
| **Hantaan, Seoul, other Hantaviruses** | Hemorrhagic fever + renal syndrome ± respiratory distress | Asia, Europe, United States | Rodent (e.g., *Apodemus*) |
| Rift Valley fever[a] | Fever, ± hemorrhage | Africa | Mosquito; ungulates |
| *Flaviviridae* (RNA) | | | |
| Yellow fever[a] | Fever, jaundice | Africa, S. America | Mosquito; monkey |
| **Dengue**[a] | Fever, ± hemorrhage | Asia, Africa Caribbean | Mosquito; human/ monkey |
| *Arenaviridae* (RNA, 2 segments) | | | |
| Junin (Argentine HF) | Fever, hemorrhage | S. America | Rodent (*Calomys musculinus*) |
| Machupo (Bolivian MF) | Fever, hemorrhage | S. America | Rodent (*Calomys callosus*) |
| Lassa fever | Fever, hemorrhage | W. Africa | Rodent (*Mastomys natalensis*) |
| *Filoviridae* (RNA) | | | |
| Marburg, Ebola | Fever, hemorrhage | Africa | Unknown, ? primate |
| *Retroviridae* (RNA + reverse transcriptase) | | | |
| **HIV** | AIDS, etc. | Worldwide | ? Primate |
| HTLV | Often asymptomatic; adult T-cell leukemia, neurological diseases (e.g., tropical spastic paraparesis) | Worldwide, with endemic foci | Human virus (? originally primate virus) |

(Adapted from Morse and Schluederberg, ref. 8, copyright 1990 by The University of Chicago. All rights reserved.)

[a]Transmitted by arthropod vector.

Underlined bold: Viruses of special concern for near future.

HF, hemorrhagic fever; HIV, Human immunodeficiency virus (types 1 and 2); HTLV, Human T-cell leukemia/lymphoma virus ( = human T-cell lymphotropic virus) (types 1 and 2).

Kerr), a more ancient disease, has enjoyed similar notoriety. In the 19th century, cholera (another bacterial infection, caused by *Vibrio cholerae*) was a major worldwide scourge (6). Of these, only smallpox has been eradicated. The others have been pushed back into the interstices of the environment by public health measures and improved-living conditions. In recent times, before human immunodeficiency virus (HIV) was known, influenza pandemics routinely swept through the world (and still do today). The most severe, the infamous pandemic of 1918–1919, has been blamed for some 25 million deaths (7). Although none of the subsequent influenza pandemics in this century (1957, 1968, and 1977) have approached the earlier one in mortality, nevertheless, they have caused considerable illness and death. While not as extensive as pandemics, ordinary epidemics of influenza still involve many millions of people in each epidemic.

TABLE 2. *Possible sources of "emerging" viruses*

**De novo evolution**
> New virus evolves (new variant; see Chapter 1, Table 2, p. 19)

**Viral traffic**
> *Interspecies transfer from the "zoonotic pool"*
> • Mutation or recombination confers expanded host range
> > Influenza
> > Canine parvovirus 2 (in dogs; probable origin, feline parvovirus)
> • Changing conditions put humans in contact with a previously rare or unfamiliar virus of another species
> > Hantaan (Korean hemorrhagic fever), other Hanta viruses ("Four Corners" Hantavirus in U.S., 1993)
> > Junin (Argentine hemorrhagic fever)
> > Lassa fever
> > Ebola, Marburg
> > Monkeypox
> • Exposure to vector in new environment, or conditions favoring increase in vector (Tick-borne or mosquito-borne diseases)
> > Rift Valley fever
> > Yellow fever
> *Dissemination of disease from a geographically localized subpopulation*
> *(Existing but previously restricted virus gains access to new host population)*
> > HIV (after introduction into a pilot human population)
> > HTLV
> > Dengue

These, then, are some of the notable epidemic diseases of the past and present. Some diseases, those that appear most suddenly and show indications of persisting, can be termed "emerging" (8). *Emerging* viruses (or, more generally, emerging diseases) are those that have recently increased their incidence and appear likely to continue increasing. In less formal terms, they are viruses that have newly appeared in the human population or are rapidly expanding their range, with a corresponding increase in cases of disease. Some examples are shown in Table 1.

What in fact are the sources of new viruses? As shown in Table 2, possible sources include new variants (de novo evolution) and existing viruses introduced into new host species or new populations. Given this background, we might consider the relative importance of viral evolution versus transfer of viruses to new host populations (a process I have termed *viral traffic*), in the emergence of new viral diseases.

It has often been assumed that new diseases imply *new*, that is, newly evolved, viruses. However, there are few examples to substantiate this, suggesting that this has rarely, if ever, been the case (for a listing of some examples, see Table 2 in Chapter 1). Critical examination of those historical examples that are available (including those in the recent past) indicates that the overwhelming majority of such instances can be accounted for by viral traffic. To summarize the conclusions from these analyses (discussed later): Most examples of viral emergence are independent of mutation, at least initially, and are more likely to be simply preexisting human or animal viruses in changed circumstances. (These processes are not necessarily mutually exclusive. It should also be cautioned that selection bias in examples cannot be ruled out, as these may be the examples people are most likely to identify. Nevertheless, the contrast is still striking.) This conclusion also readily follows from ordinary Darwinian premises, which would require that new organisms are descended from existing species (Darwin himself referred to his theory as "descent with modification"). From a Darwinian perspective, viruses would appear to be no exception.

The acquired immunodeficiency syndrome is the most novel disease of recent years and, in fact, the only new mechanism of viral pathogenesis to be identified recently. But the discovery, after the initial identification of HIV, that other lentiviruses, including some simian immunodeficiency viruses, are close relatives of HIV and may cause similar diseases in other species (see Chapter 10 by Myers and Korber) suggests that the novelty of disease, in this case, is likely to be due more to our lack of knowledge about these viruses because of their recent recognition, than to the recent development of an evolutionary novelty, although some evolutionary events (geographic diversification, perhaps specific adaptation to the human host) were also important.

If evolution is not the engine of emerging diseases, at least at this historical moment, then what is? From analyzing available examples, it would appear that most new disease-associated viruses derive from existing viruses that expand their range or acquire new hosts. Many are zoonotic (originating from animal sources). I will mention a few instances here, and will revisit this question later when I discuss factors in emergence. Considering the differences between pandemic influenza strains and those causing ordinary epidemics is instructive, as influenza is one of the paradigms of viral evolution at work. Indeed, regular annual epidemics are caused by antigenic drift in a previously circulating influenza strain. *Antigenic drift* is a change in an antigenic site of a surface protein, usually the H (hemagglutinin) protein, allowing the new strain to reinfect previously infected individuals because the altered antigen is not immediately recognized by the immune system. Unlike antigenic drift, which involves the accumulation of point mutations, pandemic strains are the result of *antigenic shift*, in which a new gene (usually for a different H protein) is introduced into a mammalian influenza strain by reassortment (9; see also Chapter 12 by Kilbourne). Pandemic influenza can be considered a zoonotic introduction, as the new gene usually originates in waterfowl and reassorts in domestic pigs (10,11).

The hantaviruses (Ha) and arenaviruses causing hemorrhagic fevers in humans, all of which are naturally occurring viruses of rodents, are classic examples of zoonotic introductions. Rodents appear to be particularly important reservoirs of zoonotic viruses, possibly because they are numerous and biologically diverse and are also among the mammals most likely to come into contact with humans. Another example of this dynamic is monkeypox in Africa. Monkeypox is a smallpox-like disease that was first described in captive monkeys newly imported from Africa. Humans can also become infected. The name, however, is a misnomer, as it is actually a natural virus of rodents (squirrels). Both people and monkeys become infected by contact with infected rodents in the forest. This example suggests that some viruses identified in nonhuman primates could actually be maintained in rodents.

Why zoonotic? Although a new and surprising variant is always possible, the existing viruses in nature constitute a large pool of viruses already relatively well adapted to a host. One can, in some sense, see them as the successful "variants of the past." In the short run, these would predominate, although we should not discount the possibility of surprises. These zoonotic viruses also possess the advantage of already being adapted to a host; hence, they are, at least in some sense, biologically preselected for viability. They will also continue to be maintained in their natural reservoir hosts if they fail to establish themselves successfully in new hosts. Viruses, hitherto established in an isolated human population, have already evolved some degree of adaptedness for human infection, and they constitute the other major source. Many of these latter (endogenous retroviruses perhaps expected) probably began as zoonotic viruses that were introduced into the human species at some time past. Although this is still uncertain, present evidence would indicate that we are now seeing this process occurring with HIV (discussed by Myers and Korber, Chapter 12).

From this analysis, we might consider the emergence of new viruses as a two-step process (perhaps a parallel not unlike Mayr's two-step process of natural selection), the first step being introduction of the virus into a human population, and the second step dissemination with the population. Emergence usually involves the introduction of a virus into a human population, typically from a zoonotic (animal) reservoir, followed by dissemination. Therefore, emphasis should be placed on understanding the conditions that affect each of these steps. We might speculate that evolution may be of special importance in the transition from step 1 to step 2, where adaptedness to the new host may be critical. Viral traffic also seems important in both steps, as shown by the examples in Table 3. Viral traffic can serve to introduce new zoonotic viruses into a human population or to spread a previously isolated virus to new hosts (12,13).

Human activities, such as occur in agriculture, during deforestation, and following population migrations to cities, can be major causes of viral traffic and, hence, often contribute to the success of newly emerged viruses (see Table 3). Many of these activities are relatively easily recognized; consequently, potentially manageable (14). For example, proliferation of water-storage containers and of other small containers that trap water, increases the incidence of dengue or fellow fever, especially in cities, by providing breeding places for *Aedes aegypti*, the major urban vector for these viruses, and for other vectors. Reducing access of mosquitoes to water would reduce dengue. As another example, various types of changes in farming practices may favor a rodent reservoir of a previously unrecognized virus, abruptly placing humans in contact with a pathogen to which they were not previously exposed (Junin, Machupo, Lassa fever, monkeypox). Other practices may increase opportunities for genetic reassortment, leading to new strains (integrated duck–pig farming may allow influenza viruses to exchange genes). Human migration can spread a previously localized virus, and commerce can disseminate viruses or vectors (yellow fever and the *Aedes aegypti* mosquito probably spread through the African slave trade; recently, another mosquito, *Aedes albopictus*, which is a vector for several viruses, including dengue, in its native Asia, was introduced into the United States, as well as Brazil and South Africa, in imported tires from Asia).

**TABLE 3.** *Human intervention in viral emergence*

| Factor | Example | Probable mechanism[a] |
|---|---|---|
| Agriculture | Influenza (pandemic reassortants) | Integrated pig–duck farming has been suggested |
| | Hantaan | Contact with rodents during rice harvest |
| | Argentine hemorrhagic fever | Agriculture may favor natural rodent host; contact with rodent host during harvest |
| | Bolivian hemorrhagic fever | Contact with rodent host during harvest |
| | Oropouche | Cacao hulls encourage breeding of insect vector |
| | Monkeypox | Subsistence agriculture and hunting at forest, increasing contact with natural rodent host |
| Tropical deforestation | Kyasanur Forest | Tick vector increased as forest land replaced with sheep grazing |
| Open water storage | Dengue, dengue hemorrhagic fever | Water containers encourage breeding of mosquito vector |
| | Yellow fever | Water containers encourage breeding of mosquito vector |
| Dams, irrigation | Venezuelan equine encephalomyelitis | Conditions favoring increase in vector (Panama) |
| | Rift Valley fever | Conditions favoring increase in vector |
| Commerce | Dengue | Disseminated by travel and migration |
| | Yellow fever | Both virus and major vector (*Aedes aegypti*) believed to have been disseminated by slave trade |
| | Seoul-like viruses (in United States) | Infected rats may have been carried to United States on ships |
| Medical technology | Lassa fever | Secondary cases in health care workers from contact with infected individuals (primary cases mostly through contact with infected rodent host) |
| | Marburg/Ebola | Secondary cases through contaminated hypodermic needles |
| | HIV | Transfusions, contaminated hypodermic needles[b] |
| | HTLV | Transfusions, contaminated hypodermic needles[b] |
| | Hepatitis B | Transfusions, organ transplants, contaminated hypodermic needles |
| | Hepatitis C | Transfusions, contaminated hypodermic needles[b] |

[a] These are the major factors, where known, but other factors may also be involved.

[b] Other important factors also involved, including sexual transmission (with all three viruses, but most notably HIV-1); HTLV has also been widespread and long-established in some human populations. With hepatitis C, additional causes are believed to exist, but have not yet been identified.

War is also a time-honored means of disseminating both microbes and their vectors (15).

This discussion of emerging viruses has necessarily been a brief overview. Additional details, further reading, and suggestions for actions that might help anticipate or prevent emergence, can be found in other references (12,14,16,17, and the various chapters in 18). However, a few words about possible actions to anticipate or prevent emergence are in order. In addition to consistent application and improvement of standard public health measures worldwide, one recommended action, long advocated by D. A. Henderson (19) as well as others, is global surveillance for emerging infectious diseases (17,19,20). Knowledge of the factors underlying disease emergence can help focus resources in the key situations and areas worldwide (14,20).

From the foregoing discussion, it would appear that viral emergence is often precipitated by environmental factors and by opportunities for dissemination into new populations (viral traffic). The role of human actions in shaping these also needs to be better appreciated to prevent the many unanticipated effects that have been seen throughout history and continue, even are increasing, today. Another major ally of the microbes has been human complacency, the feeling that infectious diseases are bygone relics of a less-enlightened past, with its consequent failure to invest in and consistently pursue appropriate public health measures.

Similar causes are often involved in the emergence of new infections caused by other classes of organisms. The appearance of new strains or variants of bacterial pathogens is driven by several factors, including traffic factors similar to those discussed here (12,16,17,21). A recent epidemic of cholera in South America probably began with the introduction of the organism in contaminated bilge water released by a freighter into a Peruvian harbor, from whence the organism spread into local shellfish and later disseminated through contaminated water supplies (22). A major factor in emergence of bacterial pathogens is selection for antimicrobial drug-resistance, driven by the wide and sometimes inappropriate use of antimicrobials in a variety of applications (23–25). Although resistance to antiviral agents has been described in several viruses, this has not become such a dominant factor as it has in bacterial and some parasitic infections, probably because present antiviral agents are few and rarely find extensive use. The potential is there, however, and the emergence of antibiotic-resistant bacteria as a result of the ubiquity of antimicrobials in the environment is an evolutionary lesson that one hopes will be heeded as more antiviral agents become available in the future. It also demonstrates the power of natural selection, as resistant strains appear and flourish under the selective pressures exerted by the presence of these antibiotics in the environment.

To sum up, then: When we ask about the viruses of the future, in a very real sense many of them (as well as the ancestors of others that may yet evolve) are here now, in other species and in isolated populations, awaiting their opportunities (one other possible source, the rarer, but occasional, possibility that some are also lurking in the genomes of our own or other species, is briefly discussed in the first chapter of this volume). Most of these conclusions would be familiar to Smith, Burnet, and others in the ecological tradition (26). In the conclusion to *Virus as Organism* (27), Burnet states:

> We have stressed throughout the importance of the transfer of virus infection from one host to another, and have given examples to show that such transfers from animal reservoirs to man are taking place at present; we have also postulated that in the past all or nearly all the known virus diseases of man must have been similarly derived from animal infections. The process by which such transfer to new hosts can, by the selective survival of suitable mutants, result in a new mode of indefinite survival of the virus species has also been discussed. . . . For those who look for progress and design in evolution, the picture that has been built up of the evolution and natural history of virus disease must be an almost intolerable one. For the viruses there has been no steady increase in complexity and emergence of new powers but, as they have evolved, an increasing simplicity of structure and a greater dependence on the host cell for nutrition. . . . In view of the extensive natural reservoirs of viruses in arthropod, bird, and mammal that have been uncovered by investigations into disease of man and domestic animals, it is only natural to believe that many others may be discovered. From these, by some suitable combination of chance and circumstance, new virus diseases of man may well arise in the future.

Chance, circumstance, and evolutionary potential contribute to the fate of a new virus as it appears for the first time. After a virus is introduced into a new host, there are several possible outcomes. Infection may be abortive or unsuccessful; this may be a common outcome, sometimes perhaps after a dramatic outbreak. Or a virus may remain geographically isolated for some time before chance and circumstance give it the opportunity to spread further, as may well have happened with HIV. Or, more rarely, like pandemic influenza strains, it may spread rapidly. Even if evolution is not necessarily the engine of new diseases, it is still a process of profound and sometimes incalculable biological importance. A virus may be too virulent in its new host and cause outbreaks of sporadic severe disease and death without spreading beyond its original hosts, as with Ebola. Eventually, given sufficient opportunity, a virus may evolve variants with increased adaptedness to the new host and, thereby, have the potential to remain in the host species and await opportunities to spread further.

These evolutionary processes occurring in the past have also led to the diversity of viruses existing today. Viral variation is itself of great importance in a number of contexts. As Domingo and Holland discuss, variation must be taken into account in the development of vaccines, antiviral agents, and diagnostic tests. At times, variants arising in an individual during infection may cause alterations in pathogenesis, with clinical consequences in that individual. An example of this is fulminant hepatitis caused by a hepatitis B variant arising during infection (discussed in the first chapter). Geographic variants may also differ somewhat in various properties, even in virulence, making them both important to track and worthy of study for the insights they may offer. Finally, lest we become complacent about the often conservative nature of evolution, variation, even if rarely, can sometimes lead to novelty, even to greatly increased virulence, as in the example of the highly virulent H5 avian influenza strain that appeared among chickens in Pennsylvania in 1983 (28), discussed in the first chapter.

Although variation itself appears beyond prediction with our present capabilities and more data will be needed, I still think that evidence of preferred mutations (29,30) and of other constraints on genetic variation (31) suggest that the process is not totally random (although it likely varies with different organisms or under different conditions), and that, in principle, it should be possible to develop mathematical methods to predict the most likely variations. To do so is an interesting intellectual challenge in any event. One approach that could serve as a basis was one that was recently developed by Benkovic and colleagues (32,32a).

Several authors have recently made ingenious suggestions for other ways to bring evolutionary biology, and especially the consequences of natural selection, to bear on other biological and biomedical questions (33,34). I will briefly mention one example. The correlation between transmissibility and virulence with infectious organisms was mentioned earlier in this book (see first chapter). Several scientists (34–36) analyzing the events following the introduction of myxoma virus in Australia as a way to control the rabbit population (see Chapter 13 by Fenner and Kerr), have noted that the level of virulence eventually attained represented a tradeoff between virulence and transmissibility; myxoma is readily transmissible, and the level of virulence was relatively high. Ewald has carried the consequences of this analysis a step further, suggesting that transmission rate can drive virulence in either direction through selection (34). Thus, he suggests that circumstances favoring rapid or simultaneous wide transmission (such as waterborne or vector-borne diseases or sexual transmission) should select for increased virulence; conversely, he suggests that slowing down or interrupting transmission would necessary select for less virulent variants (since with few opportunities for transmission the more virulent forms would kill their hosts before they could be transmitted) (34,37). By this analysis, Ewald argues that the virulence of HIV was increased by the rapid cycling that occurred during the AIDS epidemic, and could be decreased by preventing rapid transmission. Others have recently begun considering evolutionary explanations for the diversity of genes involved in the immune response (38,39), and Wills and colleagues have suggested a model based on the coevolution of host and parasite (Wills C, Green DR, personal communication). These questions are still being debated, but suggest that unexpected new insights can come from analyses grounded in evolutionary biology.

In the more immediate future, such applications of evolutionary biology as phylogenetic analysis and geographic tracking will help in identifying disease agents, both old and new, and in assessing their actual spread and their potential for spreading further. Knowledge of their biology and ecology and a better understanding of host interactions will also be useful in these assessments. Phylogenetic analysis has, to date, been based largely on molecular sequence data. Recently, structural biology has also begun to contribute here (see, for example, Chapter 5 by Heringa and Argos).

The explanation of epidemic disease and their causes need no longer rely on special creationism. Calisher quipped that, although creationism offers certainty without data, evolution offers data without certainty (40). Since then, the gap may have closed a little. Certainty is never possible in science; however, in recent years not only the data, but also the ability to make useful inferences from these data, have increased enormously, and continue to expand. The application of molecular methodology promises to bring new depth to studying ecological and coevolution-ary relationships. The molecular techniques that had previously revolutionized biomedical research are being fruitfully applied to evolution and ecology. Some of the fruits are demonstrated by the contributions in this book. The cross-fertilization of these fields has led to some promising developments in "molecular epidemiology" (41; see Chapter 10 by Myers and Korber), as well as to calls for "evolutionary epidemiology" (42). Chance and circumstance are, one hopes, propitious at last for the integration of evolutionary biology and the biomedical sciences. As scientists working in evolution and biomedical science begin to find themselves on an expanding common meeting ground, with tools of increasing resolving power, evolutionary approaches will have increasingly valuable lessons to offer biomedical researchers. In return, just as viruses have provided important insights into the molecular processes of life, they also have the potential to offer exciting insights into evolution. This two-way traffic should yield a valuable scientific harvest.

## ACKNOWLEDGMENTS

I am supported by grant RR 03121, and also in part by grant RR 01180, from the National Institutes of Health, US Department of Health and Human Services. My work on emerging viruses and their evolution was supported by the Division of Microbiology and Infectious Diseases (DMID), National Institute of Allergy and Infectious Diseases, National Institutes of Health, and by the Fogarty International Center of NIH. I thank Dr. John R. La Montagne, Director, Division of Microbiology and Infectious Diseases (DMID), National Institute of Allergy and Infectious Diseases, National Institutes of Health, and Dr. Ann Schluederberg, Virology Branch Chief.

## REFERENCES

1. Lederberg J. Viruses and humankind: intracellular symbiosis and evolutionary competition. In: Morse SS, ed. *Emerging viruses*. New York: Oxford University Press; 1993:3–9.
2. Grmek MD. *History of AIDS. Emergence and origin of a modern pandemic*. Princeton: Princeton University Press; 1990. Maulitz RC, Duffin J, translators.
3. Grmek MD. *Diseases in the ancient Greek world*. Baltimore: Johns Hopkins University Press; 1989. Muellner M, Muellner L, translators.
4. McNeill WH. *Plagues and peoples*. New York: Anchor Press/Doubleday; 1976.
5. Gottfried RS. *The Black Death*. New York: Free Press; 1983.
6. Rosenberg CE. *The cholera years*. Chicago: University of Chicago Press; 1962.
7. Crosby AW. *America's forgotten pandemic. The influenza of 1918*. Cambridge: Cambridge University Press; 1989.
8. Morse SS, Schluederberg A. Emerging viruses: the evolution of viruses and viral diseases. *J Infect Dis* 1990;162:1–7.
9. Webster RG, Bean WJ, Gorman OT, Chambers TM, Kawaoka Y. Evolution and ecology of influenza A viruses. *Microbiol Rev* 1992;56:152–179.
10. Scholtissek C, Naylor E. Fish farming and influenza pandemics. *Nature* 1988;331:215.
11. Kida H, Shortridge KF, Webster RG. Origin of the hemagglutinin gene of H3N2 influenza viruses from pigs in China. *Virology* 1988;162:160–166.
11a. LeDuc JW, Childs JE, Glass GE. The hantaviruses, etiologic agents of hemorrhagic fever with renal syndrome: a possible cause of hypertension and chronic renal disease in the United States. *Ann Rev Publ Health* 1992;13:79–98.

12. Morse SS. Emerging viruses: defining the rules for viral traffic. *Perspect Biol Med* 1991;34:387–409.
13. Shope RE, Evans AL. Assessing geographic and transport factors, and recognition of new viruses. In: Morse SS, ed. *Emerging viruses*. New York: Oxford University Press; 1993:109–119.
14. Morse SS. Regulating viral traffic. *Issues Sci Technol* 1990;7:81–84.
15. Zinsser H. *Rats, lice and history*. Boston: Little, Brown; 1935 [Reprinted: New York: Bantam Books; 1965].
16. Morse SS. Examining the origins of emerging viruses. In: Morse SS, ed. *Emerging viruses*. New York: Oxford University Press; 1993:10–28.
17. Lederberg J, Shope RE, Oaks SC Jr, eds. *Emerging infections: microbial threats to health in the United States*. Washington, DC: National Academy Press; 1992.
18. Morse SS, ed. *Emerging viruses*. New York: Oxford University Press, 1993.
19. Henderson DA. Surveillance systems and intergovernmental cooperation. In: Morse SS, ed. *Emerging viruses*. New York: Oxford University Press; 1993:283–289.
20. Morse SS. Global microbial traffic and the interchange of disease. *Am J Publ Health* 1992;82:1326–1327.
21. Krause RM. The origin of plagues: old and new. *Science* 1992;257:1073–1078.
22. Anderson C. Cholera epidemic traced to risk miscalculation [News]. *Nature* 1991;354:255.
23. Cohen ML. Epidemiology of drug resistance: implications for a post-antimicrobial era. *Science* 1992;257:1050–1055.
24. Bloom BR, Murray CJL. Tuberculosis: commentary on a reemergent killer. *Science* 1992;257:1055–1064.
25. Neu HC. The crisis in antibiotic resistance. *Science* 1992;257:1064–1072.
26. Fiennes RNT-W. Zoonoses and the origins and ecology of human disease. New York: Academic Press; 1978.
27. Burnet FM. *Virus as organism. Evolutionary and ecological aspects of some human viral diseases*. [Dunham Lectures, Harvard University; 1944]. Cambridge, Mass.: Harvard University Press; 1945.
28. Kawaoka Y, Webster RG. Molecular mechanism of acquisition of virulence in influenza virus in nature. *Microb Pathogen* 1988;5:311–318.
29. Goodenough M, Huet T, Saurin W, Kwok S, Sninsky J, Wain-Hobson S. HIV-1 isolates are rapidly evolving quasispecies: evidence for viral mixtures and preferred nucleotide substitutions. *J AIDS* 1989;2:344–352.
30. Pathak V, Temin HM. Broad spectrum of in vivo forward mutations, hypermutations, and mutational hotspots in a retroviral shuttle vector after a single replication cycle: substitutions, frameshifts, and hypermutations. *Proc Natl Acad Sci USA* 1990;87:6019–6023.
31. Fitch WM, Leiter JME, Li X, Palese P. Positive Darwinian evolution in human influenza A viruses. *Proc Natl Acad Sci USA* 1991;88:4270–4274.
32. Egger BT, Benkovic SJ. Minimal kinetic mechanism for misincorporation by DNA polymerase I (Klenow fragment). *Biochemistry* 1992;31:9227–9236.
32a. Kuchta RD, Benkovic P, Benkovic SJ. Kinetic mechanism whereby DNA polymerase I (Klenow) replicates with DNA with high fidelity. *Biochemistry* 1988;27:6716–6725.
33. Williams GC, Nesse RM. The dawn of Darwinian medicine. *Q Rev Biol* 1991;66:1–22.
34. Ewald PW. The evolution of virulence. *Sci Am* 1993;268:86–93.
35. Levin SA, Pimentel D. Selection for intermediate rates of increase in parasite-host systems. *Am Nat* 1981;117:308–315.
36. Anderson RM, May RM. Coevolution of hosts and parasites. *Parasitology* 1982;85:411–426.
37. Ewald P. *The evolution of infectious diseases*. New York: Oxford University Press; 1993.
38. Hill AV, Allsopp CE, Kwiatkowski D, Antsey NM, Twumasi P, Rowe PA, Bennett S, Brewster D, McMichael AJ, Greenwood BM. Common west African HLA antigens are associated with protection from severe malaria. *Nature* 1991;352:595–600.
39. Hill AV, Elvin J, Willis AC, Aidoo M, Allsopp CE, Gotch FM, Gao XM, Takiguchi M, Greenwood BM, Townsend AR, et al. Molecular analysis of the association of HLA-B53 and resistance to severe malaria. *Nature* 1992;360:434–439.
40. Calisher CH. Evolutionary significance of the taxonomic data regarding bunyaviruses of the family Bunyaviridae. *Intervirology* 1988;29:268–276.
41. Ou C-Y, Ciesielski CA, Myers G, et al. Molecular epidemiology of HIV transmission in a dental practice. *Science* 1992;256:1165–1171.
42. Ewald PW. Cultural vectors, virulence, and the emergence of evolutionary epidemiology. In: Harvey PH, Partridge L, eds. *Oxford surveys in evolutionary biology*, vol 5. Oxford: Oxford University Press; 1988:215–245.

# A Bibliographic Note

For readers who may desire some additional background in those areas that are unfamiliar, standard texts and references in some of the fields represented in this volume are listed below.

## EVOLUTIONARY BIOLOGY

Futuyma DJ. *Evolutionary biology*, 2nd ed. Sunderland, Mass.: Sinauer Associates; 1986.

## MOLECULAR EVOLUTION

Li W-H, Graur D. *Fundamentals of molecular evolution*. Sunderland, Mass.: Sinauer Associates; 1991. (*Introductory text.*)
Nei M. *Molecular evolutionary genetics*. New York: Columbia University Press; 1987. (*More advanced.*)

## MOLECULAR GENETICS

There are many textbooks in genetics, of which these titles are a selection.

Berg P, Singer M. *Dealing with genes: The language of heredity*. Mill Valley, Calif.: University Science Books; 1992. (*A more basic introduction than the same authors'* Genes & genomes.)
Griffiths AJF, Miller JH, Suzuki DT, Lewontin RC, Gelbart WM. *An introduction to genetic analysis*, 5th ed. New York: W.H. Freeman; 1993.
Lewin B. *Genes IV*. New York: Oxford University Press/Cell Press; 1990.
Rothwell NV. *Understanding genetics: A molecular approach*. New York: Wiley-Liss; 1993.
Singer M, Berg P. *Genes & genomes: A changing perspective*. Mill Valley, Calif.: University Science Books; 1991.
Watson JD, Hopkins NH, Roberts JW, Steitz JA, Weiner AM. *Molecular biology of the gene*, 4th ed. Menlo Park, Calif.: Benjamin/Cummings; 1987.

## METHODOLOGY FOR PHYLOGENETIC ANALYSIS
### (*see also Software, below*)

The volume by Hillis and Moritz gives general overviews of the molecular methods currently used, and includes laboratory protocols for handling and analyzing the biological samples that might be used. The other titles are for readers desiring detailed discussion of the theory and practice of constructing phylogenies.

*337*

Doolittle RF, ed. *Molecular evolution: Computer analysis of protein and nucleic acid sequences* (Methods in Enzymology, vol. 183). San Diego: Academic Press; 1990.

Hillis DM, Moritz C, eds. *Molecular systematics*. Sunderland, Mass.: Sinauer Associates; 1991.

Miyamoto MM, Cracraft J, eds. *Phylogenetic analysis of DNA sequences*. New York: Oxford University Press; 1991.

## POPULATION GENETICS

Falconer DS. *Introduction to quantitative genetics*, 3rd ed. London: Longman; 1989.

Hartl DL, Clark AG. *Principles of population genetics*, 2nd ed. Sunderland, Mass.: Sinauer Associates; 1989.

Hartl DL. *A primer of population genetics*, 2nd ed. Sunderland, Mass.: Sinauer Associates; 1988. (*A shorter version of the more detailed* Principles.)

Maynard Smith J. *Evolutionary genetics*. Oxford and New York: Oxford University Press; 1989. (*A text that combines subject matter in population genetics and molecular evolution.*)

Wallace B. *Basic population genetics*. New York: Columbia University Press; 1981.

## VIROLOGY

### Elementary:

Levine AJ. *Viruses*. New York: Scientific American Library, 1992. (*Intended as an elementary introduction for the general reader.*)

Morse SS, ed. *Emerging viruses*. New York and Oxford: Oxford University Press; 1993. (*A more specific focus than this text, but a number of the viruses in this book are included; some of the chapters may provide additional background for material in this volume.*)

### Intermediate:

Davis BD, Dulbecco R, Eisen HN, Ginsberg HS, eds. *Microbiology*, 4th ed. Philadelphia: JB Lippincott; 1990. (*Also includes immunology.*)

Joklik WK, Willett HP, Amos DB, Wilfert CM, eds. *Zinsser microbiology*, 20th ed. Norwalk, Conn.: Appleton & Lange; 1992. (*Also includes immunology.*)

Fenner FJ, Gibbs EP, Murphy FA, Rott R, Studdert MJ, White DO. *Veterinary virology*, 2nd ed. San Diego: Academic Press; 1993.

### Advanced:

Fields BN, Knipe DM, et al., eds. *Fields virology*, 2nd ed. New York: Raven Press; 1990.

Fields BN, Knipe DM, et al., eds. *Fundamental virology*, 2nd ed. New York: Raven Press; 1990. (*Selected chapters from the complete Fields, including most of the material, such as molecular virology, that will be of greatest interest to readers of this volume.*)

For information on specific animal viruses, the work by Fields et al. is a standard reference. For brief capsule descriptions, the following are useful:

Porterfield JS, ed. *Andrewes' viruses of vertebrates*, 5th ed. London and Philadelphia: Bailliere Tindall/ W.B. Saunders; 1989.

Francki RIB, Fauquet CM, Knudson DL, Brown F, eds. *Classification and nomenclature of viruses: Fifth report of the International Committee on Taxonomy of Viruses* [Arch Virol, Suppl. 2]. Vienna: Springer-Verlag; 1991.

## JOURNALS

The reader may get a sense of the most relevant journals by consulting the references cited in the chapters of this volume. A selected list follows. Many other journals, including *Nature*, *Science*, *Cell*, *Proceedings of the National Academy of Sciences (U.S.A.)*, etc., periodically publish papers of interest in evolutionary virology. As protein and nucleic acid sequences now represent an important source of data, material of interest also can be found at times in journals dealing with these subjects (such as *Nucleic Acids Research*), and in journals covering molecular genetics (e.g., *Genetics*). Data bases such as GenBank (Software, below) are indispensable for original work.

All virology journals publish at least some relevant material. The *Journal of Virology* recently has been publishing a number of highly relevant papers, and the *Journal of General Virology* has had a long-standing interest in this area. *Intervirology* has published the proceedings of the International Committee on Taxonomy of Viruses, with features dealing with viral taxonomy and related evolutionary questions. Useful reviews can be found in *Microbiological Reviews* (quarterly), *Advances in Virus Research* (Maramorosch K, Murphy FA, Shatkin AJ, eds., Academic Press, hardbound; generally several volumes annually), and *Annual Review of Microbiology*.

For HIV and related viruses, the *Journal of Acquired Immune Deficiency Syndromes*, *AIDS Research and Human Retroviruses*, and (perhaps to a lesser degree) *AIDS* often have published papers of interest.

In the field of molecular evolution, *Molecular Biology and Evolution* has published much relevant material; also the *Journal of Molecular Evolution*.

Major general journals in evolutionary biology include the *Quarterly Review of Biology* (with longer review articles), *Evolution*, and *The American Naturalist*. The first two occasionally have papers relating to viruses; the *Quarterly Review of Biology* also carries a wide variety of book and software reviews. The journal *Trends in Ecology and Evolution* ("TREE") publishes brief overviews and opinions covering the entire field of evolution.

## SOFTWARE

### Phylogenetic Analysis Software

Addresses for the software packages discussed by Andrew Leigh Brown (Chapter 4):

**PHYLIP:**
Dr. Joseph Felsenstein
Dept. of Genetics, SK-50
University of Washington
Seattle, WA 98195

e-mail: joe@genetics.washington.edu (Internet), or felsenst@uwavm (Bitnet). (*A number of platforms are supported, including IBM and compatibles, Macintosh, and Unix systems.*)

**PAUP:**
Dr. David L. Swofford
Illinois Natural History Survey
Natural Resources Building
607 East Peabody Drive
Champaign, IL 61820. (*Version 3 is now available for Macintosh; a PAUP 3 program to run on IBM-compatible computers and other platforms is under development and is expected to be available by the time this book is published.*]

**MacClade** (for Macintosh):
Version 3 of this program, by Wayne P. Maddison and David R. Maddison, is now distributed by Sinauer Associates, North Main Street, Sunderland, MA 01375; telephone, (413) 665-3722. The manual is also available separately from the same publisher.

**Clustal:**
Dr. Desmond (Des) Higgins
European Molecular Biology Laboratory
Postfach 10.2209
Meyerhofstrasse 1
6900 Heidelberg, Germany
e-mail: higgins@EMBL-Heidelberg.DE (Internet).

### Genetic and Protein Sequence Data

GenBank, the U.S. resource for DNA sequence data, was moved to the National Center for Biotechnology Information (NCBI) of the National Library of Medicine (NLM) a few years ago (those who wish to submit sequence data to GenBank should also contact NCBI). NLM makes its data bases available in a number of formats. A CD-ROM version called "Entrez" is available at low cost and can be used on both Macintosh computers and IBM compatibles using Microsoft Windows. NCBI also supports a number of other services. For further information:

NCBI
National Library of Medicine
Bldg. 38A, Room 8N-803
Bethesda, MD 20894
Tel.: (301) 496-2475
FAX: (301) 480-9241
e-mail: info@ncbi.nlm.nih.gov

The European Molecular Biology Laboratory (EMBL) in Heidelberg offers comparable on-line services and access to sequence data bases for users in Europe.

# Subject Index

Note: Page numbers followed by *t* refer to tables; page numbers followed by *f* refer to figures.

## A

Acyclovir, resistance to, 266
Adaptation, 30–31, 37, 45
  nongenetic, as constraint on selection, 39
  viral, to new host, 253–255
Adaptive radiation, 215
  in parasite evolution, 306–309
*Aedes* species, viral transmission by, 297, 308, 329
African hare fibroma virus
  geographic range of, 278*t*
  hosts, 278*t*
Alphavirus(es)
  cell receptors for, 295
  diversification, evolutionary concepts of, 305–306
  diversity, 299–304
  evolution, 243–244, 293–324
    adaptive landscape model, 313–315
    ecological concepts, 304–305
    factors regulating, 304–319
    by individual versus interdemic (group) selection, 318–319
    molecular and cellular factors, 311–316
    summary of, 319
  genomes of, 295
  mutation frequencies, 311–313
  phylogenetic analysis of, 299–304
  replication, 295
  RNA species, 295
  serology of, 294
  speciation, and geographic isolation, 309
  transmission cycles, 295–297
  vectors, dispersal of, 307–308
  virulence, for mosquito vectors, 309–311
Alphavirus-like supergroup, 107*f*, 111–113
  genetic interrelationship of, 111, 112*f*
  genomic variation among, 111–113
  members of, 106*t*
Amantadine, resistance to, 266
Amino acid sequence
  nonsynonymous substitutions in, 78, 172–173
  synonymous substitutions in, 78, 172–173

  synonymous versus nonsynonymous substitutions in, in HIV-1, 219–221, 220*f*
AMV. *See* Avian myeloblastosis virus
Antibiotic resistance, emergence, human intervention in, 331
Antigenic drift, 67–68, 328
  in canine parvovirus, 258
  in influenza A virus, 240–241
  interaction with host immunity, 68–70
Antigenic shift, 240–241, 328
Antigenic sin, and HIV variants, 194, 196
Antigenic variation, 67–71. *See also specific virus*
Antiviral drug(s)
  effect on viral evolution, 266
  resistance to, 70–71, 266
  and RNA virus heterogeneity, 174–175
Arbovirus(es), 8, 294
  evolution, phylogenetic analysis of, 297–304
  and hosts, coevolution of, 308–309
Arenavirus(es), genetic maps of, 113, 114*f*
Argentine hemorrhagic fever, emergence, human intervention in, 330*t*
Autoimmune disease(s), possible viral etiology of, 173
Avian myeloblastosis virus, base misincorporation in vitro, 188
Avipoxvirus
  biogeography of, 274*t*, 275
  genome of, 276
Avirulence, evolution of, 58

## B

Bacterial pathogens, emergence, human intervention in, 331
Bacteriophage f6, purifying-selection, 242–243
Bacteriophage K11, RNA polymerase, sequence similarities with other polymerases, 97*f*
Bacteriophage latency, 4
Bacteriophage Qb, 9
  mutation rate and frequency, 164*t*, 165
  purifying-selection, 241–243